EMERGENT BEHAVIOR IN COMPLEX SYSTEMS ENGINEERING

EMERGENT BEHAVIOR IN COMPLEX SYSTEMS ENGINEERING

A Modeling and Simulation Approach

Edited by

SAURABH MITTAL

SAIKOU DIALLO

ANDREAS TOLK

This edition first published 2018

Registered Office
John Wiley & Sons, Inc., 111 River Street, Hoboken, NJ 07030, USA

Editorial Office
111 River Street, Hoboken, NJ 07030, USA

For details of our global editorial offices, customer services, and more information about Wiley products visit us at www.wiley.com.

Wiley also publishes its books in a variety of electronic formats and by print-on-demand. Some content that appears in standard print versions of this book may not be available in other formats.

Library of Congress Cataloging-in-Publication Data
Names: Mittal, Saurabh, editor. | Diallo, Saikou Y., editor. | Tolk, Andreas,
    editor.
Title: Emergent behavior in complex systems engineering : a modeling and
    simulation approach / edited By Saurabh Mittal, Saikou Diallo, Andreas
    Tolk.
Description: 1st edition. | Hoboken, NJ : John Wiley & Sons, 2018. | Series:
    Stevens Institute series on complex systems and enterprises | Includes
    bibliographical references and index. |
Identifiers: LCCN 2017052148 (print) | LCCN 2018000725 (ebook) | ISBN
    9781119378853 (pdf) | ISBN 9781119378938 (epub) | ISBN 9781119378860
    (cloth : alk. paper)
Subjects: LCSH: Systems engineering. | System design.
Classification: LCC TA168 (ebook) | LCC TA168 .E53 2018 (print) | DDC
    620.001/1–dc23
LC record available at https://lccn.loc.gov/2017052148

Cover Design: Wiley
Cover Images: (Background image) © archibald1221/Gettyimages; (Sphere) © v_alex/Gettyimages

Set in size of 10/12pt and TimesLTStd by Spi Global Ltd, India, Chennai

Printed in United States of America

10 9 8 7 6 5 4 3 2 1

To my parents, all my teachers, and mentors, who lit the path for me,

To my wife and kids, my loving companions in this fulfilling journey,

To the Almighty, who widens my horizons and fill my heart with joy abound

Saurabh Mittal

To all who have contributed a pebble to my foundation, teachers, and family and especially my wife Faby and my son EJA, whose teachings are as entertaining and they are enlightening

Saikou Diallo

To all my mentors who helped me to become who I am, and all my colleagues who continue what I tried to start!

Andreas Tolk

CONTENTS

FOREWORD

The etymological roots for the word *Emergence* are in the Latin words "emergere" or "emergo," which mean to arise or to bring into the light: something that was covered or hidden becomes visible. Only in the recent decades has the term has been used in science and philosophy to reference the observation of some new properties that are exposed by natural and engineered systems without having been explicitly created. Such an emergent property of a system is usually discovered at the macro-level of the behavior of the system and cannot be immediately traced back to the specifications of the components, whose interplay produce this emergence. In natural systems, this may reflect a lack of depth of understanding of the phenomena of interest; for engineered systems, this tends to reflect a lack of understanding of the implications of design decisions.

The confusion with the term emergence is nearly Babylonian. The term is used in many different ways in science and philosophy, and its definition is a substantive research question itself. Researchers are not sure if their observations are domain specific, or if they must contribute in multi-, trans-, and interdisciplinary research endeavors to new insights into the bigger, general challenges of system thinking. Philosophy discusses the differences between ontological and epistemological emergence. Scientists are applying new methods, many from the field of modeling and simulation, to generate emergent behaviors and gain new insight from the study of the dynamic representation of such systems that can produce emergence. But who within these communities holds the Holy Grail?

In his book, *The Tao of Physics* (1975), Fritjof Carpa, the founding director of the Center for Ecoliteracy in Berkeley, California, writes in the epilogue: "Mystics understand the roots of the Tao but not its branches; scientists understand its branches but not its roots. Science does not need mysticism and mysticism does not need science; but man needs both." Such a view drove us to design and develop this book. Complexity has led us to understand the limits of reductionism. Such findings may not only be true for individual disciplines, but generally even more so for multi-, trans-, and interdisciplinary research. The very first step to enable such cooperation is to get to know each other. The resulting mix of invited experts in this volume therefore exposes a high degree of diversity, embracing many different views, definitions, and interpretations to show the many facets that collectively contribute to the bigger picture, hoping that we be able to reach a similar conclusion as Carpa, who states in the same source referenced above: "The mystic and the physicist arrive at the same conclusion; one starting from the inner realm, the other from the outer world. The harmony between their views confirms the ancient Indian wisdom that Brahman, the ultimate reality without, is identical to Atman, the reality within." We need experts highly educated and experienced in their facets who are willing to talk and listen to each other.

Our underlying guidance to all authors was to think about how their contribution can make a difference for those who are designing, developing, managing, operating, and maintaining systems, including system of systems, in helping them to better detect, understand, and hopefully manage emergence to reap the benefits, for example, of innovations, and avoid the dangers, for example, of unfortunate consequences. What can system scientists and engineers contribute? Can we construct simulation systems that reproduce natural systems closely enough to gain insights about emergent behavior? How should our management and governance of complex systems look? Can we validate emergence? Is emergent always repeatable, or is it path dependent? Can we apply higher principles, such as entropy, to gain more insight? What are the computational and epistemological constraints we must be aware of? A much broader approach that involves experts from many domains is needed.

Simulation always has been a melting pot of expertise coming from many disciplines interested in many different application domains. The state-of-the-art presented in this book about methods and technologies that aim to understand emergent behavior in complex systems engineering in various scientific, social, economic, and multidisciplinary systems engineering disciplines is defining the new frontiers for humankind. The insights elaborated here have broader, ongoing consequences than expected, as we are witnessing a closely related evolution in science: the increasing use of computational means to support research. Computational science emerges – pun intended – in many disciplines: computational physics, biology, social science, systems engineering, finance, to name just a few. In order to use computational means, these disciplines first have to build models about the phenomena of interest and then build algorithms to make them executable on computers: in other words, they are constructing models and simulations. The same limits and constraints on validity of simulation approaches to complexity research are therefore applicable to computational science dealing with complex systems as well.

Complexity introduces a set of challenges to engineers, scientists, and managers in many real-world applications that affect our daily life. A better understanding of emergence in such systems, including possible limits and constraints of what we can do with current methods and tools, will enable making best use of such systems to serve society better. This book is not a solution book, but a foundations book, addressing the fields that have to contribute to address the research questions that have to be answered to better detect, understand, and manage emergence and complexity.

WILLIAM ROUSE
STEVENS INSTITUTE OF TECHNOLOGY
ALEXANDER CROMBIE HUMPHREYS PROFESSOR
DIRECTOR OF THE CENTER FOR COMPLEX SYSTEMS AND ENTERPRISES
ANDREAS TOLK
THE MITRE CORPORATION
FELLOW OF THE SOCIETY FOR MODELING AND SIMULATION
SEPTEMBER 15, 2017

PREFACE

We are surrounded by emergence.

Human civilization transformed through significant periods starting from the hunter-gatherer era, through the agricultural period, to the industrial age, and now the information and digital age. Each period emerges from the previous over time not only through technological advances and economic progress but also through conflicts, war, and transformative political and social changes. What qualifies a period in history as an "era?" How does an era start, and why does it end? Among the many reasons we have listed above, it is important to emphasize the impact of technology on society and the role technological revolutions (Industrial revolution, Internet, etc.) play in shaping the direction of Humanity. Having said that, we are not completely sure how era-changing technologies come into being and are mostly unable to predict which technologies will change civilization, and which will go unnoticed. We can only observe that when a new technology appears, it is sometimes met with skepticism, mockery, ridicule, and denial. Such reactions are often due to the lack of understanding of the technology and its implications. However, some technologies – once created – add tremendous knowledge and insight while spawning new industries, disciplines, and ecosystems that generate new professions and a new workforce, thus bringing about a new societal structure that can cope with the new technology. Some technologies are so disruptive and life changing that they mark the beginning of a new era. Would not it be desirable to better understand technologies that have the potential for such large-scale emergence or maybe even able to predict and manage the consequent emergence? Might Isaac Asimov's vision of Hari Seldon's *Psychohistory* become a reality? Are we on the cusp of the emergence of a new era?

Beyond societal emergence, engineered systems capable of displaying emergent behavior are entering our daily routines at a high rate. For instance, there is currently an increasing number of unmanned system technology being applied in a wide variety of domains. Robots are conducting surgeries; we see self-driving cars maturing; packages are delivered by drones, and unmanned systems show up on the battlefield. These unmanned systems observe their environment, compare the perceived situation with their goals, and then follow rules set to achieve their objectives. Even relatively simple rules can lead to very complex swarm behavior, exposing emergent behavior beyond the intention of the designers. If this behavior is helpful in reaching their planned objective, all is good, but where is the threshold for such behavior to become dangerous or even harmful? How can we better recognize unintended consequences, which may easily be magnified due to the many and often nonlinear connections between the components? How can we ensure that such unmanned solutions evolve into a favorable direction and not like James Cameron's *Skynet* into an existential threat for society?

It is such questions and ideas that have motivated us to work on this book. We want to understand the world as a complex system and to gain some semblance of control as we inject more and more engineered systems in this existing complex system. We want to answer questions such as: Is emergence systemic, or can we reduce or even eliminate it as we gain enough knowledge about the system, its components, and relations? Do we need better tools and methods to study emergence? We strive to bring together the discipline of complex systems engineering that needs to incorporate the element of complexity, inherent in the very structure of a system, and the elements of emergent behavior that complex system engineering could never design in the first place but still needs to account for.

To this end, we are particularly interested in exploring the subject of emergence through the lens of Modeling and Simulation (M&S). Modeling is the art of simplification and abstraction, taking only "so much" from reality to answer the question put forth at the right abstraction level. Simulation is the increasingly computerized execution of a model over time to understand its dynamic behavior. Such computational means are potent tools that allow scientists and engineers to hypothesize, design, analyze, and even theorize about a particular phenomenon. Can we recreate emergence in such artificial systems in a way that helps us understand emergence in the real system of interest better? What are the limits of such M&S support? Furthermore, M&S supports scientist in social sciences with powerful tools, such as agent-based simulation systems that are increasingly used in support of computational social science. How can we gain insight regarding the natural system by evaluation of such simulations? Can we explore all types of emergence currently discussed by philosophers as well as engineers, or are there limitations and constraints computational scientists need to be aware of?

The goal of this book is to provide an overview of the current discussions on complexity and emergence, and how systems engineering methods in general and simulation methods in particular can help in gaining new insight and support users of complex systems in providing better governance. The book is organized into 16 invited chapters in four sections, providing an overview of philosophical, model engineering, computational methods using simulation, and research specific viewpoints.

The topics addressed in the chapters reflect the different viewpoints on emergence and discuss why we should not rule it out, whether complex systems can be engineered, whether all complex systems can be reduced to complicated systems if we increase our knowledge, how simulation can help to better understand and manage emergence, and what role can system thinking play in understanding emergence? The authors provide a wide variety of approaches to studying emergence ranging from formal system specification that account for emergence, deriving factors from observations of emergence in physics and chemistry, the emergence of language between two hominid agents in a resource-constrained system, and looking at emergence in complex enterprises. The editors conclude the book with observations on a possible research agenda to address some of the grand challenges the complex systems engineering community must consider.

This book is a diverse collection of contributions from a broad background of recognized experts in their field highlighting aspects of complexity and emergence important from their viewpoint. By bringing them together in one compendium, we hope to spawn a discussion on new methods and tools needed to address the challenges of complexity that obviously go beyond the limits of traditional approaches.

Saurabh Mittal, Herndon, VA
Saikou Diallo, Suffolk, VA
Andreas Tolk, Hampton, VA
September 2017

ABOUT THE EDITORS

Saurabh Mittal is the lead systems engineer/scientist in the Modeling, Simulation, Experimentation, and Analytics (MSEA) Technical Center of the MITRE Corporation. He is also affiliated with Dunip Technologies, LLC, and Society of Computer Simulation (SCS) International. He currently serves as associate editor-in-chief for Transactions of SCS and editor-in-chief of Enterprise Architecture Body of Knowledge (EABOK) Consortium. He received his M.S. and Ph.D. degrees in electrical and computer engineering from the University of Arizona. Previously, he was a scientist and architect at National Renewable Energy Laboratory, Department of Energy at Golden, Colorado, where he contributed to complex energy systems and co-simulation environments. He also worked at L3 Link Simulation & Training at 711 HPW, US Air Force Research Lab at Wright-Patterson Air Force Base, Ohio, where he contributed to integrating artificial agents and various cognitive architectures in Live, Virtual and Constructive (LVC) environments using formal systems theoretical model-based engineering approaches. He was a research assistant professor at the Department of Electrical and Computer Engineering at the University of Arizona. Dr. Mittal served as general chair of Springsim'17 and SummerSim'15, vice general chair for SpringSim'16 and SummerSim'14, and program chair for SpringSim'15. He is the founding chair for M&S and Complexity in Intelligent, Adaptive and Autonomous (MSCIAAS) Symposium offered in Springsim, Summersim, and Winter Simulation Conferences. He is a recipient of US Department of Defense (DoD) highest civilian contraction recognition: Golden Eagle award (2006) and SCS's Outstanding Service (2016) and Professional Contribution (2017) award. He has coauthored over 80 articles in various international conferences and journals, including books titled "Netcentric System of Systems Engineering with DEVS Unified Process" and "Guide to Simulation-based disciplines: Advancing our computational future" that serves the areas of executable architectures; service-oriented distributed simulation; formal Systems M&S; system of systems engineering; multiplatform modeling; intelligence-based, complex, adaptive, and autonomous systems; and large-scale M&S integration and interoperability.

Saikou Diallo is a research associate professor at the Virginia Modeling, Analysis and Simulation Center, and an adjunct professor at Old Dominion University. Dr. Diallo has studied the concepts of interoperability of simulations and composability of models for over 15 years. He is VMASC's lead researcher in Simulated Empathy where he focuses on applying modeling and simulation to study how people connect with one another and experience their environment and creations. He currently has a grant to conduct research into modeling religion, culture, and civilizations. He is also involved in developing cloud-based simulation engines and user interfaces in order to promote the use of simulation outside of traditional engineering fields. Dr. Diallo graduated with a M.S. degree in engineering in 2006 and a Ph.D. in modeling and simulation in 2010 both from Old Dominion University. He is the vice president in charge of conferences and a member of the Board of Directors for the Society for Modeling and Simulation International (SCS). Dr. Diallo has over one hundred publications in peer-reviewed conferences, journals, and books.

Andreas Tolk is technology integrator in the Modeling, Simulation, Experimentation, and Analytics (MSEA) Technical Center of the MITRE Corporation. He is also adjunct full professor of engineering management and systems engineering and modeling, simulation, and visualization engineering at Old Dominion University in Norfolk, Virginia. He holds an M.S. and a Ph.D. degree in computer science from the University of the Federal Armed Forces in Munich, Germany. He published more than 200 contributions to journals, book chapters, and conference proceedings and edited several books on Modeling & Simulation and Systems Engineering. He received the Excellence in Research Award from the Frank Batten College of Engineering and Technology in 2008, the Technical Merit Award from the Simulation Interoperability Standards Organization (SISO) in 2010, and the Outstanding Professional Contributions Award from the Society for Modeling and Simulation (SCS) in 2012, and the Distinguished Achievement Award from SCS in 2014. He is a fellow of SCS and a senior member of ACM and IEEE.

LIST OF CONTRIBUTORS

Lachlan Birdsey
School of Computer Science
The University of Adelaide
Adelaide, SA 5005
Australia

Chih-Chun Chen
Department of Engineering
University of Cambridge
Cambridge CB2 1PZ
UK

Steven Corns
Department of Engineering Management
and Systems Engineering
Missouri University of Science and
Technology
Rolla, MO 65401
USA

Nathan Crilly
Department of Engineering
University of Cambridge
Cambridge CB2 1PZ
UK

Saikou Diallo
Virginia Modeling, Analysis &
Simulation Center
Old Dominion University
Suffolk, VA
USA

Umut Durak
German Aerospace Center
Cologne
Germany

David C. Earnest
Department of Political Science
University of South Dakota
Vermillion, SD 57069
USA

Erika Frydenlund
Virginia Modeling, Analysis and
    Simulation Center
Old Dominion University
Suffolk, VA 23435
USA

Ross Gore
Virginia Modeling, Analysis and
    Simulation Center
Old Dominion University
Norfolk, VA 23529
USA

John J. Johnson IV
Systems Thinking & Solutions
Ashburn, VA 20148
USA

Matthew T.K. Koehler
The MITRE Corporation
Bedford, MA
USA

Justin E. Lane
Institute of Cognitive and Evolutionary
    Anthropology
Department of Anthropology
University of Oxford
64 Banbury Road, Oxford OX2 6PN
UK

and

LEVYNA, Ústav religionistiky
Masaryk University
Veveří 28, Brno 602 00
Czech Republic

Suzanna Long
Department of Engineering Management
    and Systems Engineering
Missouri University of Science and
    Technology
Rolla, MO 65401
USA

Saurabh Mittal
The MITRE Corporation
McLean, VA
USA

Michael D. Norman
The MITRE Corporation
Bedford, MA
USA

Akhilesh Ojha
Department of Engineering Management
    and Systems Engineering
Missouri University of Science and
    Technology
Rolla, MO 65401
USA

Tuncer Ören
School of Electrical Engineering and
    Computer Science
University of Ottawa
Ottawa
Canada

Jose J. Padilla
Virginia Modeling Analysis and
    Simulation Center
Old Dominion University
Suffolk, VA
USA

Robert Pitsko
The MITRE Corporation
McLean, VA
USA

Ruwen Qin
Department of Engineering Management
  and Systems Engineering
Missouri University of Science and
  Technology
Rolla, MO 65401
USA

William B. Rouse
Center for Complex Systems and
  Enterprises
Stevens Institute of Technology
1 Castle Point Terrace, Hoboken,
  NJ 07030
USA

Tom Shoberg
U.S. Geological Survey
CEGIS
Rolla, MO 65409
USA

F. LeRon Shults
Institute for Religion, Philosophy and
  History
University of Agder
Kristiansand 4604
Norway

Andres Sousa-Poza
Engineering Management & System
  Engineering
Old Dominion University
Norfolk, VA 23529
USA

John Symons
Department of Philosophy
The University of Kansas
Lawrence, KS 66045
USA

Claudia Szabo
School of Computer Science
The University of Adelaide
Adelaide, SA 5005
Australia

Andreas Tolk
The MITRE Corporation
Hampton, VA
USA

Wesley J. Wildman
School of Theology
Boston University
Boston, MA 02215
USA

and

Center for Mind and Culture
Boston, MA 02215
USA

Levent Yilmaz
Department of Computer Science and
  Software Engineering, Samuel Ginn
  College of Engineering
Auburn University
Auburn, AL 36849
USA

Bernard P. Zeigler
RTSync Corporation
University of Arizona
Tucson, AZ
USA

SECTION I

EMERGENT BEHAVIOR IN COMPLEX SYSTEMS

1

METAPHYSICAL AND SCIENTIFIC ACCOUNTS OF EMERGENCE: VARIETIES OF FUNDAMENTALITY AND THEORETICAL COMPLETENESS

John Symons

Department of Philosophy, The University of Kansas, Lawrence, KS 66045, USA

SUMMARY

The concept of emergence figures prominently in contemporary science. It has roots in philosophical reflection on the nature of fundamentality and novelty that took place in the early decades of the twentieth century. Although it is no longer necessary to offer philosophical defenses of the science of emergent properties, attention to basic metaphysical questions remains important for engineering and scientific purposes. Most importantly, this chapter argues for precision with respect to what scientists and engineers take to count as fundamental for the sake of their uses of the concept of emergence.

INTRODUCTION

Two defining characteristics, novelty and naturalness, mark the concept of emergence. When emergent properties are first instantiated, they are said to be novel in some difficult to specify, but presumably nontrivial, sense. Although every moment of natural history is new in the sense of being at least at a different time from what came before, the kind of novelty that is associated with emergent properties is understood

Emergent Behavior in Complex Systems Engineering: A Modeling and Simulation Approach,
First Edition. Edited by Saurabh Mittal, Saikou Diallo, and Andreas Tolk.

to constitute a metaphysically significant difference. What might that significance amount to? Very roughly, we can say that if an emergent property appears, there is a new *kind* of entity or property on the scene. Not just more of the same. To claim that a property, say for example a property like transparency, liquidity, or consciousness, is emergent is to make a judgment about the way it relates to more fundamental features of the world. The emergent property or entity differs in kind from that which preexisted it or is more fundamental than it. The first task of this chapter is to explore what it might mean for emergent properties to relate or fail to be related to more fundamental properties.

The discussion of emergent properties in scientific and philosophical research has emphasized discontinuities and differences between the emergent property and the prior or more fundamental properties from which it arises. However, emergent properties are not just discontinuous with what came before. They are also thought to be part of the natural order in some intelligible sense. According to most contemporary proponents, emergent properties are not unnaturally or supernaturally new (their appearance is not miraculous) but instead can be understood scientifically insofar as they are intelligibly connected with parts of the natural world and in particular with other properties that are prior or more fundamental.

The scientific problem of emergence involves understanding the relations between the emergent property and the more fundamental or prior properties. The practical payoff of this understanding is improved levels of prediction and control over those emergent properties and entities that concern us most.

TO EXPLAIN IS NOT TO ELIMINATE

How could there be an intelligible connection between metaphysically distinct kinds? In one sense, this is a question only a philosopher would bother asking. There are plenty of simple examples. Take Putnam's (1975, 295–298) famous example of the explanation for why a square peg fails to pass through a round hole. The rigidity of the pegs and the rigidity of the walls of the holes are dependent on their physical structure. However, the property of being able to pass through a hole of a particular size and shape is a different kind of property than the properties governing the physical constituents of the peg. Geometrical facts about the sizes of the cross section of the peg and the hole are sufficient to explain the facts about the pegs being able to pass through. An attempt to account for this higher level property in terms of the physics governing the particles in the peg would result in an unexplanatory, albeit a true and very long, description of the particular case. The geometrical explanation, by contrast is simple, provides clear guidance for interaction in the system and generalizes beyond this particular peg and hole to all pegs, all holes, and beyond.

The geometrical explanation explains many things at various scales, including why we have round manhole covers rather than square ones. Manhole covers have the property of not falling dangerously on people working in the sewers below because of the circular (rather than, say, rectangular) shape of the covers. This is one example of how we can intelligibly connect distinct kinds of properties. The *microphysical properties* of this particular peg, its particular material instantiation, can be connected with the macro-level property of *passing through this particular hole* via a geometrical explanation. That geometrical explanation has the virtue of being applicable beyond this particular case. The property of being a hole, being able to pass through, having a particular stable shape in space, having the particular microphysics that this peg has, and so on, are connected in the explanation in a way that satisfies our demand for explanation in this context perfectly.

Putnam intended this to be an example of a *non-reductive* explanation, as, he thought, the material constitution of the peg is almost completely irrelevant to the explanation of its fitting or failing to fit. His use of this example was meant to indicate the role of explanations that are not simply descriptions of physical microstates of systems. There is more going on in nature, he argued, than merely the microphysical.

Philosophers in the 1960s and 1970s were very concerned with the distinction between what they saw as reductive and non-reductive explanation. They fixated on the distinction between reductionist and anti-reductionist explanations because of their concern for the ontological implications of explanations. For some, the *threat* of reductionism is that we are encouraged to believe that one kind of object simply does not exist insofar as it can be described in terms of some more basic kind of object. This is an ontological concern. Notice that it involves a judgment that is independent of the process of explanation: We might decide that the existence of certain kinds of explanation license ontologies with fewer things in them. Thus, given the fact that we can explain traffic jams on the highway in terms of the interactions of individual vehicles, we might be tempted to draw the ontological conclusion that there is no traffic jam. Notice that if one decided to take this strategy with respect to one's ontology, it is a step beyond what the explanation of the traffic jam by itself tells us. In fact, I would argue, one needs to justify the step from a successful reductive scientific explanation to the claim that because of this successful explanation we can therefore eliminate the thing that has been explained from our ontology. Furthermore, in paradigmatically reductionist explanations, we see examples of intelligible relations being discovered between distinct kinds of properties. For example, subatomic particles are not the kinds of things that have properties like rigidity or wetness. A structural explanation of the subatomic constituents of a diamond goes some way toward explaining why the diamond in the engagement ring is rigid. There is an intelligible relation between the macro-properties of the diamond and the micro-properties of its constituents that adverts to the structure of the diamond crystal. Properties like hardness or rigidity are manifest only on some scales and result from interactions of large numbers of molecules. There is simply no non-relational explanation of why diamond molecules give rise to hardness. These relations, like the geometrical properties of Putnam's pegs, are not built into their relata.

The concern among philosophers is inspired by the concern that giving an explanation is equivalent to explaining away. Philosophers sometimes argue, following Carnap and Quine, that "explication is elimination" in natural science as well as in mathematics. This is due to a mistaken conflation of kinds of explanations and the diverse theoretical goals motivating explanatory projects. Quine's arguments concerning eliminativism were drawn from purely mathematical contexts. He was moved, primarily by his understanding of the history of analysis in nineteenth century mathematics. The infinitesimal is a puzzling artifact of early calculus that (according to popular opinion) we no longer need to include in lessons to high school students thanks to the work of Weierstrass, Dedekind, and Cantor. As the story goes, Weierstrass gave us the means to eliminate the infinitesimal, Dedekind and Cantor helped to finish the job. Quine strongly approved of this story and built his account of explication as elimination upon it.[1] He proposed a view that began by individuating metaphysically puzzling notions in mathematics, like the infinitesimal or the ordered pair, via the mathematical roles that they play. Insofar, as they are *"prima facie* useful to theory and, at the same time, troublesome," Quine recommended that we simply find other ways to perform their theoretical role. Once we find these other ways, we can stop worrying about those concepts. Like the infinitesimal, they are eliminated (1960a, 266).

The explanatory project that motivates complexity science or other studies of emergent properties is not the same as that which motivated Quine's approach to philosophical analysis. For Quine, the method of philosophical analysis is to "fix on the particular functions of the unclear expression that make is worth troubling about, and then devise a substitute, clear, and couched in terms of our liking that fills those functions" (1960b, 258–259). By contrast, the goal of research in the natural sciences is the discovery of novel objects and relations in the world. The explanatory goal is understanding rather than the rearticulation, in more parsimonious terms, of functions that make the phenomenon worth troubling about.

In scientific and engineering research more generally, this kind of elimination is simply not a goal. Insofar as things like traffic jams or epidemics are troublesome, that trouble is not eliminated by defining ways that other, less troublesome things, cause delays and illness. A traffic jam or an epidemic is not a "troublesome" theoretical entity in Quine's sense of being what he calls a "defective noun" that we wish to do without in the interest of ontological parsimony. Instead, the very goal of scientific investigation presupposes the reality of the object to be understood. If there were no epidemics or traffic jams, they would not pose any real practical problem. Defining the hurricane away will not solve our hurricane-related challenges.

Emergentism was a view that was articulated before the rise of concerns about explanation, reduction, and elimination discussed above. Since the decline of so-called British Emergentism in the 1930s, philosophers have worried about the anti-scientific connotations of the term "emergence" and have been concerned that emergentism involves an attachment to mystery, or at least the belief in limits to the power of scientific explanation. For the two most important figures in the British Emergentist tradition, Samuel Alexander and C.D. Broad, emergent properties resisted mechanistic or reductive explanation. They held somewhat different views

---

[1]Błaszczyk *et al.* (2013) raise a number of credible objections to the standard histories of analysis that Quine relies upon. Their general line of criticism focuses on what they see as the unwarranted drive for ontological minimalism motivating some prominent histories of mathematics. Prominent among their specific criticisms is what they see as the mistaken identification of the continuum with a single number system. Whether Quine had the history right is an interesting question, but it is independent of the present argument.

on the nature of explanation.[2] However, the most important aspects of their views are the following: For Broad, nothing about the laws of physics would allow an ideal epistemic agent (what Broad called a mathematical archangel) to predict all aspects of emergent properties ahead of time. Broad mentioned the smell of a chemical compound as one of the properties that the archangel would have been unable to predict (1923, 71). For Alexander, the appearance of emergent properties should be accepted as a brute fact, accepted, as he said "with natural piety" (see Alexander, 1920, 46–47). The emergentists saw the distinctive properties of, for example, the chemical, biological, or psychological levels as simply being brute facts. They argued that these distinctively non-physical aspects of reality, the smell of sulfur, the price of bread, the chemical properties of gold, the effects of crowds on individual psychology, the nature of life, consciousness, and countless other examples can be integrated intelligibly into our understanding of reality without being eliminated from our ontologies. In my view, the British Emergentists should be read as insisting that there are distinct kinds of properties or phenomena and that this distinctness cannot be explained away. At this point in the history of science and philosophy, we can have plenty of explanations that connect distinct kinds of phenomena or properties in ways that allow for understanding without assuming that such understanding entails eliminating one of these kinds.

EMERGENT PROPERTIES AND MORE FUNDAMENTAL PROPERTIES

Fundamentality is the central conceptual component of discussions concerning the emergence. Most obviously, contemporary uses of the term "emergence" vary according to their users' views of fundamentality. Varying positions with respect to emergence usually differ with respect to either (i) what their proponents take to be fundamental or (ii) whether they see emergence as a purely epistemic matter. This is evident, for example, when we compare the divergent scientific and philosophical careers of the concept of emergence. In general, contemporary scientists talk relatively freely about emergent phenomena and properties while being non-committal (beyond gesturing approvingly to fundamental physics) about what counts as genuinely fundamental.[3] Instead, scientists tend to emphasize notions like predictability, surprise, and control. This is not to say that these epistemic *seeming* concepts are unrelated to metaphysical questions. As the philosopher of mathematics Alan Baker has pointed out, being a weakly emergent property does not entail any necessary relation to the epistemic or cognitive limitations of particular agents (2010, 2.3).[4] What Baker is pointing to is that weak emergence is an objective

---

[2] See Symons (forthcoming) for a detailed discussion of the view of explanation held by the British Emergentists.
[3] See, for example, the representative papers and articles from scientists in Bedau and Humphreys (2008).
[4] A weakly emergent sequence or pattern, for example, would be one that can be computed, but cannot be compressed informationally. Every step in the pattern or sequence must be cranked out by the simulation. There can be no shortcuts and no abbreviating recipes/algorithms. Bedau (1997) is the most well-articulated source for the idea of weak emergence.

feature of certain kinds of systems in the same way that mathematical features of systems are independent of the epistemic and cognitive capacities of agents.

Since the middle of the twentieth century, most analytic philosophers have been more wary of the term "emergence" than our colleagues in science and engineering. The most central feature of the resistance to the concept of emergence in the second half of the twentieth century was been the philosophical community's attachment to the doctrine of physicalism. Physicalism is the view that physics provides us with our fundamental ontology. Ontology is our theory of what counts as real. The tendency among philosophers had been to see physics as our means of understanding the ultimate nature of being.

In recent decades, the grip of physicalism has loosened and, perhaps because of this, philosophers are again engaging with the philosophical problem of emergence. However, because of our decades' old practice of outsourcing fundamental ontology to physics, there had been relatively little philosophical engagement with the question of what counts as fundamental until very recently. Thus, philosophical debates concerning emergence have taken place in a context where physicalism dominated discussions.

Although physicalism has dominated the conversation about emergence, the problem of emergence can be articulated independently of the kind of fundamental ontology one holds. This is good news as it means that it is possible to think about the concept of emergence without having settled all other metaphysical questions ahead of time. We can begin with a common sense account of emergence as the concept operates in philosophical and scientific discourse. Then we can proceed to get clearer on the implications of the concept of emergence for matters that concern us in scientific practice. Specifically, we are concerned with the relationship between emergent properties and the challenge of scientific modeling.

Philosophical usage of the term "emergence" usually marks a single problem that can be stated very simply:

> Do genuinely novel entities or properties come into being over the course of natural history or are all facts determined by the basic facts so as to be explainable (at least in principle) in terms of those basic facts?

Although the question is easy to pose, providing a well-justified answer has proven to be a persistent conceptual challenge.[5] This question is of practical relevance to scientists and engineers in settings where we encounter complexity.[6] One reason that this is such a difficult problem involves the clash between ordinary common sense and what we can call scientifically informed common sense. Part 1 explains the conflict between these two ways of thinking.

Scientific interest in emergence is driven by the assumption that emergent properties and phenomena are real and relevant. For philosophers steeped in the doctrine of

---

[5]For a detailed discussion of the conceptual problem see Kim (1999), Symons (2002), and the philosophical (rather than the scientific) papers collected in Bedau and Humphreys (2008).

[6]For a clear and helpful account of the meaning of the term "complexity" see Mitchell (2009).

physicalism, that assumption is precisely the point of contention. In the second half of the twentieth century, most philosophers simply denied that there are really emergent properties. Until recently, philosophical reflection on emergence has focused on the challenge of understanding how one can simultaneously believe that physics provides the fundamental and maximally general story concerning the nature of being while at the same time believing that emergent properties are genuinely real. Scientific investigation of the problem of emergence from the 1990s to the present has, for the most part, pragmatically sidestepped the metaphysical questions focusing instead on explaining or modeling relationships between putatively emergent properties and their predecessors.[7] Although pragmatism might be a sensible strategy in short- to medium-term scientific research, it leaves the deeper and more basic philosophical questions unaddressed. In the sciences, as in philosophy in the late twentieth century, the implicit attitude was to defer the deepest questions to the physicists.

WHERE DOES THE PHILOSOPHICAL PROBLEM OF EMERGENCE COME FROM?

Common sense is not free from ontological questions. In ordinary experience, we puzzle over the ontological status of holes, shadows, reflections, and the like, and as we try to organize the inventories of our lives, we might wonder how to classify and count the objects that are of interest to us. Even in commercial contexts, considerable energy is expended thinking through ontological questions. The multinational oil company Shell, for example, was forced to develop its own ontological system in order to understand its own complex organization and to avoid inconsistency and waste.[8]

Common sense encourages us to believe in things like oil rigs, pipelines, dogs, minds, countries, and economies. We are inclined to think that a world without such things would have a smaller inventory of real objects than the actual world. If we imagine a scenario in which all dogs died yesterday, we tend to think that such a world would contain fewer things than actually are. Even though all the mass and energy that made up our dog will still be present in its corpse and local environment, we still believe that his death is a loss. In what sense is a dog something more than the sum of its mass and energy? Perhaps, we want to say, dog-like organization or structure is something real, over and above fundamental matter and energy. In ordinary life, it is natural for us to think of dogs as real. If pressed, we might qualify our view a bit,

---

[7] Take for example Herbert Simon's position on the metaphysics of emergence. He argues that "By adopting this weak interpretation of emergence, we can adhere (and I will adhere) to reductionism in principle even though it is not easy (often not even computationally feasible) to infer rigorously the properties of the whole from knowledge of the properties of the parts. In this pragmatic way, we can build nearly independent theories for each successive level of complexity ... " (1996, 27).

[8] Matthew West was one of the leading figures in the development of Shell's ontology. His work on the kinds of questions that it was necessary to answer as part to the management of the Shell's highly complex business demonstrate the direct role that philosophical ontology plays in the most hard-headed business contexts. See, for example, West (2009, 2011).

insisting instead that dog patterns or dog information is real. But we also recognize the difficulty of grasping the ontological status of a dog-like arrangement of parts? Would the arrangement or pattern exist independently of our ways of knowing and thinking? Do dogs make a difference to the world over and above the difference that dog parts make? Is the dog the same as its fur? If so, which fur? Presumably not the fur that he has shed.

The reason we feel that we can dismiss questions like these is because we believe that the genuinely real stuff in the universe is the matter–energy that physics studies. Thus, in spite of ordinary common sense, we are hesitant to state categorically that the animals that veterinary science and zoology studies are as real as quarks and gluons. According to physicalists, physics is the science that tells us what exists; it is the source of (most of) our ontological commitments.[9] On this view, all the other sciences, from chemistry all the way up to psychology and economics, derive what ontological legitimacy they have in virtue of the derivability of their ontologies from the fundamental physical story.

Recent decades have shown that it is extremely difficult to be fully committed to physicalism. Extreme versions of physicalism face three kinds of problem. The first, known as Hempel's dilemma, challenges the idea that our grasp of physics is good enough to serve as the source of our ontological commitments in the first place.[10] Hempel pointed out that we are certainly not committed to the ontology of the physics of the past as we believe that past physics contained numerous falsehoods. By induction, we can be reasonably sure that the physics of the present day is not perfect and that it will undergo correction. Thus, it would be unwise to look to present-day physics for our finished account of what exists. Presumably future physics, let us call it the ideal finished physics, contains the correct ontology. However, the problem with future physics is that we do not know what its ontology contains. It might be the case that future physics contains elements, like qualitative experience, for example, that current physicalists would reject. At the very least, it seems pointless for the physicalist to commit herself to the ontology of the ideal finished physics when she is unable to know what it is.[11]

The second problem for physicalism derives from the physicist's need for some kind of mathematics and the puzzle that the ontology of mathematics poses.[12]

---

[9]Even the most die-hard physicalist will have a hard time avoiding ontological commitments to selves and sets. For example, Quine (1960a) famously endorses an ontological commitment to sets. Precisely what is involved in ontological commitment can be understood in a variety of ways. The simplest way to understand the ontological commitment's that come with holding some theory is as the kinds of entities that are required in order for the sentences of a theory to be true. For an overview of the difficulties associated with this way of thinking about ontological commitment, see Rayo (2007).

[10]See Hempel (1969).

[11]It is possible to commit oneself to the answers provided by a process that one trusts. For example, I will buy roughly as many donuts as the number of people who RSVP for the breakfast meeting. However, the fact that one might be committed to believing the results of future scientific inquiry is independent of one's position with respect to current ontological disputes. One cannot preclude the possibility that physics will settle on an ontological framework that supports one's opponents.

[12]See Benacerraf (1965) for the classic presentation of the challenge of providing both a normal semantics for mathematical statements and a causal theory of mathematical knowledge.

A third is the role of subjectivity, specifically, the challenge of reconciling conscious experience with the physicalist worldview. Most of us believe that our minds are real and that we and some other animals have conscious experiences. We can entertain a thought experiment wherein we imagine a possible world that features the kinds of brains and behavior that we have in the actual world, but whose inhabitants lack any qualitative experience accompanying their brains' processes and structures, we are inclined to say that these possible people are missing something that we regard as genuinely real.[13] But if consciousness – or more precisely the qualitative dimension of phenomenal experience – makes no difference in the causal economy of the physical world, how can we be so confident that it exists?[14] Either consciousness does make a causal difference somehow, or the idea of unique causal powers as a criterion for reality is incorrect, or we are simply deluded about our own conscious experience.

In the face of considerations like those sketched above, Daniel Stoljar concludes that it is not possible to formulate a coherent and nontrivial version of physicalism. "The bad news is that the skeptics about the formulation of physicalism are right: physicalism has no formulation on which it is both true and deserving of the name. The good news is that this does not have the catastrophic effects on philosophy that it is often portrayed as having in the literature" (2010, 9).

Our pragmatic impulse is to be inclusive when deciding what kinds of things are real. "Of course dogs are real!" says the exasperated voice of common sense. Even if we are not committed physicalists, in our ontological judgments, ordinary common sense is opposed by another kind of common sense. What we might call *scientifically informed common sense* is not precisely identical with physicalism, but they share a common deference to scientific practice. Rather than being a clearly articulated philosophical thesis, of the kind that physicalists hoped for, scientifically informed common sense can be seen as a set of methodological commitments. It involves preference for reductive explanations, anti-supernaturalism, and some rough metaphysical commitments concerning causation and individuation that are drawn from conservation principles in physics. We can think of it as a disposition toward certain kinds of ontological commitments and explanations rather than a clearly defined philosophical position. A rough list of kinds of claims that scientifically informed common sense endorses runs as follows:

- there are no non-physical causes
- the physical world is all there is (more or less ... maybe (parts of) mathematics are real too)
- the current contents of the physical world are nothing more than a rearrangement of the stuff which existed during the big bang
- to be real means to make a real (and unique) causal difference to the way the world is

[13]This is David Chalmers' famous zombie argument against the identification of conscious experience with brain states (Chalmers, 1996).
[14]One answer here is that we are simply deluded about actually having conscious experiences. Dennett provides the most sophisticated account of what it could mean for subjects to tell themselves that they are conscious (Dennett, 1991).

It is important to note that the notions of causation, completeness, and reality that undergird scientifically informed common sense are not defined or articulated in detail. These three interlocking concepts are not straightforwardly scientific in nature, but are, instead, metaphysical, or at least conceptual. Currently, opposition to emergentism is not posed by physicalism per se, but rather by this more nebulous and poorly defined set of commitments that I am calling scientifically informed common sense.

INCOMPLETENESS

Although physicalism does not present a viable alternative to emergentism, this does not mean that emergentists can declare victory. The challenge to their view involves articulating the relationship between fundamental and emergent properties no matter what the account of fundamentality that we settle on.

What do we mean by fundamentality? We can begin with a sample of kinds of fundamentality relations that we might consider. Three familiar candidates are composition, governance, and determination.[15] When we think about parts and composition, the fundamentals would be the basic micro-constituents and their possible composition relations. Let us call this Part-fundamentality, or P-fundamentality for short. Philosophical atomism is the most familiar kind of P-fundamentality. For atomists, the basic components of nature have a fixed character. Although atoms are not themselves subject to change, all change can be explained in terms of the changing relations among the atoms. On this view, atomic participation in new mereological sums explains everything.

Another kind of fundamentality concerns laws and governance. In this case, we list the laws that govern nature and specify their relative authority with respect to one another. Here, the fundamentals would be the laws that govern our system with maximal generality. Notice that maximally general laws need not be the laws governing the P-fundamental parts alone. If we did posit the maximally general laws as governing the P-fundamentals, those laws would be equivalent to what are sometimes called micro-governing laws (Huttemann, 2004). However, there are other ways of understanding maximal generality that do not involve P-fundamentality at all. If we claim, for instance, that it is a law of nature that nothing travels faster than the speed of light, this law applies equally to Volkswagens as to quarks. Similarly, what Clifford Hooker has called basic and derived structural laws can be maximally general without being solely micro-governing, insofar as the scope of the quantifiers in these laws is not restricted to the micro-constituents, but can include, for instance,

[15]For the purposes of this paper I will not discuss the currently popular notion that fundamentals stand in a *grounding* relation to other properties. This is because, as Wilson (2014) argues, the idea that there is a grounding relation that does something other than standard relations of composition, governance, determination, or perhaps reduction has not been established. More importantly, I think that no matter where one settles with respect to the grounding relation, the question of the relationship between fundamental and emergent properties will remain. This is a topic for another paper though!

the entire universe (Hooker, 2004). Let us call fundamentality that is concerned with governance, L-fundamentality.[16]

Given the varieties of fundamentalism and given that one's view of what counts as emergent will be relative to one's view of fundamentality, it is prudent to maintain a provisional stance with respect to what counts as emergent. In fact, we can allow our inventory of the factors determining the basic level to discoverable a posteriori. We can make the case that our story about what counts as fundamental will change over time depending on the outcome of inquiry. Most obviously, we should leave room for a range of views about micro-determining factors. Micro-determination could take place in a variety of ways that we have not yet considered. By including a third, non-specific placeholder kind of fundamentality (we can call it H-fundamentality), we preempt the objection that the present analysis does not allow for a posteriori discoveries with respect to micro-determination.

At this point, we can present the combinations of types of fundamentality along the following lines: We could let P_N stand for those properties that are novel with respect to the composition relation operating over the basic parts of our ontology. Let P_C stand for those properties that are not novel with respect to P-fundamentality. Let L_C, L_N, H_C, and H_N be defined in the same manner. We can see that with these three fundamentality predicates in place, we can form combinations in a way that already has some useful expressive power. We could then begin to match positions with respect to emergence with the table of eight possible combinations of the three kinds of fundamentalism, as shown in Figure 1.1. The first place on the table is equivalent to an eliminativist view where all novel features are denied. The denial of novelty in the first place would be an implausibly strong kind of anti-emergentism. This would reflect,

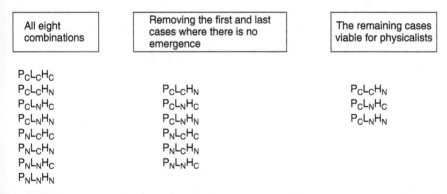

All eight combinations	Removing the first and last cases where there is no emergence	The remaining cases viable for physicalists
$P_C L_C H_C$		
$P_C L_C H_N$	$P_C L_C H_N$	$P_C L_C H_N$
$P_C L_N H_C$	$P_C L_N H_C$	$P_C L_N H_C$
$P_C L_N H_N$	$P_C L_N H_N$	$P_C L_N H_N$
$P_N L_C H_C$	$P_N L_C H_C$	
$P_N L_C H_N$	$P_N L_C H_N$	
$P_N L_N H_C$	$P_N L_N H_C$	
$P_N L_N H_N$		

FIGURE 1.1 Combinations of novelty and naturalness with respect to kinds of fundamental property.

---

[16]Neither of these standard views captures the account of structural fundamentality that Sider has recently defended. I won't have time to discuss this in detail, but suffice to say that Sider sees structure as equivalent to nature's joints. The structural characteristics of nature, on his view, are accessible to us through common sense, mathematics, and physics. We could call his view S-fundamentality. Since Sider's account is new enough not to have figured in these debates, we can leave it out for now.

for example, the position that Quine takes with respect to most ontological questions (see the discussion above). However, the last place on the table where $P_N L_N H_N$ is also a denial of emergence since, although it embraces novelty, it denies naturalness, asserting the existence of properties with no connection to the fundamentals whatsoever.

In the case of the remaining six spots on our table, each represents a combination of novelty and naturalness with respect to the three fundamentals listed. Note also that some of the combinations in a list like this will be either self-undermining or will be judged to be implausible metaphysical positions for some reason or another. So, for example, most contemporary physicalists will want to hold onto micro-constitution and will stick to the top half of the table. Given the three types of fundamentality presented here and given the usual physicalist qualms, there will be only three options left for the emergentist; the second, third, and fourth slots on the table.

From the remaining three slots on the table, for example, a $P_C L_C H_N$ property would be one that obeys micro-laws, is constituted by micro-constituents, but is determined by non-micro-level factors. For instance, when complexity theorists talk about emergence, they often suggest that "Systems with a higher level of complexity emerge from lower level entities in new structural configurations," here the usual reductionist composition relations and micro-governing laws are still in effect, and yet such phenomena are thought to be novel by virtue of their novel structural properties. Specifically, the new structure is taken to be the factor that determines the appearance of the emergent property.

Of the remaining three accounts of "emergent" properties, the first is that emergent properties obeys micro-laws and are constituted by micro-constituents but are determined by factors other than those at the micro-level ($P_C L_C H_N$). The second is that laws of basic physics do not govern the emergent properties but those properties are determined by the micro-constitutents and micro-factors ($P_C L_N H_C$). Finally, in the third case, an emergentist who is also a physicalist could assert that emergent properties are constituted by the microphysical constituents but are neither determined nor governed by microstructural laws or determiners ($P_C L_N H_N$).

This is an exercise in a kind of conceptual bookkeeping with respect to fundamentals that invites us to reconsider the problem of emergence in terms of clearly articulated accounts of what it is that we think metaphysical fundamentals are doing in our theories. If there are other candidate kinds of fundamentality or relations to fundamental properties, they could be added, generating a new table of positions along the lines described here.

Completeness is essential to arguments against the possibility of strongly emergent properties. There are a variety of ways one could imagine carving up the question of completeness for a metaphysical system. One might regard completeness as equivalent to capturing all true causal judgments, all facts more generally, all the right ways of carving up the world, the capturing grounding relations, and so on. At the very least, to determine completeness, we need to decide on the relevant set of truths that we hope to capture and we need to specify as precisely as possible the set of objects, relations, laws, and so on that the metaphysical position proposes. This is an unwieldy *interpretive* task for most nontrivial metaphysical systems. However, as I have suggested elsewhere, for the purposes of understanding completeness, we can treat metaphysical frameworks abstractly by recasting them as *generative fundamentals* (Symons, 2015). Generative fundamentals are the set of total states of a world and the list of possible transformations on that set. They should be thought of as being a set of unexplained explainers. Elsewhere, I explain in detail how to think of a metaphysical framework in terms of generative fundamentals for the purpose of determining whether it is complete (Symons, 2015).

From the anti-emergentist perspective, fundamental theories provide a total account of the behavior of a system. For example, Lewisians argue for the view that all facts (the entire actual factual landscape) are entailed by the most basic facts (1986, 15). On this view, there exists a determination relation of some kind between the basic facts and all the other facts.

In these discussions, it is worth noting the burden that is being assigned to the fundamentals and how easily we can slip from a claim about supervenience or necessitation at an instant to the claim that all past, present, and future facts are packed into the facts about the fundamentals at any instant.

One way to understand the behavior of a system is to specify the possible states it can occupy and to provide some account of how the system changes from state to state. If the anti-emergentist is right, listing all the fundamental facts would suffice for generating such an account. With this in mind, we can reconsider the problem of emergence in terms of the relationship of the putatively emergent property to some specified set of states and transformations. Different kinds of fundamentality will result in different sets of states and transformations, whether a property is emergent will be determined relative to that set of states and transformations. A complete set of states and fundamentals will have no emergent properties associated with it.

When it comes to the problem of determining completeness, there are two significant challenges. First, there is the challenge of excluding the kinds of interactions between states and transformations that lead to truths about the system that are not derivable from the generative fundamentals. Prospects for finding a principled way to accomplish such an exclusion for metaphysical systems of sufficient complexity to be of interest are bleak.

One could simply define the space of relevant facts recursively in the following way:

- If a fact is fundamental, it is a true reachable fact.
- If a fact can be obtained from true reachable facts by means of the rules of the generative fundamentals, it is a true reachable fact.
- The set of true reachable facts is the smallest set of facts satisfying these conditions.

In any suitably interesting or complex set of generative fundamentals, there will be at least one non-reachable fact. This is probably the case for arithmetical facts as Gödel's proof of the incompleteness of the system of *Principia Mathematica* demonstrated (1931). The anti-emergentist might decide to exclude facts about mathematics from his claims about the domain of reachable facts. At that point, one could imagine trying to exclude the possibility of genuinely novel properties using two general strategies. First, one could exclude unanticipated results of interaction by positing a system that anticipates (or claims to anticipate) all possible interactions ahead of time (the Leibnizian way) or by claiming that one's system does not have potentially problematic interactions between states of the system (the Humean way).[17] Both ways are unattractive for reasons I discuss in Symons (2015). In that paper, I argue that any approach that attempts to rule out the existence of emergence a priori is simply unscientific.

Unanticipated interactions are difficult to prevent completely without creating systems that are so trivially simple as to generate little to no insight. In Boschetti and Symons (2011), we attempt to demonstrate how even a trivially simple computational system can generate unanticipated and novel interactions.

In addition to preventing or anticipating interaction, the proponent of metaphysical completeness faces a stranger problem, namely the problem of transients. A transient is defined as a state or sequence of states or a subset of that sequence of states that has a first member.[18] This way of understanding transients is similar to the concept of transients as they appear in a Markov chain analysis. In a Markov chain, if there is some non-zero probability that the system will never return to a state, we say that this state is transient. If a state is not transient, it is recurrent.[19] For any generative fundamentals, F the possibility of transients entails that F might have resulted from some other generative fundamentals $F*$.[20]

There will be some cases where $F*$ is epistemically inaccessible from the perspective of agents in some system governed by F. More intuitively, for any system or universe that we imagine completely captured by some generative fundamentals, we cannot exclude the possibility that the set of fundamentals themselves are the result of some non-repeating process – a transient – that is not part of that set. One could imagine a simple series of states in some oscillating universe, for example, where the denizens live between a Big Bang and a Big Crunch. They might have developed a cosmological theory that correctly predicts all the truths of the future of their universe and perhaps does a good job retrodicting the past states of the universe as well. However, the apparent completeness of this account is threatened by the possibility of a transient that was part of the history of the universe, but not part of the cycle of bang and crunch into which their universe has settled.[21]

Properties in some system governed by F can be such that, relative to the successor or predecessor system, they can be called emergent. The kind of emergence exhibited by these systems can be called strongly emergent insofar as the novel system's generative fundamentals differ from the system that preceded it. In this sense, apparent completeness at the level of generative fundamentals governing the later system would not be sufficient to account for all the metaphysically basic features of reality.

[17] I discuss these alternatives in detail in Symons (2015).
[18] Booth (1967).
[19] For an overview of Markov chains see Booth (1967).
[20] This section repeats some of the discussion of transients in Symons (2015).
[21] See Symons (2015) for more elaboration on the relationship between emergence and the idea of incompleteness for generative fundamentals.

The purpose of the argument from transients is simply to note a limitation on attempts to use the completeness of some set of generative fundamentals as the basis for an argument against emergence.

At this point, the advocate of fundamentalist metaphysics might respond that one can opt for an a posteriori view of the fundamentals such that whatever this additional extra emergent something is, it can simply be added to the proposed list of fundamentals in order to ensure completeness.[22] As argued by Hempel (1969) and later by Symons (2015), an ad hoc strategy of adding to the list of fundamentals as required by new evidence is insufficient if one's goal is to defend something like physicalism or any other metaphysically complete account of the fundamentals against the possibility of emergence. Given the possibility of transients, one's metaphysics can fail with respect to the project of generating a complete list of fundamentals even when we allow our account of the fundamentals to be modified a posteriori. The possibility of an incomplete fundamental metaphysics turns out to be unavoidable and cannot be remedied by the addition of extra principles or categories. This is because, as we have seen even in cases where the present and future states of the natural world appear to have been completely captured by some set of fundamental principles, the possibility that these principles themselves are the result of the process of emergence cannot be excluded. Of course, what this means is that the possibility of emergent properties is simply an indication that the ambitions of theorists hoping for a fundamental theory can always be dashed.

MODELS, EMERGENCE, AND FUNDAMENTALITY

As we have seen, emergent properties are not indicators of trouble with respect to scientific explanation. However, they run counter to the ambition of metaphysical fundamentalism. In the engineering and modeling context, when we refer to emergent properties of systems, we are often referring to those that were not originally intended as part of the functional characterization of the system. The unplanned behaviors of a system or the unintended uses to which these systems are put are referred to as emergent.[23] This is a straightforward analog to the more abstract treatment of properties that are not (and perhaps could not be) anticipated given what I called the generative fundamentals alone. Building on work by John Holland, 1992 and others, Das et al. (1994) defines emergent computation as "the appearance in a system's temporal behavior of information-processing capabilities that are neither explicitly represented in the system's elementary components or their couplings nor in the system's initial and boundary conditions." There are a wide variety of accounts of emergence in the scientific literature. The key to all of them is the role of either limitations on the explanatory power of some existing system or the role of interactions as the source of novelty.

---

[22]Or the fundamentals can be modified in some other way in order to ensure completeness.
[23]Following Crutchfield and Mitchell some computer scientists have proposed actively exploiting features of emergent computing via the use of genetic algorithms.

Scientific models are intended to provide guidance with respect to explanations and predictions of emergent properties or to offer possible interventions that would allow control over those properties. The relation between these models and the philosophical problem of emergence is multifaceted and challenging. As we have seen, our desire for theoretical completeness faces a set of hard limits. The fact that these limits characterize of attempts to build complete systems in general is directly relevant to the engineering and modeling community. Discussions of novelty and emergence in complex systems science and computer science often take place in a way that obscures the central problem of defining the boundaries of the systems under consideration. In Boschetti and Symons (2011), we argue that the conceptual features of the problem of interaction can be characterized in a straightforward and non-question begging way. We argue that in scientific and engineering contexts, the specification of system or model boundaries determines our commitments with respect to the nature of interaction and the possibility of genuinely novel causal powers for emergent properties.

The science of emergence involves understanding the role of system boundaries and interactions. Part of this task involves being explicit about what counts as fundamental in one's model. With respect to what properties are the putatively emergent properties emergent? Properties that are called emergent will be understood as such relative to those features of the model taken to be fundamental and different scientific domains will understand this differently. The conceptual relationship between the emergent and the fundamental will be the same across different scientific domains, but they will vary with respect to what is taken to be fundamental and what is understood to count as a system boundary.

This chapter discussed the limitations of one prominent account of ontological fundamentality, physicalism. We went onto provide a general characterization of fundamentality, explaining the challenges faced by the anti-emergentist versions of fundamentalism. Anti-emergentist approaches must claim completeness, must block potentially novel interactions, and must make claims about epistemically inaccessible pasts. These are significant challenges for the anti-emergentist that force him into the unscientific position of having to exclude unwanted possibilities a priori. We concluded by asserting that scientific uses of the term emergence depend on stipulating as precisely as possible the nature of system boundaries and being as clear as possible about what one takes to be fundamental in the systems in question.

ACKNOWLEDGMENTS

I would like to thank the editors and referees for very valuable criticism.

REFERENCES

Alexander, S. (1920) *Space, Time, and Deity: The Gifford Lectures at Glasgow, 1916–1918*, vol. **2**, Macmillan.

Baker, A. (2010) Simulation-based definitions of emergence. *Journal of Artificial Societies and Social Simulation*, **13** (1), 9.

Bedau, M.A. (1997) Weak emergence. *Noûs*, **31** (s11), 375–399.

Bedau, M. and Humphreys, P. (2008) *Emergence: Contemporary Readings in Philosophy and Science*, MIT Press.

Benacerraf, P. (1965) What numbers could not be. *The Philosophical Review*, **74** (1), 47–73.

Błaszczyk, P., Katz, M.G., and Sherry, D. (2013) Ten misconceptions from the history of analysis and their debunking. *Foundations of Science*, **18** (1), 43–74.

Booth, T.L. (1967) *Sequential Machines and Automata Theory*, vol. **3**, Wiley, New York.

Boschetti, F. and Symons, J. (2011) Novel properties generated by interacting computational systems. *Complex Systems*, **20** (2), 151.

Broad, C.D. (1923) *The Mind and its Place in Nature*, Kegan Paul, London.

Chalmers, D.J. (1996) *The Conscious Mind: In Search of a Fundamental Theory*, Oxford University Press.

Das, R., Mitchell, M., and Crutchfield, J.P. (1994) A genetic algorithm discovers particle-based computation in cellular automata, in *International Conference on Parallel Problem Solving from Nature*, Springer, Berlin, Heidelberg, pp. 344–353.

Dennett, D.C. (1991) *Consciousness Explained*, Little, Brown.

Gödel, K. (1931) Über formal unentscheidbare Sätze der Principia Mathematica und verwandter Systeme, I. *Monatshefte für Mathematik und Physik*, **38**, 173–198.

Hempel, C. (1969) Reduction: ontological and linguistic facets, in *Essays in Honor of Ernest Nagel* (eds S. Morgenbesser *et al.*), St Martin's Press, New York.

Holland, J.H. (1992) Genetic algorithms. *Scientific American*, **267** (1), 66–73.

Hooker, C.A. (2004) Asymptotics, reduction and emergence. *The British Journal for the Philosophy of Science*, **55** (3), 435–479.

Huttemann, A. (2004) *What's Wrong with Microphysicalism?* Routledge.

Kim, J. (1999) Making sense of emergence. *Philosophical Studies*, **95** (1), 3–36.

Lewis, D. (1986) *On the Plurality of Worlds*, Oxford.

Mitchell, M. (2009) *Complexity: A Guided Tour*, Oxford University Press.

Putnam, H. (1975) *Philosophical Papers: Volume 2, Mind, Language and Reality*, vol. **2**, Cambridge University Press.

Quine, W.V. (1960a) Carnap and logical truth. *Synthese*, **12** (4), 350–374.

Quine, W.V. (1960b) *Word and Object*, MIT Press.

Rayo, A. (2007) Ontological commitment. *Philosophy Compass*, **2** (3), 428–444.

Simon, H.A. (1996) *The Sciences of the Artificial*, MIT Press.

Stoljar, D. (2010) *Physicalism*, Routledge.

Symons, J. (2002) Emergence and reflexive downward. *Principia*, **6** (1), 183.

Symons, J. (2015) Physicalism, scientific respectability, and strongly emergent properties, in *Cognitive Sciences: An Interdisciplinary Approach* (eds D. Tudorel and L. Mihaela), Pro Universitaria, Bucharest, pp. 14–37.

Symons, J. (forthcoming) Brute facts about emergence, in *Brute Facts* (ed. E. Vintiadis), Oxford, Oxford University Press.

West, M. (2009) Ontology meets business, in *Complex Systems in Knowledge-Based Environments: Theory, Models and Applications* (eds A. Tolk and L.C. Lain), pp. 229–260.

West, M. (2011) *Developing High Quality Data Models*, Elsevier.

Wilson, J.M. (2014) No work for a theory of grounding. *Inquiry*, **57** (5–6), 535–579.

2

EMERGENCE: WHAT DOES IT MEAN AND HOW IS IT RELEVANT TO COMPUTER ENGINEERING?

Wesley J. Wildman[1,2] and F. LeRon Shults[3]

[1] *School of Theology, Boston University, Boston, MA 02215, USA*
[2] *Center for Mind and Culture, Boston, MA 02215, USA*
[3] *Institute for Religion, Philosophy and History, University of Agder, Kristiansand 4604, Norway*

SUMMARY

This chapter begins with a primer on basic philosophical distinctions pertaining to emergence: weak emergence, two types of strong emergence depending on affirmation of supervenience, and several types of reductionism. These distinctions are then related to the mathematics of complex dynamical systems and to the concerns and accomplishments of computer engineers, particularly in the domain of modeling and simulation. We contend that computer engineers are (perhaps unintentionally) effecting a transformation in philosophical disputes about emergence, steadily promoting confidence in theories of weak emergence and undermining the plausibility of both types of strong emergence. This is occurring through the systematic production of probable explanatory reductions using computer models capable of simulating real-world complex systems.

INTRODUCTION

Very few philosophers are interested both in analyzing the concept of emergence and in understanding what computer engineers are up to. Some of our heroes, all of whom

Emergent Behavior in Complex Systems Engineering: A Modeling and Simulation Approach,
First Edition. Edited by Saurabh Mittal, Saikou Diallo, and Andreas Tolk.
© 2018 John Wiley & Sons, Inc. Published 2018 by John Wiley & Sons, Inc.

are better philosophers than us, fall into this category. However, we were the only two philosophers that the editors of this volume could convince to write a chapter in just 2 weeks while organizing conferences, traveling around the globe, and preparing classes for a new semester. Despite these crazy circumstances, we are happy for the opportunity to share our thoughts and hope they prove useful for computer engineers who may turn to this volume for insight into the conceptual puzzles surrounding emergence.

We are delighted that this team of authors has decided to take on this subject. In some fundamental sense, computer engineering is deeply entangled with the problems of complexity that the term *emergence* was coined to describe. And unlike the mathematical study of complexity in nonlinear dynamical systems, computer engineering is about real-world tractability, which gives it a refreshingly practical dimension. Despite being philosophers, we like being practical, and we are personally quite fond of a fair number of engineers, so we are not overstating the case when we say we are delighted at the opportunity to contribute to this volume.

We begin by outlining some of the varied meanings of the heavily contested term *emergence*. Analysis of the philosophical discussions around this term might help us penetrate more deeply into the conundrums that human beings face as they try to make sense of themselves and the world around them. Sketching some of the most common definitional debates and their implications may help computer engineers locate themselves in the wider context of these discussions.

We then address the mutual relevance of emergence and computer engineering. To repeat, we are not computer engineers, and so we initially felt a little hesitant to suggest lines of relevance in either direction. But, clearly, we overcame these feelings. We have enough experience working with computer engineers, especially in the tasks of building computer simulations, to think we have an inkling of the kind of connections that may prove meaningful. We also think that this sort of dialog among computer scientists and scholars from other fields can yield rewarding insights and surface new multidisciplinary methodologies (Wildman *et al.*, 2017). So, yes, we are going there – where more angelic (or less foolish) philosophers might fear to tread.

We conclude with remarks on some possible future directions for scientific research that focuses on the sort of complex systems where emergent phenomena are prominent. It is probable that there is a hard limit on the extent to which any type of research can decisively resolve metaphysical disputes in interpretations of emergence. Nevertheless, we suspect that sustained research will probably wind up supporting one conception of emergence as more plausible than competitor interpretations. After that, poorly supported interpretations might still be affirmed by some scholars, but only through special pleading, and for reasons that are unrelated to the prodigious achievements of computer engineering.

Emergence and related words such as supervenience are terms of art in philosophy, especially Anglo-American analytic philosophy. There are types and subtypes of emergence, various degrees of "strength" within the types, and other differentiations that depend on modal operators, possible worlds semantics, or other auxiliary concepts employed to render a conceptually precise definition.[1] The term *emergence* points to a conceptual tangle that has been keeping analytical philosophers busy for decades.

Computer engineers are very likely going to demand a reason for taking the time to elaborate such a host of definitions and distinctions. In good pragmatic manner, the engineer needs to know what difference such fine distinctions could possibly make. Fair enough.

Philosophers develop arrays of definitions and distinctions to clarify concepts. That can lead to applying duly clarified concepts to the more practical task of describing the world, but often enough philosophers content themselves with fine-grained debates over concepts and meanings. For their part, engineers are prone to embrace any readily intelligible definition and then immediately start applying it regardless of the conceptual chaos that may result. A few engineers, such as those editing this volume, may later decide that they need to tidy up their conceptual house and take on the task of persuading a larger number of engineers to take their housekeeping efforts seriously.

So, let us grant at the outset that we all – including engineers – can benefit from *some* of what philosophers feel driven to do in their special kind of conceptual policing. And let us admit that most philosophers do not build or fix anything and that if organizing human life was entirely up to them, we would likely not have very developed civilizations, although we would be extremely thoughtful and conceptually clear about whatever we did have. In other words, computer engineers (and everyone else) probably do not need to devote a lot of attention to *everything* that philosophers have to say about emergence.

With that in mind, our aim here is to introduce the philosophical concept of emergence in a way that focuses on the possible applications of the concept in computer engineering.

MAPPING MEANINGS OF EMERGENCE

Emergence is an old word with a complex array of meanings. Its Latin root is *emergere*, which is a compound of *e-* (out) and *mergere* (dip), suggesting arising of a hidden object out of some fluid – the opposite of submerge. Those origins reach back in the deep past of Indo-European languages. In English, usage dates back to the seventeenth century. A spin-off meaning is the arising of something unexpected requiring urgent action, which is the line of usage yielding the English word *emergency*. The basic idea is that we cannot see the hidden thing coming but it gradually manifests itself, demanding attention.

---

[1] The most sweeping overviews are found in the Stanford Encyclopedia of Philosophy; see the discussion of emergence in "Emergent Properties" (https://plato.stanford.edu/entries/properties-emergent/) and the article on Supervenience (https://plato.stanford.edu/entries/supervenience/). See also Bedau and Humphreys (2008).

In the nineteenth century, scientists applied this basic idea specifically to physical systems, first in medicine, then in geology, and then in biology. By the beginning of the twentieth century, philosophers were using the idea to describe a captivating and powerful evolutionary view of the world, which charts a course all the way from inanimate matter to self-conscious beings with moral codes, institutional commitments, and intricate cultures – with lots of other things emerging at various levels along the way.

These early philosophical discussions sought to distinguish between products that *result* from causes, where we can spell out the causal story, and products that *emerge* from causes, where no amount of cleverness leads to a clear causal story. Philosophers quickly found themselves resorting to the concept of levels as they sought to integrate scientific analyses of different physical systems into a unified scientific account of nature. Instead of being baffled as to how self-awareness emerges from the physical material and energy of the earlier universe, which is a very big question, we can break down our bafflement into bite-sized pieces that ask how one level of reality emerges from another.

These days philosophers often express the intractable big question of emergence in a series of more tractable problems, organized by levels: How do atoms emerge from the early universe? How do galaxial structures emerge from scattered atoms? How do stars and planetary systems emerge from galaxial clouds? How do life-friendly ecosystems emerge from the hostile conditions of early planetary bodies? How do large molecules emerge from soupy fluids? How do replicating cells and organelles emerge from large molecules? How do multicellular organisms emerge from single cells? How do intelligent brains emerge from brainless bodies? How does self-consciousness emerge from brain-equipped bodies? And how do traffic jams emerge from self-aware culture creators? We are so committed to the concept of levels that we break our scientific endeavors into disciplines organized by levels and grapple with questions of emergence in micro-form, slowly building up a detailed story of how traffic jams emerge from the early universe.

What precisely do we mean by *levels* as a way to break down the deep puzzle of emergence into a series of smaller puzzles? Levels of what? Evidently, there is some quantitative conception underlying our use of levels. The best candidate concepts to make sense of levels are "complexity" and "information," and ultimately the two probably yield the same metric. Levels rise with increasing complexity and denser information. Thus, emergence can be framed as the arising at one level of something unexpected, demanding explanation, from another, lower level. The emergent "thing" will have distinctive properties that do not apply to the lower level, as when we say that wetness is a property of many water molecules at an appropriate temperature but is not a property of the individual molecules. The Australian-born British philosopher Samuel Alexander (1859–1938) had already come this far in his thinking by the early twentieth century (Alexander, 1920). We have to pay attention to the intellectual contributions emerging from that gigantic island with so few people and so many dangerous creatures!

Fairly early in twentieth century philosophical debates over emergence, two large and still unresolved problems came to the fore (not to say they emerged). The first problem is already evident in Alexander's writings and bears on the idea of the "strength" of different types of emergence. The second problem takes a while longer to show up, but when it does, it transforms the debate: this is the philosophical encounter with nonlinear dynamical systems in mathematics. We discuss both in what follows.

Strong and Weak Emergence

When we ponder the scientific reduction of the unmanageably large question of emergence into a series of smaller questions sorted by levels of complexity, we quickly notice differences in degrees of intractability. These differences in the difficulty of micro-problems of emergence have inspired, and still inspire, energetic debates about the *strength* of the concept of emergence needed to make sense of what we encounter in nature.

To illustrate, consider the emergence of atoms from the early universe. In this case, scientists can understand the emergence process in fine detail. In particular, we have serviceable mathematical models of symmetry breaking for the prismatic separation of fundamental forces, for the stabilization around low-energy first-generation of matter, and for the phase transition from a radiation-energy-dominated universe to a matter-energy-dominated universe. There are plenty of details yet to be sorted out (dark matter, dark energy, surplus of matter over anti-matter, the status of a multiverse, etc.), and the whole scheme may change in some sense, but much of the mystery has dissolved, to be replaced with wonder at the beautiful implacability of the process.

Again, consider the gas laws of thermodynamics, particularly the law pertaining to a gas in a closed container stating that the product of pressure and volume is proportional to temperature. Pressure, volume, and temperature are properties of a higher level system and somehow emerge from the lower level molecular components of the enclosed gas. In this case, we have a comprehensive reduction of the higher level properties, characterized by the gas laws, to the behavior of molecules by way of statistical mechanics. That is, just by considering how gas molecules bounce around in a container, we can derive the gas laws. Here again, the mystery dissolves, to be replaced by wonder at the intricate complexity of emergent properties.

In both cases, the micro-problem of emergence is tractable, has clear mathematical models that yield testable predictions, and leads to an account of emergent properties that requires no new forces or powers. All of the causation occurs at the lower level and it is the *organization* or *form* of those causes that yields higher level properties. Two-and-a-half millennia ago, the Greek philosopher Plato and his student Aristotle saw the need to introduce the concept of form to explain the emergence of complex things from lower level components. It is literally an ancient idea. These guys were not Australian, but they were pretty smart just the same, so apparently we need to keep our eye on the Greeks, too.

Now, to establish the contrast, let us consider the building of a great ship. The ship is conceived, designed, funded, and built by many conscious beings. The piecing together of the material components of the ship never happens without an underlying intention and imagination, and the vast object that eventually sails the seas permanently bears the marks of those imaginative intentions. The properties of the ship, such as the host of safety systems, can be analyzed in terms of the organization of matter, which is a nontrivial exercise as any engineer will tell you, but there is a lot more to be said. We cannot leave conscious design out of the account of the ship in a way that we can for the gas laws. That makes the account of emergence extremely complicated. To begin with, the relevant causes are not just at the low level of material components making up the ship; human intentions are an ineliminable aspect of the causal story.

Again, consider consciousness. Famously, consciousness in human beings has subjective features (*qualia*) that we know about or "experience," that we attribute to other beings like us, and that we have no idea how to explain. Most scientists and philosophers assume that consciousness, with its functional and subjective properties, emerges somehow from the physicality of brains, but we are nowhere near any consensus on how to explain the emergence process in detail. The degree of mystery here is very high, and many scholars think that it may be permanently intractable.

In both of these latter two cases, the micro-problem of emergence is *not* tractable, in a rather dramatic way. Causation appears to be both bottom-up, as in the gas example, and top-down. Correspondingly, there appears to be some new causal power introduced in the process of emergence: the power to initiate causal chains from the top downwards through a complex material system. We intend to make the ship; we intend to take a back route to evade the traffic jam; we intend to put gas in a jar to study it. No amount of knowledge of brain states could account for the qualitative features; in fact, neuroscience appears to be the wrong *type* of knowledge for the job. *What kind of emergence is this*?! Certainly not the kind of emergence we employed to explain the first two examples above.

Thus, we arrive at the basic philosophical distinction between weak emergence and strong emergence. *Weak emergence* interprets emergent properties as the organization of causes among (or, equivalently, the form of) lower level components in a complex system. *Strong emergence* interprets at least some emergent properties as novel causal forces capable of shaping complex systems in a top-down way.

Strong emergence is worth a moment's further thought. The concept of strong emergence enters the philosophical debate because it appears to be needed to give an empirically responsible account of nature. But it also throws a wrench into the mechanics of science as it seems to propose qualitative differences in the sorts of things that exist in the world – some tractable for physical science and some not. We can use psychology to study human intentions, but we can never reduce those intentions to the accumulation and organization of lower level causes.

Proponents of strong emergence have to make a crucial choice at this point. Either they affirm the claim that there are no differences in emergent properties without matching differences in the organization of lower level components or they reject this claim. This claim is known as *supervenience* and it establishes a strong connection between higher level properties and lower level components. Affirming supervenience keeps strong emergence in touch with scientific methods while still allowing for qualitative difference in the nature of what reality contains (e.g., objects with material causes and consciousness with mental causes). Rejecting supervenience

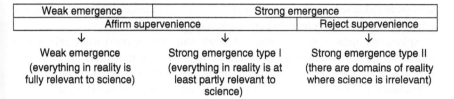

Weak emergence	Strong emergence	
Affirm supervenience		Reject supervenience
↓	↓	↓
Weak emergence (everything in reality is fully relevant to science)	Strong emergence type I (everything in reality is at least partly relevant to science)	Strong emergence type II (there are domains of reality where science is irrelevant)

FIGURE 2.1 Types of emergence.

divorces philosophical accounts of strong emergence from the scientific process and leads to a view of the world of causal relationships that often simply never corresponds to material causal processes. Of course, weak emergence necessarily affirms supervenience. The result is three fundamental views of emergence, depicted in Figure 2.1.

The term *reductionism* is as complicated and contested as emergence, which is not surprising given the entanglement of the two concepts. We need to distinguish several types of reductionism to grasp the meaning of the three types of emergence just sketched. We will employ the idea of levels again as a framing conception for making sense of reductionism.

- *Ontological reductionism*: Emergent realities at a higher level just are lower level components organized in a particular way; there are no novel higher level forces or causal powers.

- *Explanatory reductionism*: we can exhaustively explain higher level emergent phenomena in terms of lower level components. This is sometimes because we have a theoretical reduction, as we do with thermodynamics, in which the gas laws are theoretically reducible to molecular movements via statistical mechanics. Often we do not have the theoretical reduction, but we sense no absolute impediment to explaining an emergent property, so we bet on an explanatory reduction.

- *Value reductionism*: Emergent realities at a higher level are *nothing but* lower level components organized in a particular way and possess no special value beyond what pertains to the value of the components, also known as *nothing-buttery*.

This three-way distinction helps us further elaborate the three types of emergence in Figure 2.1. To begin with, value reductionism is a moral mistake of a very high order. This is equivalent to treating human beings as no more valuable than the material constituents of human bodies. The way low-level constituents are organized in high-level emergent things often adds value, *form matters*. Nothing-buttery must be firmly resisted for moral as well as rational reasons, and all the three types of emergence have the resources to resist value reductionism in resolute manner.

Ontological reductionism is straightforward: weak emergence affirms it, whereas both types of strong emergence reject it. The strong emergence views reject ontological reductionism because both affirm emergent causal powers or forces that were not present in the constituent components.

Explanatory reductionism is more difficult and (we suspect) more relevant to engineering concerns. Importantly, in most cases, *we do not actually have in hand the explanatory reductions that science invites us to think must exist.* In the thermodynamics of gasses, the existing theoretical reduction guarantees an explanatory reduction. In the story of the universe, we have a patchwork of pretty good theories that imply an explanatory reduction, and that is good enough for scientific theorizing. But in the domain of consciousness, we do not have much in the way of theoretical reduction and there is a real question about whether any explanatory reduction will ever be possible – thus the disagreement between proponents of weak and strong emergence.

Practically speaking, rejecting value reductionism is a moral imperative. We cannot easily figure out whether ontological reductionism is correct or not; as usual with metaphysical issues, we can affirm either position more or less indefinitely. But we think that engineers play a key role (perhaps unintentionally) in making the affirmation of ontological reductionism more plausible than its rejection, thereby supporting weak emergence over both types of strong emergence. The practical importance of explanatory reductionism is that it is the means by which we decide which kinds of reductionism and which kinds of emergence are actually present in the world around us. We will return to these practical questions in the second part of this chapter, with a special focus on computer simulations of complex systems. Before we get there, however, we need to discuss the second major development in the history of the development of the concept of emergence: its encounter with the mathematics of complexity and especially nonlinear dynamical systems.

Complexity in the Mathematics of Non-linear Dynamical Systems

The mathematical treatment of nonlinear dynamical systems has a history of its own, running parallel with the history of the concept of emergence. Importantly, the two lines of thinking merged in the twentieth century as complexity theory came into its own. A good place to begin the short version of this story is with Isaac Newton (1642–1727).

After Newton had demonstrated the stability of the two-body system (e.g., Sun and Earth), he turned his attention to the three-body system (e.g., Sun, Earth, and Earth's Moon). He was unable to prove the stability of this slightly more complex system and, based on his shockingly confident expectation that he should be able to prove anything provable, Newton concluded that the three-body system was in fact not stable. This conjured the specter of planetary collisions, thereby frustrating the divine providential purpose. As that was unthinkable, Newton's solution was to suggest that God must reach into nature and periodically adjust planetary orbits to avoid those collisions.

We know today why Newton could not prove the stability of the three-body system: it is a complex dynamical system capable of (mathematically speaking) chaotic behavior. We also know that Newton overreacted. In fact, Pierre-Simon

Laplace (1749–1827) famously used perturbation theory to prove the stability of the Sun-Earth-Moon system. In a well-known (although possibly apocryphal) story about Laplace, he is said to have responded to a question from then emperor Napoleon about Newton's theory of divine intervention and the place of God in managing the solar system, by replying that he had "no need of that hypothesis." If Laplace actually made this remark, it was probably not explicitly meant as a rejection of theism but intended as a comment on complex systems and the introduction into mathematics of techniques for coping with the bizarre behavior that complex systems sometimes produce.

In the early nineteenth century, the logistic mapping was introduced to explain longitudinal variations in the population of animal species. The population in year $n + 1$, x_{n+1}, will be proportional to the population in the previous year, x_n, due to reproduction, and also proportional to $(1 - x_n)$, due to the pressure on resources and ease of predation when the population is too large. Thus, $x_{n+1} = Kx_n(1 - x_n)$, where K is a constant. This highly intuitive equation proved useful for modeling animal populations. So long as K is not too big (i.e., stays between 0 and 3), the sequence x_0, $x_1, x_2, x_3, \ldots$, called an orbit, matches real-world numbers tolerably well. When K goes above 3, however, very strange behavior emerges, including periodic and chaotic orbits. This actually helped explain periodic animal populations, bouncing between a high level 1 year and a low level the next, then back again – when $3 < K < 3.45$. But the logistic map seemed inapplicable to the real world for $K > 3.45$.

By the last two decades of the twentieth century, mathematicians had analyzed the heck out of the logistic mapping and a bunch of other recursive nonlinear equations, discovering a ton of absolutely fascinating results in the process. For our purposes, what matters is a key property of the formal notion of mathematical chaos that emerges from this process. Chaotic orbits are *eventually unpredictable* in the sense that, no matter how close two starting points might be (i.e., the values of x_0), eventually the two orbits emanating from these two starting points diverge and show no relationship whatsoever. The 2 orbits might closely match for 10 iterations, or a 100, but eventually they look nothing like one another. This is called *sensitive dependence* on initial conditions, or the *butterfly effect*. Eventual unpredictability in nonlinear dynamical mathematical systems has important implications for modeling complex systems using both mathematics and computer systems. Also, the emergence of higher order behaviors in (mathematical) nonlinear dynamical systems speaks to emergence.

It was once thought that unpredictability was evidence of non-determinism in nature and a sign of the impossibility of mathematical modeling. Those days are long gone. We now know that nonlinear dynamical systems can display unpredictability (technically, *eventual* unpredictability) and yet are completely deterministic recursive equations. This extends the explanatory reach of the metaphysical hypothesis of determinism, allowing many more systems, including unpredictable systems, to be explained using deterministic nonlinear dynamical systems. But eventual unpredictability also shows that we can never decisively prove determinism using mathematical modeling: the data from a real-world system will eventually diverge wildly from any theoretical model no matter how precise our knowledge of initial conditions.

So the metaphysical hypothesis of determinism is both greatly strengthened and permanently checked by the same set of mathematical discoveries.

Complex systems in nature routinely display emergent properties, some of which seem highly non-predictable. The mathematical analysis of nonlinear dynamical systems has rightly increased our confidence in our capacity to explain such systems – and this has paid off in dozens of domains of inquiry from irregular heart rhythms to turbulence in airflow over aircraft wings. This strengthens the hand of explanatory reductionism, which in turn strengthens the position of weak emergence and leads us to say, with Laplace, that we have no need of the hypothesis of novel forces and emergent causal powers.

This story of the coalescing of emergence and nonlinear dynamical systems in complexity theory has been very brief, but we hope it has been adequate to prepare the reader for the next phase of our discussion: how all of this affects the kinds of issues that computer engineers worry about (to the extent that we philosophers appreciate those issues – and we cannot say often enough that we know we are not engineers).

EMERGENCE IN COMPUTER ENGINEERING

Computer engineers are a diverse group. We know most about computer modeling and simulation (M&S), so that is where we will focus our attention. Most of our work has been on computational models involving the simulation of psychological and social systems (Wildman and Sosis, 2011; Shults *et al.*, 2017a,b; Shults and Wildman, 2018). But it is worth pointing out in passing that the combination of periodic and chaotic behavior just described has applications in semiconductor manufacturing, in network management, and in analysis of interacting signals from several electronic components.

M&S can be understood as the use of virtual complex systems to model real-world complex systems. If the low-level interactions within the virtual system can be validated against the low-level behavior of the real-world system, then the emergent properties of the virtual system can be meaningfully compared to the real-world system. When there is no match, the system does not generate the desired insights. When there is a match, we have a strong case that we should be able to generate insights into the causal dynamics of the real-world system by examining the causal architecture of the virtual system. For example, sensitivity analyses can show us which parameters in a model are most salient for producing particular types of emergent properties. We can also interfere with the virtual system in a variety of ways to produce emergent effects we would like to see matched in the real-world system. That takes us straight into design and safety considerations for bridges and freeways, into efficiency and productivity considerations for work flows and communication or transportation networks, and into policy considerations when we are using computer models to ease social or economic problems.

As Manuel Delanda points out, the advent of computer simulations is "partly responsible for the restoration of the legitimacy of the concept of emergence [in science and philosophy] because they can stage interactions between virtual entities from which properties, tendencies, and capacities actually emerge" (DeLanda, 2011). In other words, simulations can help us elucidate the mechanisms at one level through which properties (such as wetness, temperature, or density) "emerge" at another level. Intensive differences in populations (e.g., atoms, molecules, chemical compounds, or even human individuals) can have the capacity to act as energy storage devices, which Delanda calls "gradients." The latter can play a role in the self-organization of the moving parts of a population into a larger whole. Computer models attempt to identify and simulate the mechanisms by which the relevant gradients lead to the "emergence" of properties in the real-world at a variety of levels (thermodynamic systems, ancient organisms, social networks, human societies, etc.). The goal is to generate or "grow" the relevant emergent properties, capacities, and tendencies of these somewhat stable, larger (simulated) "wholes" from the local interaction of "lower" level parts using computational tools such as cellular automata, genetic algorithms, neural nets, or multi-agent artificial intelligence systems (Epstein and Axtell, 1996).

Of course, even when we have successfully validated a computer model against a real-world system, both at the low level of component interactions and at the high level of emergent behaviors, we really do not *know* that the causal dynamics of the virtual system match the causal dynamics of the real-world system. In fact, sometimes we know they do not match, as in the very different approaches to chess playing adopted by then world champion Gary Kasparov (intuition and analysis) in his famous 1997 match with IBM's Deep Blue (position scanning). But often we are just not sure how closely the causal dynamics match. The better and more comprehensive validation is, at both the low and high levels and anywhere else we can get empirical traction, the more confidence we are entitled to have that the model is shedding light on real-world dynamics.

Suppose, then, that we have a richly validated computer model that generates useful insights into some real-world system – how to manipulate it, how to control it, how to optimize it, how to protect it, how to crash it, and so on. *Does this explanation count as an explanatory reduction of the emergent properties of the real-world system?* This is a philosophical question of some importance. To say definitely that we *do not* have an explanatory reduction is to ask for certainty, whereas the study of nonlinear dynamical systems shows that certainty about the operations of complex systems is typically impossible, as noted above. To say definitely that we *do* have an explanatory reduction is to claim too much for the ability of virtual computer simulations to match real-world complexity; the best we can hope for is usefully close approximations. Thus, computer engineers find themselves in a middle position where all they can claim is that they *probably* have an explanatory reduction – maybe very probably, depending on how good the model proves to be. There is no certainty here, just likelihood.

This sort of claim might not seem that impressive to scholars invested in absolute certainty. But there are not very many of the latter in engineering, and not even in philosophy anymore. When it comes to estimating the contribution of computer engineering to the ongoing philosophical disputes about emergence, this outcome is rather impressive. The fact that we can build virtual complex systems that *probably* explain the operation of real-world complex systems lends powerful support to the weak emergence position and to the view that we have no need for the hypothesis of novel forces and causal powers embedded in Strong Emergence Type I. Again, there is no proof here. But we think developments in computational modeling can help one discern an implacable movement of knowledge toward greater explanatory power without assuming top-down causes, whole-part causes, intentional mental causes, mystical entelechies, or supernatural beings. Who knew computer engineering could be so important to philosophers?!

CONCLUSION: FUTURE DIRECTIONS

The basic message of this chapter is that computer engineers are effecting a transformation (perhaps unintentionally) in philosophical disputes about emergence, steadily promoting confidence in theories of weak emergence and undermining the plausibility of both types of strong emergence. This is occurring through the systematic production of probable explanatory reductions using computer models capable of simulating real-world complex systems.

But computer engineers are not getting everything their own way. The reason strong emergence was proposed in the first place was to deal with a sharp empirical problem: nobody could explain the subjective features (qualia) of consciousness or mental causation as weakly emergent properties of a complex bodily system, so strong emergence seemed to be the only credible option. No amount of computer modeling seems likely to help with this problem directly.[2] Eliminative materialist philosophers "solve" this ongoing problem by proposing (way too) simply that there are no such things as qualia or mental causes, treating them as misunderstandings to which we are prone as biological creatures, but having no independent reality. Understandably, this is not a very popular position – but at least such proposals take the problem seriously. To us, it appears that philosophers such as Alfred North Whitehead (1861–1947), Charles Sanders Peirce (1839–1914), John Dewey (1858–1952), William James (1842–1910), and Gilles Deleuze (1925–1995) had it right: the reason we cannot explain the subjective qualities of consciousness is that we have a defective understanding of the fundamental constituents of reality. Fix that problem, and we can stay with weak emergence, explain the empirical embarrassment of qualia without pretending they do not exist, and relegate the varieties of strong emergence to the pages of history.

But have we really failed (so far) to understand the fundamental constituents of reality? Surely quantum mechanics has demonstrated that we do not *know* what matter, energy, or even causation *are*. All we have are philosophical gestures in certain directions, all too often shaped by folk-physics intuitions that should have been wiped out by quantum physics a century ago. For that matter, we are struggling mightily with the ideas of space, time, and force, as well. We have solid mathematically voiced theories about all of these concepts, but no intuitive sense of how to follow the hints embedded in such mathematical expressions to articulate and explain their entailment

---

[2]Note that some proponents of the "new mechanism" in philosophy of science might argue that they could indirectly contribute to solving the problem. See, for example, Bechtel (2008, 2009).

relations in the material world. As many philosophers of physics have pointed out, we often navigate our way around and through the mathematical theories at our disposal, changing narratives as needed without being sure we are making real progress. Meanwhile, predictions and technological applications do advance, progressively, if not always safely or wisely or justly.

Telling stories about invisible causal agents or other causal powers, as required by strong emergence, is a failing approach: we have argued that computer engineers and others are advancing explanations that are steadily strengthening the hypothesis of weak emergence and undermining both versions of strong emergence. But there might still be a way to tell stories about the fundamental constituents of reality that makes sense of qualia of consciousness (and what we call mental causation) without winding up on the wrong side of advancing explanations about the way the world works. We think it is worth trying to construct this sort of story, following philosophers such as those mentioned above. One option is to treat the fundamental constituents of nature as dipolar in character – that is, as both matter-energy and mentality (with Whitehead), or as both matter-energy and value (as with Peirce, James, and Dewey) – or even as both matter-energy and form (as with Plato and Aristotle). Another option is to develop a univocal ontology (as with Deleuze), in which the fundamental constituents of nature are events of pure becoming, emissions of singularities whose intensive differences mark the birth of thought.

This is where we expect most computer engineers to stop listening. We get it. This is the sort of thing we philosophers love to argue about even if no one listens. We are among the growing number of philosophers who have learned from listening to computer engineers, whose work increasingly shows how desperate the research programs of strong emergence really are. The explanatory reach of weak emergence will continue to expand as computer engineers keep building virtual complex systems, which dissolves reasons to resort to special forces and causal powers as demanded by strong emergence. There is nothing computer engineers can do to aid strong emergence, save the cataclysmic failure of their modeling aspirations, which is the best that proponents of strong emergence can hope for. The only constructive movement engineers can make is in support of weak emergence. In the final analysis, weak emergence might not be correct – strong emergence might eventually win the day. But every new success in virtual complex systems achieved by computer engineers only makes this reversal of fortunes less likely. We live in the era of the decline of strong emergence. A consensus around weak emergence is rapidly consolidating. And perhaps more than any other group, computer engineers are to blame. Thanks for that. For those of you who made it this far, we hope we have convinced you that it might be worth your while to take the time (occasionally) to chat with us philosopher types to learn more about the havoc you are wreaking in our disciplines.

BIBLIOGRAPHY

Alexander, S. (1920) *Space, Time, and Deity*, Macmillan, London.

Bechtel, W. (2009) Looking down, around, and up: mechanistic explanation in psychology. *Philosophical Psychology*, **22** (5), 543–564.

Bechtel, W. (2008) *Mental Mechanisms: Philosophical Perspectives on Cognitive Neuroscience*, Taylor and Francis, London.

DeLanda, M. (2011) *Philosophy and Simulation: The Emergence of Synthetic Reason*, Continuum, London.

Epstein, J.M. (2006) *Generative Social Science: Studies in Agent-Based Computational Modeling*, Princeton University Press, Princeton, NJ.

Epstein, J.M. and Axtell, R. (1996) *Growing Artificial Societies: Social Science from the Bottom up*, Brookings Institution Press, New York.

Bedau, M. and Humphreys, P. (eds) (2008) *Emergence: Contemporary Readings in Philosophy and Science*, MIT Press, Cambridge, MA.

Shults, F.L., Gore, R., Wildman, W.J., Lane, J.E. *et al.* (2017a) Mutually escalating religious violence: a generative and predictive computational model. SSC Conference Proceedings.

Shults, F.L., Lane, J.E, Wildman, W.J., Diallo, S. *et al.* Modeling terror management theory: computer simulations of the impact of mortality salience on religiosity. *Religion, Brain & Behavior*, **8** (1), 77–100.

Shults, F.L. and Wildman, W.J. (2018) Modeling Çatalhöyük: simulating religious entanglement in a neolithic town, in *Religion, History and Place in the Origin of Settled Life* (ed. I. Hodder), University of Colorado Press, Denver, CO.

Wildman, W.J., Fishwick, P.A., and Shults, F.L. (2017) Teaching at the Intersection of Simulation and the Humanities, in *Proceedings of the 2017 Winter Simulation Conference* (eds W.K.V. Chan, A. D'Ambrogio, G. Zacharewicz *et al.*), pp. 1–13.

Wildman, W.J. and Sosis, R. (2011) Stability of groups with costly beliefs and practices. *JASSS*, **14** (3).

3

SYSTEM THEORETIC FOUNDATIONS FOR EMERGENT BEHAVIOR MODELING: THE CASE OF EMERGENCE OF HUMAN LANGUAGE IN A RESOURCE-CONSTRAINED COMPLEX INTELLIGENT DYNAMICAL SYSTEM

Bernard P. Zeigler[1] and Saurabh Mittal[2]

[1]*RTSync Corporation, Rockville, MD, USA and University of Arizona, Tucson, AZ, USA*
[2]*MITRE Corporation, McLean, VA, USA*

SUMMARY

A recent paper laid a systems theoretic foundation for understanding how human language could have emerged from prelinguistic elements. The systems theoretic approach to incorporating emergence (as a construct) to understand a complex phenomenon first required the formulation of a system model of the phenomena. Having the correct formal system specification is a key to demonstrating that the obtained holistic behaviors denote the intended emergent behavior. The case study of human language is presented as an instance of a set of language-ready components that must be coupled to form a system with innovative inter-component information exchange. In this chapter, we first review the systems theoretic foundation for modeling accurate emergent behavior. We then review the activity-based monitoring paradigm and

Emergent Behavior in Complex Systems Engineering: A Modeling and Simulation Approach,
First Edition. Edited by Saurabh Mittal, Saikou Diallo, and Andreas Tolk.
© 2018 John Wiley & Sons, Inc. Published 2018 by John Wiley & Sons, Inc.

the emergent property of attention-focusing in resource-constrained activity-based complex dynamical intelligent systems (RCIDS): a class of systems that exhibit intelligent behaviors. We then pose the problem of shared attention among hominins within these paradigms. A fundamental property of stable systems may lie in their ability to obtain useful information from the world by suppressing activity unrelated to current goals and thereby to satisfy the requirements for well definition of self as system. This includes the case of nascent social systems containing humans as emerging linguistic agents.

INTRODUCTION

Following Ashby (1956) and Foo and Zeigler (1985), *emergence* is conceived as global behavior of a model that might be characterized as

$$\text{Emergent behavior} = \text{components} + \text{interactions (computation)} + \text{higher order effects}$$

where the latter can be considered as the source of emergent behaviors (Kubik, 2003; Mittal and Zeigler, 2014a,b; Szabo and Teo, 2015).

In a fundamental characterization of emergence, Mittal and Rainey (2015) contrast positive emergence with negative emergence. Positive emergence fulfills the systems of systems (SoS) purpose (Boardman and Sauser, 2006) and keeps the constituent systems operational and healthy in their optimum performance ranges. Negative emergence does not fulfill SoS purposes and manifests undesired behaviors such as load hot-swapping, cascaded failures. Some perspectives on the role of modeling and simulation in the study of emergence are presented in Mittal (2013), Mittal and Rainey (2015), Zeigler (2016a,b), Zeigler and Muzy (2016a,b), and Zeigler (2017).

It is an understated assumption that emergent behavior is manifested when there are a large number of agents or systems[1] engaged in multi-level information/physical exchanges. Consequently, it is appropriate to begin our discussion with the notion of a system. Failure to do so will lead to inaccurate exposition of emergent behavior. Systems are generally best described with General Systems Theory (GST) (see, e.g., Wymore (1967) and Ören and Zeigler (2012)). To describe a system, we first need a system boundary, described using input and output interfaces, and an input/output (I/O) function that transforms input into output. To develop a multicomponent system, many such systems are brought together and connected with explicit coupling relations using GST's composition principle. The composition specifications may yield a hierarchical system that limits complexity through containment. Additionally, if the component systems are managerially, operationally, and evolutionary independent and are geographically displaced, the overall system can be termed *SoS* in a strict sense (Maier, 1998). However, for simplicity purposes, and for the purpose of

---

[1] An agent and system are used interchangeably. Both have distinct boundaries with explicit input and output interfaces, and both transform input into output with an internal I/O function.

modeling, a multi-level system (that may or may not exhibit hierarchical structure) composed of many systems can be effectively termed as an *SoS*. The collective behavior of such an SoS that is irreducible at any subsystem level is considered as an emergent behavior, which is also a primary principle for architecting SoS (Maier, 1998).

Components and couplings in complex system models must include representation of decision-making in natural and artificial environments. Such decision-making is embedded in various I/O functions in various subsystems. The modeling of SoS needs to be specified in a layered manner using the levels of the hierarchy of systems specification framework (Zeigler *et al.*, 2000) that guides on creating these multi-level I/O pairs. Another important implicit fact in these multi-level systems is the exchange of information between two systems. In a component-oriented perspective, which we have already established with the notion of an agent of a system, such exchanges are synonymous with discrete events. For example, sending information is an event, receiving information is an event, and transformation of an information element into another can also be termed as a *logical event*. Consequently, as we subscribe to such discrete event world view in a SoS domain context, we employ formal Discrete Event System Specification (DEVS) to model the SoS and thereby formally define the resulting emergent behavior of such SoSs (Mittal, 2013).

DEVS has the universality (Zeigler *et al.*, 2000) to represent the *discrete* (for agent models), *continuous* (for natural environments), and *hybrid* (for artificial environments) formalism types needed for adequate complex system model construction. DEVS supports dynamic structure for genuine adaption and evolution (Park and Hunt, 2006; Steiniger and Uhrmacher, 2016). Strong dynamic structure capabilities are needed to specify and flexibly control the changes in components and their coupling to be able to adequately model adaptation, evolution, and emergence (Mittal, 2013) in ways that include the possibility of genuine surprise (Muzy and Zeigler, 2014).

Employing DEVS to then describe a class of intelligent systems deployed in resource-constrained dynamical environment is one of the objectives of this chapter. In our previous work (Mittal and Zeigler, 2014b), such investigation led to the modeling of attention-switching of system as an emergent property of the system. In this chapter, we capitalize on this emergent property of attention-switching toward emergence of language between two agents (as a result of positive emergence between two agents). We argue that in the absence of the ability to switch attention, language will not result (a result of negative emergence) in a resource-constrained dynamical environment. Next, after demonstrating through an example, we extend the idea toward the concept of emergence of shared attention in an SoS context.

The chapter is laid out as follows. Background section provides an overview on the DEVS formalism, the underlying architecture for positive emergence, the notion of a well-defined system, and characteristics of resource-constrained complex intelligent dynamic system (RCIDS). Emergence of Language Capabilities in Human Evolution section discusses the emergence of language capabilities in human evolution. Fundamental Systems Approach for RCIDS for Emergence of Language Example section dives into the detail of systems theoretic approach for modeling a two-agent RCIDS system. Perspectives: Evolution and System Modeling section presents the evolutionary and system perspective on the solution so obtained. RCIDS and Activity-Based Perspective on Shared Attention section follows with an activity-based perspective on the emergence of shared attention that postulates attention interrupting and redirecting mechanisms to enable the emergence of a composite RCIDS. The chapter is then summarized in Conclusions and Future Work section.

BACKGROUND

DEVS Formalism

The DEVS formalism based on systems theory provides a framework and a set of modeling and simulation tools to support systems concepts in application to SoS engineering (Zeigler et al., 2000; Mittal and Risco-Martín, 2013). A DEVS model is a system theoretic concept specifying inputs, states, and outputs, similar to a state machine. Critically different, however, is that it includes a time advance function that enables it to represent discrete event systems, as well as hybrids with continuous components, in a straightforward platform-neutral manner. DEVS provides a robust formalism for designing systems using event-driven, state-based models in which timing information is explicitly and precisely defined. Hierarchy within DEVS is supported through the specification of atomic and coupled models. Atomic models specify behavior of individual components. Coupled models specify the instances and connections between atomic models and consist of ports, atomic model instances, and port connections. The input and output ports define a model's external interface, through which models (atomic or coupled) can be connected to other models.

Based on Wymore's systems theory (Wymore, 1967), the DEVS formalism mathematically characterizes the following:

- DEVS atomic and coupled models specify Wymore systems (Ören and Zeigler, 2012).
- Composition of DEVS models: component DEVS and coupling result in a Wymore system, called the *resultant*, with structure and behavior emerging from their interaction (Mittal and Risco-Martín, 2013).
- Closure under coupling: the resultant is a well-defined DEVS just like the original components (Zeigler et al., 2000).
- Hierarchical composition: closure of coupling enables the resultant coupled models to become components in larger compositions (Oren and Yilmaz, 2015).

Table 3.1 shows the levels of system specification as implemented in DEVS formalism (Zeigler et al., 2000).

Well-Definition of Systems

Before proceeding to review the more fundamental issues, we need to start with a primitive and perhaps more intuitive notion of a system (summarizing Zeigler and

TABLE 3.1 DEVS Levels of Systems Specification

Level	DEVS Specification	Systems Specification and Agent Development
4	Coupled	Hierarchical system with coupling specification. Multi-agent system is defined at this level
3	Atomic	State space with transitions. An agent behavior and internal structure is specified at this level
2	I/O function	State space with defined initial state. Agent behavior with respect to initial state is defined at this level
1	I/O behavior	Collection of input/output frames for behavior specifications
0	I/O frame	Input/output variables with associated ports over a time base

Muzy, 2016a). Consider a concept of an input-free system (or autonomous, i.e., not responding to inputs) with outputs not considered and with states, transitions, and times associated with transitions. Then, we have a transition system

$$M = \langle S, \delta, ta \rangle, \quad \text{where } \delta \subseteq S \times S, \text{ and } ta : S \times S \to R_0^\infty$$

where S is set of states, δ is the transition function, and ta is the time advance. To further simplify, we are allowing the transition system to be *non-deterministic* so that rather than δ and ta being functions they are presented as relations.

For example, in Figure 3.1, we have

$$S = \{S1, S2, S3, S4, S5, S6, S7\},$$

$$\delta = \{(S1, S3), (S3, S4), \dots \}$$

$$ta(S1, S3) = 1, ta(S3, S4) = 1, \dots$$

For example, there are transitions from state S1 to state S3 and from S3 to S4, which each takes 1 time unit and there is a cycle of transitions involving S4 … S7 each

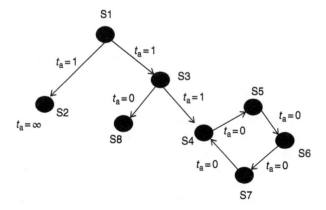

FIGURE 3.1 System with timed transitions.

of which takes zero time. There is a self-transition involving S2, which consumes an infinite amount of time (signifying that it is passive, remaining in that state forever). This is distinguished from the absence of any transitions out of S8. A state trajectory is a sequence of states following along existing transitions, for example, S1, S3, and S4 is such a trajectory.

This example gives us a quick understanding of the conditions for system existence at the fundamental level:

We say that the system is

- *not defined* at S8 because there is no trajectory emerging from it, that is, because there is no transition pair with S8 as the left member in δ
- *non-deterministic* at S1 because there are two distinct outbound transitions defined for it, that is, because S1 is a left member of two transition pairs (S1, S2) and (S1, S3)
- *deterministic* at S2 and S4 because there is only one outbound transition for each, that is, because there is only one transition pair involving each one as a left member.

We say that the system is *well defined* if it is defined at all its states. These conditions are relative to static properties, that is, they relate to states not how the states follow one another over time. In contrast, state trajectories relate to dynamic and temporal properties. When moving along a trajectory, we keep adding the time advances to get the total traversal time, for example, the time taken to go from S2 to S4 is 2. Here, a trajectory is said to be *progressive* in time if time always advances as we extend the trajectory. For example, the cycle of states S4 … S7 is not progressive because as we keep adding the time advances, the sum never increases. Conceptually, let us start a clock at 0 and, starting from a given state, we let the system evolve following existing transitions and advancing the clock according to the time advances on the transitions. If we then ask what the state of the system will be at some time later, we will always be able to answer if the system is well defined and progressive. A well-defined system that is not progressive signifies that the system gets stuck in time and after some time, it becomes impossible to ask what the state of the system is in after that time. Zeno's paradox offers a well-known metaphor where the time advances diminish so that time accumulates to a point rather than continue to progress and offers an example showing that the pathology does not necessarily involve a finite cycle. Our concept of progressiveness generalizes the concept of legitimacy for DEVS (Zeigler *et al.*, 2000) and characterizes the "zenoness" property.

Active-Passive Compositions   An interesting form of the interaction among components may take on a pattern found in numerous information technology and process control systems. In this interaction, each component system alternates between two phases, active and passive. When one system is active, the other is passive - only one can be active at any time. The active system does two actions: (i) it sends an input to the passive system that activates it (puts it into the active phase) and (ii) it transits to the passive phase to await subsequent reactivation. For example, in a Turing Machine

(TM), the TM control starts a cycle of interaction by sending a symbol and move instruction to the tape system then waiting passively for a new scanned symbol to arrive. The tape system waits passively for the symbol/move pair. When it arrives, it executes the instruction and sends the symbol now under the head to the waiting control.

Such active-passive compositions provide a class of systems from which we can draw intuition and examples for generalizations about system emergence at the fundamental level. We will employ the modeling and simulation framework based on Systems theory formulated in Zeigler *et al.* (2000), especially focusing on its concepts of iterative specification and the DEVS formalism (Zeigler *et al.*, 2000). Zeigler and Muzy (2016a) employed the pattern of active-passive compositions to illuminate the conditions that result in ill definition of deterministic, non-deterministic, and probabilistic systems. They provided sufficient conditions, meaningful especially for feedback-coupled assemblages, under which iterative system specifications can be composed to create a well-defined resultant system and that moreover can be expressed and studied within the DEVS formalism.

Attention-Switching in Resource-Constrained Complex Intelligent Dynamical Systems (RCIDS)

Mittal and Zeigler (2014a) defined a scalable RCIDS with the following properties:

1. *Resource-constrained environment*: In the modeled connectionist system, network bandwidth and computational resources available to any sensor are finite. The constraints may take the form of energy, time, knowledge, control, and so on, that are available to any processing component.
2. *Complex*: Presence of emergent behavior that is irreducible to any specific component in the system. Attention-switching is an emergent phenomenon (Tsotsos and Bruce, 2008).
3. *Intelligent*: The capacity to process sensory input from the environment and act on the sensory input by processing the information to pursue a goal-oriented behavior.
4. *Dynamical*: The behavior is temporal in nature. The system has emergent response and stabilization periods.
5. *System*: The model conforms to systems theoretical principles.

Within the context of the RCIDS, Mittal and Zeigler (2014b) presented a model capable of focusing attention to components displaying high activity and directing resources toward them so that they can accomplish their task effectively. The designed model was built using the DEVS formalism and is a proof-of-concept realization of the more general RCIDS. It illustrates the DEVS formalism's suitability for modeling complex dynamical systems. Moreover, it exploits the dynamic structure capabilities of DEVS, which allows the system to exhibit both the adaptive behavior as its environment changes and maintenance of steady state in a dynamic environment.

The architecture design can be mapped to any real-life system that is hierarchically organized working along the rule of "chains of command," which implies that the information is filtered and condensed as it travels up in the hierarchy. To demonstrate the basic concepts, the system can be conceptually mapped to geographical area distribution under the control of managers that allocate communication network resources (such as bandwidth and channel capacity) to areas under their control, which in turn distribute the resources to end users (sensors). The distribution is done intelligently with the most active component receiving the largest amount of resources. As the components pass through cycles of high and low activity, so does the assignment of the resources allocated to them. The simulation results have confirmed that the system is capable of directing and switching its focus to components that display persistent high activity during simulation and also can withdraw attention from components that are not displaying any activity. Hu and Zeigler (2013) showed that the reduction in energy in actual implemented systems (e.g., hardware) can be measured and compared to the ideal level given by a metric that measures the disparity in activity demands among components. Likewise, the attention-switching architectures saves energy by not wasting energy on components that do not need it in dynamic manner. Accordingly, the (nonfunctional) performance of these architectures can be gauged by the disparity metric – the greater the disparity, the greater the savings due to attention-switching.

Mittal and Zeigler (2014a) adapted the winter-take-all (WTA) algorithm by incorporating various sampling algorithms that determine where an activity has arisen of highest importance, thereby receiving attention from data-driven decision maker (DDM) or a resource allocation manager (RAM). The criteria for deciding an activity is "important" is based on the sensitivity of the sensor and the rate estimator (RE) threshold. This component "validates" the importance of any activity sensed by the sensor – a higher threshold represents a higher importance. The WTA model enables the system to continually shift focus and direct its attention to the most active component.

The model just described is an abstracted version of RCIDS, where the resources are directed toward a focused activity. This makes the system a generalized architecture capable of focusing attention and concentrating on the task at hand by providing more resource to the task by redistribution and reallocation. For intelligent goal-pursuing behavior by an agent, the activity (both designed and emergent) needs to be partitioned according to both bottom-up and top-down phenomena (Tsotsos and Bruce 2008). Incorporating both these phenomena results in a sensor-gateway system (Figure 3.2). It can also be construed as an agent-gateway system where an agent has the sensory and the perceptual apparatus. The implementation is attempted through RCIDS that display an emergent property of focusing attention to an agent-gateway that detects "change" in activity. RCIDS work with finite resources, and consequently, resource management is a critical aspect of such systems. The distribution is done through feedback loops from other supporting system components, with the most active component receiving the most resources. As the agent passes through cycles of high and low activity, so does the assignment of the resources allocated to it. RCIDS are equipped with programmable sensor

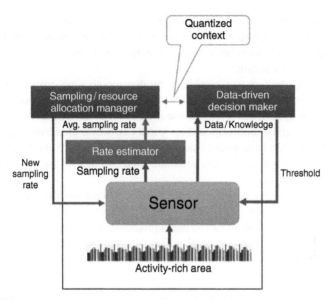

FIGURE 3.2 Sensor-gateway system model for a single-level architecture. *Source*: Reproduced with permission of Mittal and Zeigler (2014a).

gateways that can dynamically change their sampling rate and their threshold value to report the data. A fractal sensor/agent-gateway system is minimally composed of four components:

1. *Sensor/agent*: This component is attuned to the appropriate activity currency and has a sampling rate to detect a quantum change.
2. *Sampling* or *resource allocation manager (RAM)*: This component samples the activity currency in a pragmatic context using multiple computational algorithms.
3. *Rate estimator (RE)*: This component estimates the dynamical behavior of gateway and provides a smoothing function to prevent rapid system oscillations and to ensure that an activity persists "long enough."
4. *Data-driven decision maker (DDM)*: This component quantifies the information flow from sensor, sets new goal (threshold for sensor), and quantifies the context.

There exists a RAM at every level to direct focus and attention and an RE coupled to every sensory element. The RE may or may not be present at the intermediate levels in the hierarchy, but it must be at the coarsest level to deduce and validate what the sensors are witnessing. The system also allows resources and peripheral attention to the ongoing working sensors and does not inhibit or stall their operation in the pursuit of focusing attention to the important one. For different WTA mechanisms, the sensor population is met accordingly and in no case, the resources are completely withdrawn from the running sensors as it is not predictable which sensor might produce an important information the next instant. The system lets the other sensors keep working at their default settings and provide the resources for their operation and intermittently switches when an activity of high importance is encountered and advertised by any sensor.

The nature of attention was elucidated by developing a system model that validates its emergent nature. This research also validated that in a resource-constrained environment, the system has to switch attention to focus the task-at-hand and to live within the energy constraints imposed on it as a real-world phenomenon. We developed two metrics, that is, response time and stabilization time that quantify the time taken for the system to switch attention and then continue to pay attention to the selected activity region.

The component-based architecture and distributed operation of the system facilitate its deployment in the real world in terms of federates and participate in larger system of systems. The system can be made predictive and robust with more detailed modeling of the RE and WTA mechanisms implemented in the RAM, supported with efficient synchronization strategies.

EMERGENCE OF LANGUAGE CAPABILITIES IN HUMAN EVOLUTION

Speech acts involve humans speaking to one another in order to change their knowledge or behavior or to otherwise affect them. Questions concern the evolution of motivations and mechanisms and how such speech acts could have emerged in early hominins. According to Everett (2012), humans' interest in interacting with fellow humans differs from that of other hominids for reasons such as

1. the discovery of fire and management of fire technology,
2. improving the efficiency of securing and preparing food (accelerated by fire (cooking)), and
3. emerging sense of community.

Everett asserts that evolution had to prepare early hominins to develop instances of multimedia channels in order to communicate with one another laying the foundation for language, the ultimate communicative tool. The media for conveying messages could have been whistles, humming, or physical signing predating the use of the modern vocal apparatus to produce today's speech sounds. An early speech act may have been spontaneous grunting and pointing vigorously to a serendipitously appearing prey. "Theory of mind" – recognizing others as having the same mental propensities as oneself – may be prerequisite to shared symbols in messages. The range of employed speech acts is governed by general principles by which all members of the culture recognize that a speech act of a certain kind has taken place. Thus, culture sets the channel symbol size and response of senders and receivers. Although syntax concerns message formation, and semantics concerns message meaning, pragmatics concerns a speaker's ability to understand another's intended meaning (Zeigler and Hammonds, 2007). Mutual comprehension requires speaking the same language

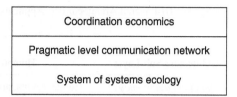

FIGURE 3.3   Tri-layered architecture for positive emergence.

with a shared ontology. The language, as it has come to be known, is a well-structured sequence of symbols with accompanying grammar at the syntactic level, with each symbol been deliberately assigned a particular meaning for its semantic value, and then used in context and understood exactly like it is by the other person in the same context for its pragmatics and mutual comprehension.

Tri-Layered Architecture

Now we lay the basis for formulating the problem of emergence of language using the emergence of communication channel as a positive emergence outcome. First, we review the tri-layered architectural framework from Zeigler (2018), shown in Figure 3.3.

The layers (from bottom) are roughly characterized as follows:

1. *System of systems ecology (Primeval SoS)*: the systems that will become component systems of the emergent SoS already exist as viable, *stable* autonomous entities in an ecology; however, left unperturbed, they would not emerge into the SoS under consideration.

2. *Network supporting pragmatic level of communication*: the ability to communicate among putative component systems of the SoS, not only at technical level, but a level that supports the coordination needed at the next level (this is the "pragmatic" level vs the underlying syntactic and semantic layers (Zeigler and Hammonds, 2007)).

3. *Coordination economics*: a compound term we introduce to refer to (i) the *coordination* required to enable the components to interact in a manner that allows emergence of the SoS with its own purposes and (ii) the economic conditions that enable emergence – the collective benefit that the SoS affords versus the cost to individuals in their own sustainability to contribute to the SoS objectives.

The architecture was illustrated with examples of trending topics in Twitter and in National Healthcare Systems (Zeigler, 2016a,b). The reader is encouraged to review the exposition of tri-layered architecture in Zeigler's (2016a,b) recent work.

In the case of language, Table 3.2 identifies the constituent agents, their mode of communication, and other considerations facilitating positive emergence. For a language to emerge, we need two parties interested in communicating an

TABLE 3.2   Positive Emergence Tri-Layered Architecture as Applicable in the Case of Human Language

Architecture Layer	Architecture Sub-Layer	Human Language
SoS ecology	*Component autonomous Systems*	Human agents, artificial agents
	Component interactions	Agent communicate an object of interest to other agent
	Basis for economic survival	Agent's value in providing meaning to a situation of interest
Pragmatic level communication network	*Communication platform at technical levels*	Hand gestures, specific sounds, facial expressions, body signals
	Communication platform at pragmatic level	Objects embedded in context. Situated agents sharing meaningful information through multiple media
Coordination economics	*Coordination required*	Agents alternating actions of "speaking" and "listening" OR "showing" and "understanding" OR "hypothesizing" and "validation"
	Favorable economic conditions	Shared understanding and shared attention of a situation

object of interest. The two parties could be both humans, both machines, or a human-machine pair. The economic benefit of two agents communicating an object of interest may have an economic value when shared understanding develops from a confused and ambiguous situation to a well-understood situation that can be effectively put to use by both the agents. This kind of positive emergence exemplified by a coordinated system or a group of intelligent agents capable of bringing meaning to an ambiguous situation is a case of strong emergence, wherein, new knowledge and new structure yield a better system that better supports their economic survival.

FUNDAMENTAL SYSTEMS APPROACH FOR RCIDS FOR EMERGENCE OF LANGUAGE EXAMPLE

To model RCIDS using systems approach as implemented in DEVS formalism, Table 3.3 presents the mapping of RCIDS characteristics with DEVS levels of system specifications. It allows us to formulate the notion of well-defined system between two agents who wish to share information, leading to emergence of language, that aids repeatability and shared understanding of the context.

TABLE 3.3   Relating RCIDS Characteristics with DEVS Levels of Systems Specifications for Emergence of Language

RCIDS Feature	Remarks on Emergence of Language in RCIDS	Level of Systems Specification	DEVS Mapping
Resource constrained	Resource is context dependent. For the language case study, resource could be medium of exchange, the object used for communicating the message, the available time to convey the message or interpret the message. Constraints may appear at both the agent level and at the environment level	Levels 0, 1, 2, and 3 at the agent level Levels 3 and 4 at the environment level	Behavior trajectories in atomic and coupled models
Complex	The system structure between the agents and between the agent and the environment may be multi-level resulting in time-sensitive interactions that are causal at multiple levels. This leads to emergent behavior that becomes causal as well	Levels 2, 3, and 4	Atomic and coupled model specifications, couplings, and variable structure systems
Intelligent	The agent-environment-agent system is a goal-oriented system where information is exchanged between the agents for a specific purpose such as communicating the presence of a threat or food	Levels 0, 1, 2, and 3	Atomic model behavior trajectories
Dynamical	The agent-environment-agent system is a time-sensitive system and each interaction has a spatiotemporal pattern. This is analogous to defining a grammar between the symbols. The presence of "pause" between a "tone" constitutes a symbol or a "word"	Levels 0, 1, and 2	Atomic model behavior specification
System	The emergence of language should be understood in the context of agent-agent system	Levels 0-4	Atomic and coupled models

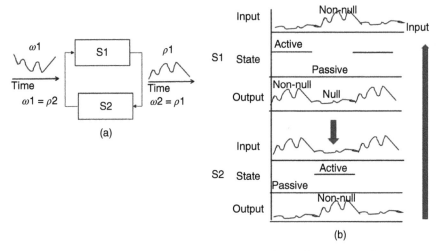

FIGURE 3.4   Interaction of two components attempting to form a composite system.

Well-Formed RCIDS for Allowing Accurate Emergent Behavior

Our approach to emergence of language starts with this background but formulates the problem as an instance of a set of components that must be coupled to form a system with the ability to communicate that does not exists initially. We note that standard languages and principles for computer agent to agent communications are being researched (Chopra *et al.*, 2016).

However, from the systems composition perspective, agent communication research assumes *existing* linguistic competence. Also, while evolutionary theory speculates on how such competence could have developed, it does not do so as an instance of emergence from a formal M&S perspective.

We examine the problem of emergence of a well-defined system from a coupling of a pair of component systems (the smallest example of the multicomponent case). As in Figure 3.4, the time base is a critical parameter to take into account on which input and output streams (functions of time) are happening. The cross-coupling imposes a pair of constraints as shown that the output time function of one component must equal the input time function of the other. The problem is given that the system components have fixed I/O behavior, how can the constraints be solved so that their composition forms a well-defined system?

We must recognize that inputs and outputs are occurring simultaneously in continuous time so that the constraints must be simultaneously satisfied at every instant. One step toward a solution is to break the time functions into parts or segments that allow restricting the problem to segments rather than complete streams. This approach was formalized within the mechanism of iterative specification of systems (Zeigler and Muzy, 2016a) and a complete review is beyond the scope of this paper. However, the way that active-passive compositions help can be explained as in Figure 3.4b.

In this interaction, each component system alternates between two phases, active and passive. When one system is active, the other is passive - only one can be active at any time. The active system does two actions:

1. it sends an input to the passive system that activates it (puts it into the active phase) and
2. it transits to the passive phase to await subsequent reactivation.

Moreover, during any one such segment, only the active component sends a non-null output segment to the passive component. The I/O constraints are satisfied as long as this output segment is accepted by the passive component and the component outputs only null segments while passive.

DEVS-Based Well-Formed RCIDS for Two Agents in Conversation

Assuming that both prelinguistic agents have the necessary basic components – motivation and theory-of-mind basis for communication and Shannon audio-visual mechanisms to create signs for encoding/decoding messages. One agent needs to get another to attend to an audio-visually encoded message about the environment and/or need for action. To do so requires setting up the channel whereby the receiver needs to decode the sender's message and interpret it as information and urge to action. In this way, a unidirectional channel emerges from one agent to another. DEVS enables formal modeling – and subsequent simulation – of this dynamic structure change. The notion of "dynamic structure" involves the first agent producing something within an environment that allows the setting of a channel for the other agent to take notice and attend to. Once the attention-switching has occurred, they form a system dynamically that is composed of two agents and a rudimentary channel that begins the unidirectional information flow.

The addition of a new component and associated couplings is mediated by dynamic structure transformation that can be induced when a model satisfies the requisite conditions (Barros, 2002; Uhrmacher, 2001). The addition of ports provides potential points of input and output information/energy flow among component models and couplings from specific output ports to input ports determine the actual flow paths. Mittal (2013) described the variable structure operating at the levels of systems specification, as reproduced in Table 3.4. It shows the critical importance of variable structure capability in adapting to the new behavior both at the agent (levels 0–3) and at the overall system (levels 3–4) levels.

The next step in emergence of language might have been the addition of a second channel allowing bidirectional communication (or the same channel with duplex capacity). Conceptually, this would be a second instance of the process. However, the establishment of a discipline or protocol for when to speak and when to listen may be more novel and problematic. The problem might be suggested by talking with a friend on the phone (perhaps a successful instance) or pundits from opposite sides of the political spectrum attempting to talk over each other (a negative instance).

TABLE 3.4 Introducing Dynamism at Various Levels of System Specifications (Mittal, 2013)

Level	Name	How Dynamism is Introduced	Outcome
4	Coupled systems	1. System substructure 2. System couplings 3. Subsystem I/O interfaces 4. Subsystem active/dormant state	1. Dynamic component structures 2. Dynamic interaction
3	I/O system	1. Addition/removal of states 2. Augmentation of transitions with constraints/guard conditions	1. Dynamic states 2. Dynamic transitions (probabilistic) 3. Dynamic outputs
2	I/O function	1. Initial state 2. Addition/removal of initial state 3. Addition/removal of I/O pairs	1. Dynamic initial state
1	I/O behavior	1. Time scale between the I/O behavior 2. I/O mapping changing the behavior itself 3. Allowed behavior 4. Addition/removal of I/O pairs	1. Dynamic I/O behavior
0	I/O frame	1. Allowed values 2. I/O to port mapping	1. Dynamic interfaces

PERSPECTIVES: EVOLUTION AND SYSTEM MODELING

We outline two perspectives – evolution domain and system modeling - and show how they may converge to provide a solution.

Evolution Domain Perspective

From the evolution domain perspective, we can recognize that two Shannon components are needed – thinking (e.g., encoding, decoding, and interpreting) and production/transduction (vocalizing and auditioning). Limited cognitive capabilities would bound the number and level of activities that could command attention simultaneously (single-tasking more likely than multitasking). Thus, agents would likely be in either speaking or listening mode - not both. Philosophically, the mind can hold only one thought at one time. It can switch rapidly across a number of thought streams, but it holds only one thought (state) at a given time instant. Conversational interaction might ensue as the talker finishes and the listener now has motivation to speak, for example, "Let's grab it" stimulates the listener to take action. The discipline of alternation between speaking and listening might then become institutionalized. Appendix A provides more evidence that apes cannot develop the required alternation of discourse discipline.

System Modeling Perspective

From the system-modeling perspective, we return to the problem, illustrated in Figure 3.4, that components with cyclic (looped) – as opposed to acyclic – coupling face in forming a well-defined system. Solutions do exist to satisfy the simultaneous I/O constraints imposed by such coupling (sufficient conditions are known for particular classes of systems) (Zeigler *et al.*, 2000). However, as mentioned before, such solutions are difficult to realize due to the concurrency of interaction and must overcome the potential to get stuck in zeno-like singularities. On the other hand, we see that the passive/active alternation interaction for a cross-coupled pair can more easily support the conditions for a well-defined system. Identifying *listening* with *passive* (although mental components are active, they do not produce output) and *speaking* with *active* (output production), we see that alternation between speaking and listening would satisfy the conditions for establishing a well-defined coupled system from a pair of hominin agents. Figure 3.5 illustrates a simplified DEVS representation of such a speak-listen active-passive system. The active agent transitions from thinking to talking and outputting a command that starts the passive agent listening. The receiving agent must detect the end of the command and commence with an associated action. Although not shown, the reverse direction would have the roles reversed with the second agent transmitting a reply and the first agent receiving it.

Evolutionary studies indicated that language should be regarded as a means to coordinate behaviors and requires management of shared attention (Appendix A). The activity-based monitoring paradigm developed by Mittal and Zeigler (2014a) and the emergent property of attention-focusing in such RCIDS (Mittal and Zeigler, 2014b) provide evidence that in a resource-constrained environment, the state of becoming "active" by an agent would allow focusing of that agent's faculties in

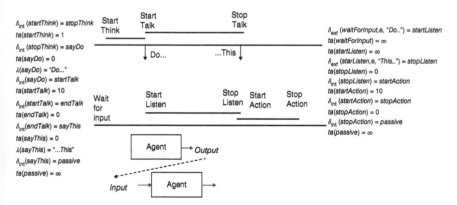

FIGURE 3.5 DEVS modeling of one-way agent-agent communication.

producing something of value for the other agent who is passive and receptive of such an exchange through its sensory apparatus. There may be a fundamental property of systems in that their ability to obtain useful information from the world is related to the ability to suppress activity unrelated to current goals and thereby to satisfy the requirements for well definition of self as system.

RCIDS AND ACTIVITY-BASED PERSPECTIVE ON SHARED ATTENTION

Having proposed a model for emergence of language in hominins, and a required protocol for participating in linguistic interaction, we now raise the problem of how the necessary attentional switching can also arise. Considering the exposition of the RCIDS in Attention Switching in Resource-Constrained Complex Intelligent Dynamical Systems (RCIDS) section, and extended into activity-based intelligent systems at-large (Mittal and Zeigler, 2014a), can they be used to explain how hominins developed a means to share attention and develop the speaker/listener protocol needed to establish linguistic communication?

Assume that each agent is a RCIDS with a closed internal attention system, for example, each agent is attending to a forage activity. Interaction and emergence of a composite RCIDS requires that

1. first, an attention redirecting mechanism (output of agent) and
2. second, that an attention interruption and diversion mechanism (input) be developed.

The first requires creating an interrupt to disrupt the normal ongoing activity of the other agent by the high-value activity (or a highly valued information message) by the agent possessing the new knowledge that is ready to share it. The second requires the ability to be interrupted and redirect the current activity toward externally given knowledge (as an input). This process is the attention-switching mechanism at the receptive agent. Successful completion of this process establishes the presence of three things: a highly valued message output stream by an agent (the one with the message), a communication channel, and a receptive listening agent that stops its current task and focuses on the new activity. This is the emergence of half duplex channel in RCIDS. Now that attention-switching has occurred in the receptive agent, its mental resources are processing the new information to come up with new outputs from his knowledge-based/mental model. When this agent is ready, it uses the symbols available at its disposal (due to the constraints of the environment and possible lack of symbology adding more to the constrained nature) to repeat the process, and on reciprocation by the first agent, they begin to share symbols and refine the content of conversation. Given sufficient time and mutually engaging behavior, the sequence of such interactions will result in shared understanding of the communicated context. This process can then be extended toward one-to-many scenario where one speaker with the message is communicating to many attentive listening agents.

CONCLUSIONS AND FUTURE WORK

DEVS has a number of features that enable the representation of systems components and couplings that can change dynamically as needed for genuine adaptation and evolution and therefore provides the concepts and formalisms needed for modeling and simulation of emergent behavior. Active-passive compositions provide a class of systems from which we can draw intuition and examples for generalizations about system emergence at the fundamental level. Here, we employed the modeling and simulation framework based on Systems Theory with particular focus on its concepts of iterative specification. Earlier development formulated sufficient conditions for feedback-coupled assemblages under which iterative system specifications can be composed to create well-defined systems that can also be represented in the DEVS formalism. Our approach to emergence of human language here has formulated the problem in terms of a set of components that must be coupled to form a system with innovative inter-component information exchange. We showed that passive/active alternation interaction of a cross-coupled pair can satisfy the required conditions for well definition. The implication is that learning to alternate between speaking and listening may have been a necessary precondition for a pair of hominin agents to establish a well-defined symbol-based information exchange system. The fundamental rule in the setup of two agents engaged in communication is that both cannot be simultaneously in active mode or listening mode. For a channel to emerge, one agent has to be in active and the other in listening mode, followed by an alternation. The overall process consists of three steps:

1. Define the system agents and their timing behaviors accurately. The timing is a decisive factor in development of symbols. The constraints present in the environment as well as the lack of available symbols in a given context adds to the emergence of new symbols as the message gets formulated. The received message is then communicated in reciprocating manner due to the characteristic property of alternation, leading to refinement of symbols.
2. Evaluate the mechanisms to add new information and structures so that they become causal for both the agents.
3. Develop protocols and grammar such that the process becomes repeatable, reproducible, and efficient.

Earlier RCIDS and ABIS research had established the capability of attention-switching in a resource-constrained environment. However, it assumed the presence of communication channels between agents. A fundamental concept brought forward in this work is the necessary property of resource constraints in the system that warrants the agents in the system to use the environment in a way that is novel and directs actions that are still goal pursuing. Repeated use of such mechanism leads to emergent behavior that either gets strengthened due to positive feedback or dissipates due to negative feedback. This chapter focused on the systems theoretic foundations for fundamental emergence of communication channel between the two agents and the high-valued messages that warrants attention-switching, to begin with. More research is required for a more complete account of how human language emerged. In this chapter, we addressed the motivation to create a language by early hominins that would help them understand their environmental situation. We provide theoretical constructs using DEVS systems specification on how such new phenomenon could emerge out of necessity from environmental conditions. More research is needed to understand the dynamic nature of the communicated audio-visual multimedia objects and various patterns these can be put into communicate the objects of interest by the two (or more) communicating agents. The structure of such patterns can be studied using ontology engineering. The behavior of such patterns needs to be understood using dynamical systems theory. Together, this needs to be integrated in the RCIDS and ABIS frameworks toward understanding the emergence of language as a whole. Another particular area is the cyber domain where the boundaries are not clearly defined between the agents and the environment. Mittal and Zeigler (2017) discussed the role of M&S in cyber environments. Exposition of RCIDS and ABIS concepts in cyber domain is an opportunity worth considering.

APPENDIX A: LIMITATIONS OF NON-HUMAN PRIMATES WITH RESPECT TO ALTERNATION OF DISCOURSE

Observations indicate that apes point only for humans and not for one another (Kinneally, 2008). Even apes trained to communicate with humans do not exhibit this behavior with each other. The reason seems to be that humans respond to apes' gestures, whereas other apes are oblivious to such gestures. Perhaps, apes do not recognize that they are being addressed (Tomasello, 2006). Kinneally (2008) relays the account of two apes that had successfully acquired many signs and used them effectively. However, when one day they were asked to converse, there resulted a sign-shouting match; *neither ape was willing to listen*.

Savage-Rumbaugh (2004) provides an analysis that strongly supports our system theoretic-based emergence approach. She says that "language coordinates behaviors between individuals by a complex process of exchanging behaviors that are punctuated by speech." At its most fundamental, language is an act of *shared attention*, and without the fundamentally human willingness to listen to what another person is saying, language would not work. Symbols like words are devices that coordinate attention, just as pointing does. They presuppose a general give-and-take that chimpanzees do not seem to have (Tomasello, 2006).

DISCLAIMER

The author's affiliation with the MITRE Corporation is provided for identification purposes only and is not intended to convey or imply MITRE's concurrence with, or support for, the positions, opinions, or viewpoints expressed by the author. Approved for Public Release, Distribution Unlimited [Case: PR_17-3254-3].

REFERENCES

Ashby, W. (1956) *An Introduction to Cybernetics*, University Paperbacks, Methuen, London.

Barros, F.J. (2002) Towards a theory of continuous flow models. *International Journal of General Systems*, **31** (1), 29–40.

Boardman, J. and Sauser, B. (2006) System of systems – the meaning of "of". IEEE/SMC International Conference on System of Systems Engineering. Los Angeles, CA: IEEE.

Chopra. A. K. et al. 2016. Research directions in agent communication, *ACM Transactions on Intelligent Systems and Technology*, V, N, **2012**, 1-27.

Everett, D.L. (2012) *Language: The Cultural Tool*, Vintage, 368 p.

Foo, N.Y. and Zeigler, B.P. (1985) Emergence and computation. *International Journal of General Systems*, **10** (2–3), 163–168.

Hu, X. and Zeigler, B.P. (2013) Linking information and energy–activity-based energy-aware information processing. *Simulation: Transactions Social Modelling & Simulation*, **89** (4), 435–450.

Kinneally, C. (2008) *The First Word: The Search for the Origins of Language*, Penguin Pub.

Kubik, A. (2003) Toward a formalization of emergence. *Artificial Life*, **9** (1), 41–65.

Maier, M. (1998) Architecting principles for system of systems. *Systems Engineering*, **1** (4), 267–284.

Mittal, S. (2013) Emergence in stigmergic and complex adaptive systems: a formal discrete event systems perspective. *Cognitive Systems Research*, **21**, 22–39.

Mittal, S. and Rainey, L. (2015) Harnessing emergence: the control and design of emergent behavior in system of systems engineering. Proceedings of the Conference on Summer Computer Simulation, SummerSim '15, Society for Computer Simulation International, San Diego, CA, USA.

Mittal, S. and Risco-Martín, J.L. (2013) *Netcentric System of Systems Engineering with DEVS Unified Process*, CRC-Taylor & Francis Series on System of Systems Engineering, Boca Raton, FL.

Mittal, S. and Zeigler, B.P. (2014a) Context and attention in activity-based intelligent systems. *ITM Web of Conferences*, **3** (1), 1–16.

Mittal, S. and Zeigler, B.P. (2014b) Modeling attention switching in resource constrained complex intelligent dynamical systems (RCIDS). Symposium on Theory of M&S/DEVS, Spring Simulation Multi-Conference, Tampa, FL.

Mittal, S. and Zeigler, B.P. (2017) Theory and practice of M&S in cyber environments, in *The Profession of Modeling and Simulation* (eds A. Tolk and T. Oren), Wiley. doi: 10.1002/9781119288091.ch12

Muzy, A. and Zeigler, B.P. (2014) Specification of dynamic structure discrete event systems using single point encapsulated control functions. *International Journal of Modeling, Simulation, and Scientific Computing*, **05** (03), 1450012.

Oren, T.I. and Yilmaz, L. (2015) Awareness-based couplings of intelligent agents and other advanced coupling concepts for M&S. SIMULTECH, Proceedings of the 5th International Conference on Simulation and Modeling, Methodologies, Technologies and Applications, Colmar, Alsace, France, 21–23 July, pp. 3–12, www.site.uottawa.ca/~oren/y/2015/D04_couplings-pres.ppsx.

Ören, T.I. and Zeigler, B.P. (2012) System Theoretic Foundations of Modeling and Simulation: A Historic Perspective and the Legacy of A Wayne Wymore. *Simulation*, **88** (9), 1033–1046.

Park, S. and Hunt, C. (2006) Coupling permutation and model migration based on dynamic and adaptive coupling mechanisms. Agent Directed Simulation Conference/Spring Simulation Multiconference, Von Braun Center, Huntsville, Alabama USA, pp. 6–15.

Savage-Rumbaugh, S. (2004) http://www.ted.com/talks/susan_savage_rumbaugh_on_apes_that_write (accessed February 2017).

Steiniger, A. and Uhrmacher, A.M. (2016) Intensional couplings in variable-structure models: an exploration based on multilevel-DEVS. *ACM Transactions on Modeling and Computer. Simulation*, **26** (2), 9:1–9:27.

Szabo, C. and Teo, Y.M. (2015) Formalization of weak emergence in multiagent systems. *ACM Transactions on Modeling and Computer. Simulation*, **26** (1), 6:1–6:25.

Tomasello, M. (2006) Why don't apes point? in *Roots of Human Sociality: Culture, Cognition and Interaction* (eds N.J. Enfield and S.C. Levinson), Berg, Oxford & New York, pp. 506–524.

Tsotsos, J.K. and Bruce, N.B. (2008) Computational foundations for attentive processes. *Scholarpedia*, **3** (12), 6545. doi: 10.4249/scholarpedia.6545

Uhrmacher, A.M. (2001) Dynamic structures in modeling and simulation: a reflective approach. *ACM Transactions on Modeling and Computer. Simulation*, **11** (2), 206–232.

Wymore, A.W. (1967) *A Mathematical Theory of Systems Engineering: The Elements*, John Wiley & Sons, New York.

Zeigler, B.P. (2016a) A note on promoting positive emergence and managing negative emergence in systems of systems. *The Journal of Defense Modeling and Simulation*, 133–136. doi: 10.1177/1548512915620580

Zeigler, BP 2016b. Contrasting emergence: in systems of systems and in social networks, *The Journal of Defense Modeling and Simulation*. **13** (3), 271-274, /1548512915620580,13,1, http://journals.sagepub.com/doi/abs/10.1177/1548512916636934

Zeigler, B.P. (2017) Emergence of human language: a DEVS-based systems approach. Proceedings of M&S of Intelligent, Adaptive and Autonomous Systems, SpringSim, Virginia Beach, VA.

Zeigler, B.P. (2018) DEVS-based modeling and simulation framework for emergence in system-of-systems, in *Engineering Emergence: A Modeling and Simulation Approach* (eds L. Rainey and M. Jamshidi), CRC Press, in press.

Zeigler, B.P. and Hammonds, P. (2007) *Modeling & Simulation-Based Data Engineering: Introducing Pragmatics into Ontologies for Net-Centric Information Exchange*, Academic Press, Boston.

Zeigler, B.P. and Muzy, A. (2016a) Iterative specification of input/output dynamic systems: emergence at the fundamental level. *Systems*, 4 (4), 34. doi: 10.3390/systems4040034

Zeigler, B.P. and Muzy, A. (2016b) Some modeling & simulation perspectives on emergence in system-of-systems. Proceedings of the Modeling and Simulation of Complexity in Intelligent, Adaptive and Autonomous Systems 2016 (MSCIAAS 2016) and Space Simulation for Planetary Space Exploration (SPACE 2016) Pasadena, California, SCS, April 3–6.

Zeigler, B.P., Praehofer, H., and Kim, T. (2000) *Theory of Modeling and Simulation: Integrating Discrete Event and Continuous Complex Dynamic Systems*, Academic Press, NY.

4

GENERATIVE PARALLAX SIMULATION: CREATIVE COGNITION MODELS OF EMERGENCE FOR SIMULATION-DRIVEN MODEL DISCOVERY

Levent Yilmaz

Department of Computer Science and Software Engineering, Samuel Ginn College of Engineering, Auburn University, Auburn, AL 36849, USA

SUMMARY

The three pillars of emergent behavior are variation, interaction, and selection. In this chapter, these three processes are characterized and their roles in emergent behavior are examined. As computational substrates for modeling creative cognition, these processes, along with systems models of creativity, provide a foundation for model discovery. The view of models as emergent phenomena allows us to rethink simulation modeling so that creativity in exploratory phases of model-based problem solving is enhanced rather than stifled. Generative parallax simulation (GPS) is introduced as a strategy and an abstract specification for its realization is presented. GPS is based on an evolving ecology of ensembles of models that aim to cope with ambiguity and uncertainty, which pervade in early phases of model-based science and engineering. GPS views novelty as an emergent phenomenon; it harnesses the principles of self-organization and cognitive coherence while drawing from the science of complexity to enable simulation technology development to enhance discovery in model-based science and engineering.

Emergent Behavior in Complex Systems Engineering: A Modeling and Simulation Approach,
First Edition. Edited by Saurabh Mittal, Saikou Diallo, and Andreas Tolk.
© 2018 John Wiley & Sons, Inc. Published 2018 by John Wiley & Sons, Inc.

INTRODUCTION

Many natural and artificial phenomena that are characterized as complex system involve large networks of individual autonomous components, which operate using specific rules of behavior with no central control (Holland, 2000; Mitchell, 2009). For systems, such as intelligent urban transportation and smart power grids, the principles of self-organization and self-adaptive behavior are increasingly becoming the operating mechanisms. However, the underlying mechanisms of these systems often result in hard-to-predict and changing patterns of behavior (Parunak and VanderBok, 1997). These systems produce and use information and signals from both components within and the external environment, and as the context evolves, they are expected to adapt to improve the quality of service through learning or evolutionary strategies. This mode of decentralized coordination and self-adaptive nature of systems call for modeling strategies that facilitate emergent discovery of mechanisms that exhibit desired behavioral regularities while providing support for characterizing and explaining emergent behavior.

Emergent behavior arises as novel and coherent patterns because of local interactions among individual elements of a system during the process of self-organization in complex systems. Among the characteristics of emergent behavior are (i) novelty, (ii) coherence in the form of patterns that arise at the macro-level, and (iii) evolution of patterns via dynamic processes. Variation provides the requisite constructs for novelty as well as adaptation. Yet, to take advantage of what has been learned, there needs to be limits in the amount of variety in the system. The key challenge is to establish a balance between variety and uniformity. Interaction is essential also because patterns of behavior arise from the interactions among the elements of the system, and the mechanisms underlying interaction can harness complexity by altering patterns of interaction (Axelrod and Cohen, 2000). Trust and cooperation among the components of a system are governed and influenced by patterns of interaction. Furthermore, the uncertainty and the evolving nature of the context require the provision of processes that amplify the success of the system through selection of components and their strategies.

The ability to influence emergent behavior and to develop confidence about a system's ability to exhibit desired behavior is a critical challenge. Such ability requires deliberate strategies for orchestrating the degree of variety, the mechanisms of interaction, and means for selecting effective components and behavioral strategies. In this chapter, the process of scientific discovery is used as a metaphor to provide a framework for modeling systems with emergent behavior and experimenting with such models to discern plausible mechanisms that influence and account for the desired behavior. Such a methodology requires (i) generating and prioritizing modeling assumptions that are ripe for exploration, (ii) enabling hypothesis-guided automated generation and execution of experiments (Yilmaz et al., 2016), and (iii) interpreting results to falsify and revise competing behavioral mechanisms.

The critical obstacles to supporting these steps include the disconnect between model discovery and experimentation, as well as consideration of justification only after models are built. We conjecture that addressing these issues requires viewing models as emergent phenomena via dynamic coupling of model building and experimentation while enabling continuous feedback between their technical spaces. Direct coupling of these spaces needs to provide sound explanatory characterization about what alternative and complementary mechanisms are plausible, whether they cohere and under what conditions. This is in sharp contrast with the current practice. Often when experiment results deviate from the expected behavior, engineers locate the cause of the problem and mitigate the imminent issue. This leads to premature convergence to an authoritative model that is not robust enough to address variability and uncertainty. Yet, the ability to retain, experiment with, and relate alternative model mechanisms is critical for developing robust solutions. This is akin to the issue in philosophy of science (Klahr and Simon, 1999) of discovering robust theories (Levins, 1966) that aim to achieve a balance between generality, precision, and accuracy.

Motivated by these observations, the objective of this chapter is to put forward a methodological basis, which aims to (i) explore the utility of viewing models as adaptive agents that mediate among engineering domain theories, data, requirements, principles, and analogies; (ii) underline the role of cognitive assistance for model discovery, experimentation, and evidence evaluation to discriminate between competing models; and (iii) examine alternative strategies for explanatory justification of plausible solutions via cognitive models that explicate coherence judgments. For this chapter, creativity is interpreted as the production of novel and useful ideas (Amabile, 1996). More specifically, creative discovery is viewed as the product of a process that involves concept (e.g., model) combination, expansion, metaphor, and analogy along with mechanisms acting in concert to expand the frontiers of knowledge and conceptualization in a domain. Hence, creative cognition can be construed as a system by which processes transform and create structures to produce results that are novel, yet rooted in existing knowledge (Ward et al., 1997). Generative parallax simulation (GPS) views novelty as an emergent phenomenon; it harnesses the principles of self-organization and draws from the science of complexity to enable simulation technology development to enhance discovery in model-based science and engineering.

Our main thesis is that model building involves coherent molding of various ingredients (i.e., domain theory, data, requirements, analogies, and principles) to meet certain quality criteria. This thesis is predicated on two sub theses: the context of discovery and the context of justification. In the context of discovery, mechanism integration plays a key role. The integration needs to take place by transforming the constituent elements by merging them into a unified mechanism. One aspect of molding is the mechanistic representation, and the other aspect is of calibration: the choice of the parameters in such a way that the model not only fits the data but also integrates other ingredients, such as constraints, domain theory, principles, and analogies. The context of justification refers to built-in justification; that is, when the set of ingredients implement the requirements, the model is expected to address, then justification is built-in.

BACKGROUND

The adaptive nature of models as highlighted above is consistent with the "models as mediating instruments" perspective (Morgan and Morrison, 1999). Per this view, a model needs to align theoretical and empirical constructs to converge to an explanation of expected behavior. The integration of the ingredients of the model toward aligning theory and data suggests the autonomous character of a model, measured as a function of the degree of coherence of the competing assumptions, mechanisms, and evidential observations. However, models should not merely mediate existing theoretical and empirical elements. New elements and constructs are generated not only to introduce mechanisms but also to integrate them in ways that are logical and consistent.

Models as Cognitive Mediators

Figure 4.1 illustrates the mediation role of models between theory and data. Theoretical principles are leveraged to construct mechanistic hypotheses in the form of behavioral rules. The initial construction of the model is followed by the experimentation process through simulation.

Targeted instrumentation of the model results in observed simulation data that forms the basis for learning about the efficacy of the hypothesized mechanistic assumptions. The dynamic and mutually recursive feedback between the model and data refers to adaptive learning. The data gathered through experimentation are analyzed to make decisions about model representation. Model revisions, if there is sufficient consensus, result in mechanism revision. Consequently, the theoretical principles induced from limited data become increasingly accurate in their explanatory and generative power.

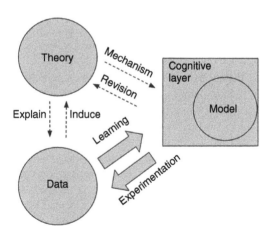

FIGURE 4.1   Models as mediators.

Similarly, Dynamic Data-Driven Application System (DDDAS) paradigm (Darema, 2004) promotes incorporation of online real-time data into simulation applications to improve the accuracy of analysis and the precision of predictions and controls. As such, DDDAS aims to enhance applications by selectively imparting newly observed data into the running application to make it congruent with its context. The ability to guide measurement and instrumentation processes is critical when the measurement space is large. By narrowing measurement within a subset of the overall observation space, the methodology reduces both the cost and the time to test, while also improving the relevance of data. Viewing coupled model experiment system dynamics as a DDDAS requires advancements in variability management, model interfaces to instrument the simulation, and facilities to incorporate data-driven inference back into the model's technical space. As such, the provision of run-time models and dynamic model updating are requisite for closing the loop.

Creative Cognition and Learning

Creative processes often involve a broad idea generation phase that is approached from different perspectives, followed by idea evaluation and selection (Amabile, 1996). Because creativity requires novel yet useful solutions to make creative leaps, appropriate tradeoffs between constraints and flexibility are needed for the models' representation of the problem. Hence, the effectiveness of simulation systems that support creative discovery will rely heavily on their ability to start behaving robustly across many hypotheses, constraints, and propositions, followed by narrowing toward a limited range of conditions that are found to be plausible in terms of an explaining extensible set of attributes. Development of such simulation systems will require progress in addressing the following:

- when does discovery involve exploitation of a problem space, and when does it involve exploring alternative problem representations?
- what are appropriate tradeoffs between constraints and flexibility in supporting incremental and iterative expansion of the model space to explain expanding sets of mechanisms and attributes underlying the phenomena of interest?
- how can models and theories of creative cognition help us rethink simulation modeling so that creativity is enhanced rather than stifled?

There has been extensive research on creativity, discovery, and innovation at different levels spanning from cognition in individuals to groups, organizations, and collective creativity in large-scale community forms of organization. Although significant progress is achieved in many disciplines, the topic is relatively new for the modeling and simulation discipline. To improve creativity and discovery in model-based science and engineering, advanced simulation technologies can be extended to provide facilities and opportunities that go beyond its conventional experimentation capabilities. Principles of creative problem solving help establish the role of evolutionary dynamics in generating emergent novelty.

- As indicated by Gero and Kazakov (1996), evolution is creative in the sense of generating surprising and innovative solutions. Analogous to creative and innovative problem solving, evolutionary mechanisms improve solutions iteratively over generations.
- Ambiguity and lack of clarity about knowledge about existing relations between the requirements for ideal solution and forms that satisfy these requirements (Rosenman, 1997) are useful opportunities for creativity.
- Exploring a search space in an effective and efficient manner and ability to explore alternative search spaces by redefining the problem representation are critical in creative problem solving. Evolutionary mechanisms that do not have considerable freedom to vary their representations are clearly not creative.
- Creativity requires transfer of knowledge and use of metaphor (Holland, 2000) and analogical reasoning across disciplines (Goldberg, 1999). Hence, evolutionary dynamics coupled with ecological perspective that favors transfer is more likely to be creative.

MODEL DISCOVERY AND EMERGENCE AS A CREATIVE COGNITION PROCESS

Bottom–up theories of creativity include generate and explore model (Smith and Blankenship, 1991), systems model of creativity (Csikszentmihalyi, 1996), and evolutionary models of creativity (Campbell, 1960; Simonton, 1999). The commonality between these models is twofold. First, they all hypothesize and substantiate that creativity is influenced by synergistic interaction with and confluence of environmental factors. Second, their underlying mechanisms are based on the principles of models of evolutionary systems. That is, creative ideas emerge through a process analogous to the natural selection process.

Systems Model of Creativity. According to the systems view of creativity, introduced by Csikszentmihalyi (1996), creativity is not the product of an individual, but rather the interaction between the individuals, context, and the problem domain. For creativity to occur, original and novel contributions submitted by individuals should be evaluated for inclusion as a new body of knowledge in the domain. Individual contributions are predicated on the practices and knowledge stored in the problem domain, so that novelty can be produced as a variation in the content of the domain. Figure 4.2 depicts the components and interactions of this model. The domain is considered as a critical component of creativity because it is impossible to introduce a variation without reference to an existing pattern specified in the domain knowledge. The technical contributions made by individuals produce creative solutions to domain-specific problems. Such technical contributions induce novel variations in the domains that constitute the context.

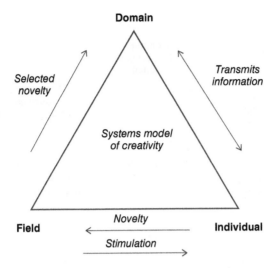

FIGURE 4.2   Systems model of creativity.

Generate and Explore Model. The view of creative problem solving as a bottom–up generative process is the basis for the generate and explore model. Introduced by Smith and Blankenship (1991), the model is based on a two-stage process. In the first stage, the individual *generates* pre-inventive forms, which are ideas and conceptual structures that might be useful for creative production. In the second phase, that primarily focuses on the *exploration* of pre-inventive structures to interpret their utility for solving specific problems. The generation phase can leverage various cognitive processes. The individual may retrieve prior learned concepts for evaluation, or transform concepts in novel ways or configure them in new forms to orient the solution in a new context. Constraints on the final product can be imposed to influence the generation and exploration phases.

Evolutionary Models of Creativity and Emergence. According to Campbell (1960), three major conditions are necessary for creative thinking, which are also requisite for emergence: (i) similar to organic evolution, there should be a mechanism to generate new ideas in the form of ideational variation; (ii) following the generation of new ideas, specific criteria must be used to select those ideas that are useful for the problem under consideration; (iii) the variations that are selected should be retained and used for reproducing new ideas. The arguments against Campbell's model of creative cognition focus on the need for incorporation of experience and learned behavior.

GENERATIVE PARALLAX SIMULATION: BASIC CONCEPTS

Development of simulation methods that support creative problem solving requires leveraging principles that explain emergence of creativity. The perspective examined

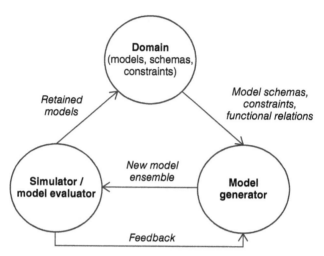

FIGURE 4.3   Generate and explore reference model.

in this work is the creative cognition world view that focuses on bottom–up idea generation and evaluation strategies that enable optimal combinations of explorative and exploitative modes of inquiry.

An Abstract Model of Creative Cognition

Examination of creative cognition models reveals three main components that interact with each other to produce useful novelty: domain, generator, and evaluator. We define a high-level reference model (see Figure 4.3) to delineate each component along with its role.

Domain: The domain embodies the ensemble of plausible models (problem formulations), hypothesized mechanisms believed to represent referent processes, constraints (e.g., experimental conditions and range of values of known variables), phenomena being explored, plus schemas (e.g., templates for hypothesized model mechanisms) used to specify analogues.

Generator: The generation phase of the creative cognition process can be based on any number of novelty generation actions. To be successful in improving creative insight into a problem, a simulation platform and its underlying mechanisms need to be aware of principles and operators underlying the process for generating creative novelty. Sawyer (2017) discusses and illustrates four major variation creation operators that often enable creative outcomes:

– *Concept elaboration* – extending existing concepts through new features and constraints to obtain more specialized concepts.

– *Concept combination* – requires integration of two or more concepts to obtain a new novel concept.

- *Concept transfer* – involves establishing a metaphor through analogy to reuse a collection of related concepts in a new context.

- *Concept creation* – refers to invention of new concepts that do not exist in the problem domain.

Evaluator: Model synthesis is a hypothesis: these components as composed become a mechanism upon execution, and that mechanism will lead to measurable phenotypic attributes that mimic prespecified, targeted, referent attributes. A more interesting model is one capable of a greater variety of phenotype mimicry, and for which the mappings from model to referent mechanisms can be concretized; conceptual mappings cannot. Improved model-to referent mappings at the mechanism level are expected to lead to deeper insight. Models capable of greater mimicry of targeted attributes are retained. Phenomimetic measures are needed to compare phenotype overlap: attribute similarity. Comparative phenometrics will depend on the relative ability of two or more models to achieve prespecified measures of similarity to referent. Included will be quantitative validation metrics as well as more qualitative measures, such as behavioral similarities. Comparative phenometrics should also consider the degree of phenotype overlap and mechanistic similarity between models. Substantial multi-attribute similarity coupled with some mechanistic divergence has the potential to catalyze creative leaps. The feedback provided back to the generator improves its effectiveness in selecting the model generation operators through a cognitive learning mechanism.

Abstract Specification of the Structure and Dynamics of GPS

To formalize GPS, we define the structure of the domain of models as a graph of ensembles, $G = (V, R)$, V is the set of nodes, and each node $v \in V$ denotes an ensemble of models, and R is the set of relations depicting affinity (e.g., similarity in terms of function and form) between the ensembles. Each ensemble E has a neighborhood $N(E)$, which refers to a connected subgraph of G containing E. For our purposes, each ensemble contains a collection of meta-objects, each one of which specifies the schema of a corresponding model. Figure 4.4 depicts the structure of the graph of ensembles. The strength of relations (e.g., $w(i, k) = w_{ik}$ or $w(k, i) = w_{ki}$) between ensembles signify the degree of similarity analogs in the source ensemble exhibit with respect to phenotypic attributes of the target ensemble's referent.

Evaluation: Given the above specification, we need to define evolutionary aspects of the ensemble. Three major factors are of interest. First, analogs that exhibit behaviors like those targeted must be favored, as they generate sufficiently valid behavior, if a similarity measure value exceeds a prespecified threshold. Second, those analogs that use divergent mechanisms yet have a sufficient level of validity may facilitate discovery of novel mechanisms. Consequently, they should be retained. Finally, those models that demonstrate success in generating behaviors and features imposed by neighboring ensembles should be favored, as they may be able to extend their usefulness and scope. Models that satisfy the constraints of multiple phenotypic attributes

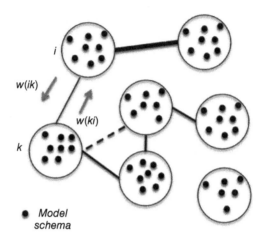

FIGURE 4.4   Model ensembles.

will relate to schemas (hypothesized mechanisms) from more than one ensemble: they are expected to have a larger impact.

Transfer: Given the set of edges, $R \subseteq V \times V$ of the ensemble graph, G, each edge (E_i, E_j), is associated with two components: w_{ij} and w_{ji}. These components, shown in Figure 4.4 as the strength of relations, are positive integers that define transfer rates from E_i to E_j and E_j to E_i, respectively, and $Q = \sum w_{ij} \leq N$. Each schema s in the ensemble has a propensity to transfer $\mu(s)$, which is a monotonic function of the change of similarity over k iterations. Initially, w_{ij} for each ensemble i is set to a low value. These transfer rates, which emulate conceptual transfer and analogy-based discovery, may change over time. Learning takes place as information about the similarity of copied and transferred models is gathered. If models that are transferred from E_i to E_j improve their average similarity, the transfer rate for migration from E_i to E_j is updated to increase the number of transfers; otherwise, the transfer rate is decreased. At each round of evaluation, for every (E_i, E_j) in the analogue ensembles graph, the population in ensemble i is scanned to locate K schemas with $\mu(s) \geq \gamma_{transferThreshold}$.

Implications of the Ecological Perspective: The generation and transfer of mechanisms update the contents and structure of the ensemble graph as an evolving ecology. The network of ensembles enables interaction between generative model ensembles. The boundaries denote separate attributes and targeted objectives in the referent. Ensembles communicate with each other and share models across their boundaries. An overall solution is discovered through continual flows of models so that ensembles in the graph can sustain themselves and improve the impact and usefulness of local solutions. Those models that do not survive after migration to other ensembles are considered as falsified. Exchanges of models in this evolving ecology of ensembles are sustained by pervasive cooperation because a model that migrates to a new ensemble contributes its traits to its new context. Furthermore, viewing a solution to complex multi-aspect and multi-attribute problem as an evolving ecology achieves stability and resilience through richness and complexity of interaction. Due to the synergistic combination of evolution and ecological perspective, the generated solution that is defined in terms of an ensemble of ensembles is flexible due to consequence of multiple feedback loops that keep the solution in a state of dynamic balance. That is, the solution is not biased toward a single targeted attribute; hence, the solution is not optimized toward one specific aspect. Rather, ensembles fluctuate around their optimal forms.

A REFERENCE ARCHITECTURE FOR MODEL ENSEMBLES AS COGNITIVE AGENTS

As an active entity, models with cognitive capabilities provide features that overlay the simulation model and augment it to support computational discovery. In this view, a model is construed as a family of models, which evolve as learning takes place. As such, models need to be designed with variability management (Metzger and Pohl, 2007) in mind to support seamless customization and to address a variety of experiment objectives, especially when the target system has multiple facets.

Figure 4.5 illustrates the building blocks of an active model that is coupled to an experimentation environment to maintain a mutually beneficial and adaptive feedback between theory and data. Theoretical constructs are characterized by the variability management layer (VML) and the models that encapsulate mechanistic hypotheses, principles, and constructs underlying the theory. A family of models is defined in terms of features (Kang *et al.*, 2002; Oliveira *et al.*, 2013) that can be configured to synthesize alternative models. Features define mandatory or optional standard building blocks of a family of models along with their interrelationships. Variability management via feature models is a common strategy in product-line engineering practice. To support such a process, a variability management language should enable the specification and weaving of variants in the host simulation modeling language.

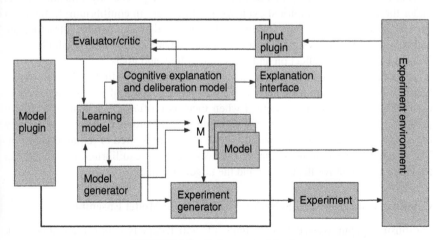

FIGURE 4.5   Reference architecture.

The selection of such variants in the form of features triggers specific actions that customize a model by adding, removing, or updating model fragments.

Extending variant management to online model tuning to support experimentation requires a run-time evaluation mechanism. To this end, an input plug-in transforms raw experiment data into a format that can be analyzed using the Evaluator/Critic component, shown in Figure 4.5. The evaluator can be as simple as a filter that abstracts the data. However, to provide cognitive assistance, a sophisticated model can facilitate coupling model's technical space to formal methods such as Probabilistic Model Checking (Kwiatkowska et al., 2011) to determine the extent to which expected behavior is supported by the simulation data. Those mechanisms that lend significant support to expected behavior are retained, whereas others are revised or declined from further consideration.

The learning model uses the results of the evaluator to update the confidence levels of competing hypotheses. For example, a Bayesian Net can revise conditional posterior probability estimates using the Bayes' rule in terms of prior probabilities and the observed evidence. Alternatively, cognitive models such as explanatory coherence (Thagard, 2013) can be used to acquire, modify, reinforce, or synthesize hypotheses to steer the model-driven discovery process. This incremental and iterative strategy is coordinated by a mediation process that governs the interaction between the technical spaces of models.

MODEL EMERGENCE VIA REFLECTIVE EQUILIBRIUM

In this section, we aim to demonstrate the feasibility of a formal model that can support attaining dynamic balance among competing hypothesized models and experiments in a model ensemble. The theory of *Reflective Equilibrium*, originally formulated by Rawls (1971) in his seminal work on the Theory of Justice, can present a view to achieving such balance. According to this theoretical framework, in the presence of conflicts among competing hypotheses, we proceed by adjusting beliefs under the presence of empirical evidence until they are in equilibrium. In the most general sense, Reflective Equilibrium can be construed as an emergent attractor state in a complex adaptive system. The attractor state emerges at the end of a deliberation process by which we reflect on and revise our beliefs and goals.

If we can view equilibrium as a stable state that brings conflicts to a level of resolution, the equilibrium state serves as a coherence account of justification, and an optimal equilibrium can be attained when there is no further inclination to revise judgments about the plausibility of competing model features because together they have the highest degree of acceptability with respect to desired or required behavior. The principles and judgments that one arrives when the equilibrium is reached can provide an account for the context and the target examined.

Cognitive Coherence as a Selection Mechanism in Model Ensembles

For illustration purposes, consider Figure 4.6, which depicts a connectionist network comprised of observed or expected system behaviors, system requirements that aim

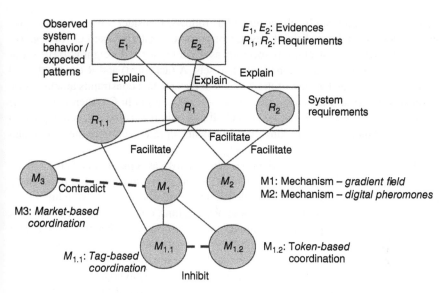

FIGURE 4.6 Connectionist constraint network.

to explain these observations, and model mechanisms that facilitate addressing the requirements at different levels of resolution. Furthermore, suppose that the intended system is a self-organizing system, and the mechanisms are well-known patterns such as gradient-field, digital pheromones, as well as low-level supportive mechanisms such as tag-based, token-based, or market-based coordination with their specific template structures. The hypothetical network shown in Figure 4.6 evolves as learning takes place through experimentation. In this example, M_1 and M_2 together facilitate R_1, and therefore they are compatible, whereas, M_3 is alternative to M_1, and hence contradicts with it. That is, they do not cohere. Similarly, tag-based coordination mechanism, $M_{1.1}$, and token-based coordination, $M_{1.2}$, are alternative strategies that contribute to the higher level mechanism, M_1. Therefore, they compete. If the network is viewed as a connectionist network with node activation values depicting the acceptability of model mechanisms (hypotheses), while the strength of links representing the degree of compatibility or coherence among nodes toward satisfying system requirements, we can start developing insight into evaluation of hypothesized mechanisms. The strategy can lend support to providing a coherence-driven justification of decisions made in model discovery and design.

If we can view equilibrium in such a network as a stable state that brings design conflicts to a level of resolution, the equilibrium state serves as a coherence account of model justification, and as indicated by (Thagard, 2013), an optimal equilibrium can be attained when there is no further inclination to revise decisions because together they have the highest degree of acceptability. The decisions that one arrives when the equilibrium is reached can provide an account for the solution under the system requirements examined. In relation to our example shown in Figure 4.6, the act of resolving conflicts among hypotheses from uncertainty to a stable coherent state of equilibrium can be characterized as follows: Consider a finite set of cognitive elements e_i and two disjoint sets, C^+ of positive constraints, and C^- of negative constraints, where a constraint is specified as a pair (e_i, e_j) and weight w_{ij}. The set of cognitive elements are partitioned into two sets, A (accepted) and R (rejected), and $w(A,R)$ is defined as the sum of the weights of the satisfied constraints among them. A satisfied constraint is defined as follows: (1) if (e_i, e_j) is in C^+, then e_i is in A if and only if e_j is in A. (2) if (e_i, e_j) is in C^- if and only if e_j is in R. When applied to practical and theoretical reasoning, these elements represent assumptions, goals, and propositions.

The coherence problem can then be viewed partly as a parallel constraint satisfaction problem, where the goal is of satisfying as many constraints as possible by propagating a wave of activations while taking the significance (i.e., weights) of the constraints into consideration. With these observations, coherence can be adopted in terms of the following premises to support model discovery: *Symmetry*: Explanatory coherence is a symmetric relation, unlike conditional probability. *Explanation*: (i) A hypothesized model mechanism coheres with what it explains/supports, which can either be evidence/requirement or another hypothesis; (ii) hypotheses that together support some other requirement cohere with each other; and (iii) the more hypotheses it takes to satisfy a requirement, the lower the degree of coherence. *Analogy*: Similar hypotheses that support similar requirements cohere. Evidence/requirement priority: Propositions that describe pieces of evidence and requirements have a degree of acceptability on their own. Contradiction: Contradictory propositions are incoherent with each other. *Competition*: If P and Q both support a proposition, and if P and Q are not explanatorily connected, then P and Q are incoherent with each other. P and Q are explanatorily connected if one explains the other or if together they explain something. *Acceptance*: The acceptability of a proposition in a system of propositions depends on its coherence with them.

Computing the Equilibrium State of the Model Ensemble via Coherence Maximization

Given the characterization of coherence, the activations of nodes in the network are analogous to the acceptability of the respective propositions. This strategy is similar to viewing the state as a unit-length vector in an N-dimensional vector space. Each node receives input from every other node that it is connected with. The inputs can then be moderated by the weights of the link from which the input arrives. The activation value of a unit is updated as a function of the weighted sum of the inputs it receives. The process continues until the activation values of all units settle. Formally, if we define the activation level of each node j as a_j, where a_j ranges from -1 (rejected) and $+1$ (accepted), the update function can be defined as follows:

$$a_j(t+1) = \begin{cases} a_j(t)(1-\theta) + \text{net}_j(M - a_j(t)), & \text{if } \text{net}_j > 0 \\ a_j(t)(1-\theta) + \text{net}_j(a_j(t) - m), & \text{otherwise} \end{cases}$$

In this rudimentary formulation, the variable θ is a decay parameter that decrements the activation level of each unit at every cycle. In the absence of input from other units, the activation level of the unit gradually decays, with m being the minimum activation, M denoting the maximum activation, and net j representing the net input to a unit, as defined by the following equation: $\sum_i w_{ij} a_i(t)$. These computations can be carried out for every node until the activation levels of elements stabilize and the network reaches an equilibrium via self-organization. Nodes with positive activation levels at the equilibrium state can be distinguished as maximally coherent hypotheses. For experimentation purposes, the design of the network can be calibrated or fine-tuned to alter the links' weights, which represent the significance of the constraints.

Consensus Formation Among Parallax Model Ensembles

Each coherence model indicates the plausibility of a model ensemble from a particular perspective as a parallax view. This perspective highlights specific target attributes and hypothetical mechanisms so that the model(s) can be evaluated with respect to the evidence or behavioral requirements pertinent to this view. A useful and robust model is one that can address multiple attributes from different perspectives. The robustness of a model is predicated on how well its mechanisms are able to account for the behavioral regularities across a wide range of contexts and situations. This requires attainment of consensus across the coherence models of individual ensembles, each one of which is focusing on a specific set of attributes that collectively define a perspective.

Model ensembles, using their coherence models, evaluate competing hypothetical mechanisms. In the previous section, we argued about how coherence can be construed as the maximization of constraint satisfaction and that it can be computed by a connectionist strategy. It is expected that when model ensembles reach different coherence-based conclusions about which mechanisms to accept and reject. Consensus is achieved when a set of model ensembles accept and reject the same set of elements that represent the behavioral mechanisms and assumptions. Consensus emerges as a result of exchanges between model ensembles to a sufficient extent that they make the same coherence judgments. The exchanges include the propositions about hypotheses, observations, experiment results, as well as pieces of evidence considered.

The process is as follows: (i) Start with a group of model ensembles, each one representing a specific perspective akin to a viewing an object from a different line-of-sight in parallax view. Each ensemble can accept and reject different hypotheses and experiments, resulting in distinct coherence judgments due to variations in requirements and explanatory mechanisms. (ii) The ensembles exchange information to indicate their coherence judgments. (iii) Repeat step 2 until all model ensembles have acquired sufficiently similar evidence and explanations so that they all accept and reject the same hypotheses, implying the achievement of consensus.

The implementation of the above process can leverage alternative communication strategies. Following the coherence computations for each ensemble, the strategy can check for network consensus. This check fails if two elements of the network are found to differ in their coherence model propositions. So, if consensus does not exist, communication starts. The communication can be peer-to-peer or a combination of peer-to-peer and multicast. In the peer-to-peer communication, two model ensembles E_1 and E_2 are selected to communicate with each other. The transfer of constraints E_1 to/from E_2 is stochastic, in that the transfer depends on a probability that is a function of the strength of the connection between two ensembles. The strength is defined as a function of similarity between propositions of respective coherence models. As such, communication that succeeds in bringing coherence models of E_1 and E_2 closer to each other will amplify the probability of future communication. Alternatively, a selected group of model ensembles can multicast their coherence model elements and constraints (e.g., a lecture mode) to other ensembles in the network. The transfer of information at this stage is still stochastic; nevertheless, the multicast/broadcast phase facilitates outreaching to a broad range of elements across the network. After this phase, peer-to-peer communication continues through stochastic local interactions.

CONCLUSIONS

Parallax models will allow us to approach a cognitive systems-inspired form of modeling derived from synergistic integration of emergence, creative cognition, and ecological perspective. The ability to instantiate, generate, transform, execute, and if necessary, evolve multiple models, in parallel, all of which take similar but slightly different perspectives on the same referent system, opens the door to the automatic generation, and selection (by falsification) of many somewhat different hypothetical, including non-intuitive mechanisms for that referent. In other words, the above ability would allow us to construct, execute, and falsify, many more hypotheses for the way a system works than can be achieved feasibly through our current sequential, iterative, modeling methods. Modeling throughput would increase exponentially. Such an exponential increase in model and hypothesis throughput would promote creative discovery and increase opportunities for creative leaps; that increase is necessary for generative simulation to begin significantly supplanting some of the current trial and error methods in domains like development of new therapeutics. Development and use of GPS have potential to advance computational science and engineering and achieve targeted objectives on at least two fronts: experimental methods and information science and theory and methodology of M&S. The outcome envisioned could in time change how computational modeling is done and how scientists and engineers are trained, while opening new territories for systems engineering.

REFERENCES

Amabile, T.M. (1996) *Creativity in Context: Update to the Social Psychology of Creativity*, Westview Press.

Axelrod, R. and Cohen, M.D. (2000) *Harnessing Complexity*, Basic Books.

Campbell, D.T. (1960) Blind variation and selective retentions in creative thought as in other knowledge processes. *Psychological Review*, **67** (6), 380.

Csikszentmihalyi, M. (1996) *Flow and the Psychology of Discovery and Invention*, Harper Collins, New York.

Darema, F. (2004) Dynamic data driven applications systems: a new paradigm for application simulations and measurements. Proceedings of the International Conference on Computational Science (ICCS 2004). Lecture Notes in Computer Science (LNCS), vol. 3038, pp. 662–669.

Gero, J.S. and Kazakov, V.A. (1996) An exploration-based evolutionary model of a generative design process. Computer-Aided Civil and Infrastructure Engineering, 11 (3), 211–218.

Goldberg, D.E. (1999) The race, the hurdle, and the sweet spot: lessons from genetic algorithms for the automation of design innovation and creativity, in Evolutionary Design by Computers (ed. P. Bentley) chapter 4, Morgan Kaufmann, San Mateo, CA, pp. 105–118.

Holland, J.H. (2000) Emergence: From Chaos to Order, Oxford University Press.

Kang, K.C., Lee, J., and Donohoe, P. (2002) Feature-oriented product line engineering. IEEE Software, 19 (4), 58–65.

Klahr, D. and Simon, H.A. (1999) Studies of scientific discovery: complementary approaches and convergent findings. Psychological Bulletin, 125 (5), 524.

Kwiatkowska, M., Norman, G. and Parker, D. (2011) PRISM 4.0: verification of probabilistic real-time systems. International Conference on Computer Aided Verification. Springer, Berlin Heidelberg, pp. 585–591.

Levins, R. (1966) The strategy of model building in population biology. American Scientist, 54 (4), 421–431.

Metzger, A. and Pohl, K. (2007) Variability management in software product line engineering. Software Engineering-Companion, 2007. ICSE 2007 Companion. 29th International Conference on. IEEE, pp. 186–187.

Mitchell, M. (2009) Complexity: A Guided Tour, Oxford University Press.

Morgan, M.S. and Morrison, M. (1999) Models as Mediators: Perspectives on Natural and Social Science, vol. 52, Cambridge University Press.

Oliveira, B.C.D.S., Van Der Storm, T., Loh, A. and Cook, W.R. (2013) Feature-oriented programming with object algebras. European Conference on Object-Oriented Programming, Springer, Berlin Heidelberg, pp. 27–51.

Parunak, H.V.D. and VanderBok, R.S. (1997) Managing emergent behavior in distributed control systems. Ann Arbor, 1001, 48106.

Rawls, J. (1971) A Theory of Justice, Harvard University Press.

Rosenman, M.A. (1997) An exploration into evolutionary models for non-routine design. Artificial Intelligence in Engineering, 11 (3), 287–293.

Sawyer, K. (2017) Group Genius: The Creative Power of Collaboration, Basic Books.

Simonton, D.K. (1999) Origins of Genius: Darwinian Perspectives on Creativity, Oxford University Press.

Smith, S.M. and Blankenship, S.E. (1991) Incubation and the persistence of fixation in problem solving. The American Journal of Psychology, 4 (1), 61–87.

Thagard, P. (2013) Emotional cognition in urban planning, in Proceedings of the 2nd Delft International Conference on Complexity, Cognition, Urban Planning and Design (eds J. Portugali and E. Stolk), Springer, Berlin, pp. 197–213.

Ward, T.B., Smith, S.M., and Vaid, J.E. (1997) Creative Thought: An Investigation of Conceptual Structures and Processes, American Psychological Association.

Yilmaz, L., Chakladar, S. and Doud, K. (2016) The goal-hypothesis-experiment framework: a generative cognitive domain architecture for simulation experiment management. Winter Simulation Conference (WSC), 2016, IEEE, pp. 1001–1012.

SECTION II

EMERGENT BEHAVIOR MODELING IN COMPLEX SYSTEMS ENGINEERING

5

COMPLEX SYSTEMS ENGINEERING AND THE CHALLENGE OF EMERGENCE

Andreas Tolk[1], Saikou Diallo[2], and Saurabh Mittal[3]

[1] *The MITRE Corporation, Hampton, VA, USA*
[2] *Virginia Modeling, Analysis & Simulation Center, Old Dominion University, Suffolk, VA, USA*
[3] *The MITRE Corporation, McLean, VA, USA*

SUMMARY

Systems engineering methods do not only help us to engineer artificial systems, but they also provide means that help to understand natural systems. Complexity and emergence add challenges that require new methods to cope with them, leading to complex systems engineering, which applies increasingly simulation solution to provide insights into the dynamic behavior of complex systems, allowing to detect and understand some classes of emergent behavior. This chapter introduces simple, complicated, and complex system definitions and shows how these system classes are related to simple, weak, strong, and spooky emergence, and which systems engineering methods can be applied to support the detection, understanding, and management of such emergence.

Emergent Behavior in Complex Systems Engineering: A Modeling and Simulation Approach,
First Edition. Edited by Saurabh Mittal, Saikou Diallo, and Andreas Tolk.
© 2018 John Wiley & Sons, Inc. Published 2018 by John Wiley & Sons, Inc.

INTRODUCTION

After we dealt with the philosophical and general scientific view on complexity and emergence in the leading chapters of this book, we turn our attention to emergence in the state of the art of system's engineering, more specifically in the context of complex systems engineering. We will address the questions of how complex systems are different from complicated systems, what new methods can be used to support complex systems engineering, and whether there are any methods of systems engineering we can apply to discover, understand, and manage emergence? We will also show how modeling and simulation methods are increasingly playing a special role in this context.

Recent research on system of systems (SoSs) has shown that if engineered systems are coupled within a new context, it is likely that (i) unintended consequence will occur, or (ii) new behavior results out of the interactions of the systems with each other. And even if the systems themselves are fully specified and understood, the moment they interact with a changing environment, we are facing the same issues again. Are we still a "watchmaker" who focuses on every detail down to the tiniest component of our systems? Or do we have to restrain ourselves to the role of a "gardener" who uses the characteristics of the various plants to create a garden as he wants it to be, but who constantly has to work on individual plants, ripping out the weed, and making sure that no spots evolve into an unwanted status? How can we use modeling and simulation to learn more about such systems, and where are our boundaries of learning? Can we at least identify regions in which our systems behave like we expect them to do? Is it eventually possible to take advantage of emergence?

Within the chapter, we will first consider the purpose of systems engineering, as a discipline that allows us to apply scientific principles and engineering methods to cope with the challenges of complexity and emergences. We will then consider several simulation methods that can help to better understand and manage emergence, as well as the limits of such approaches. The two topics emphasized in this chapter are the design and engineering of complexity and emergence, and the use of systems engineering and modeling and simulation methods to better understand and manage complexity and emergence.

SYSTEMS ENGINEERING AND EMERGENCE

Complexity is a multivariable and multidimensional phenomenon within the space–time continuum. Multivariable implies an often-large number of variables with sometimes incomplete knowledge about their interdependencies. Multidimensional implies possibly multiple vantage points that are dependent on the frame of reference when observing the phenomenon. These vantage points are not mutually exclusive, but allow to deal with different facets. Furthermore, the phenomenon may manifest over time or over space in the space–time continuum, requiring methods allowing for spaciotemporal analysis instead of exclusively looking at local snapshots.

One of the characteristic and defining property of complex systems is the manifestation of emergent behavior by the system, which we usually understand in the context of system engineering as a macro-level behavior that dynamically arise from the spaciotemporal interactions between the parts at the micro-level. In traditional systems engineering, the aim is to engineer a closed system that is predictable. For natural systems, the situation is a bit different as they are open systems and replete with irreducible macro-level emergent behaviors at the micro-level. This novel irreducible macro-behavior adds further complexity, specifically in adaptive systems when it becomes causal at the micro-level leading to a fundamental shift in the behaviors at multiple levels of the systems. The system is no longer behaving as originally specified, but it adapted to a new environment by developing new multi-level interactions and positive and negative feedback loops. How can such self-modification and adaption be unified with the views of systems engineers, who want to design a system to fulfill certain requirements by providing necessary functionality via well-defined interfaces? Are we still able to design such complex adaptive systems, if they modify themselves in often surprising ways?

The existence of emergence and its influence on our systems, including the fact that emergence can and will occur in our engineered systems, is well recognized. Ignoring emergence is no longer an option, as the coupling of systems into a SoSs or the interaction with a complex environment will result in unintended consequences or even emergent behavior. As described in the earlier chapters, emergence is studied extensively in many disciplines. In this chapter, our work has been shaped by the four schools of thought described by Wolf and Holvoet (2005):

1. complex adaptive systems theory: macro-level patterns arising from interacting agents
2. nonlinear dynamical systems theory and chaos theory: concept of attractors that guide systems behavior
3. synergistic school: macro-order parameter that influences macro-level phenomena
4. far-from-equilibrium thermodynamics: dissipative structures and dynamical systems arising from far-from-equilibrium conditions

Emergence in the context of systems engineering has been addressed in the context of SoS, among others by Maier (1998), Mittal and Rainey (2015), and Rainey and Tolk (2015), as well as complex sociotechnical systems (Mittal, 2014). Maier (2015) extended his earlier work to postulate the following four types of emergence in simulations.

- *Simple emergence*: The emergent property or behavior is predictable by simplified models of the system's components.
- *Weak emergence*: The emergent property is readily and consistently reproduced in simulation of the system but not in reduced complexity non-simulation models of the system, that is, simulation is necessary to reproduce it.

- *Strong emergence*: The emergent property is consistent with the known properties but, even in simulation, is inconsistently reproduced without any justification of its manifestation.

- *Spooky emergence*: The emergent property is inconsistent with the known properties of the system and is impossible to reproduce in a simulation of a model of equal complexity as the real system.

We also apply Maier's (2015) definition of a SoSs, which proposes that we have operationally and managerially independence, evolutionary development, and geographical distribution of comprised systems that will result in emergence as described above.

Traditional Systems Engineering

The systems engineering community did not embrace these new "emerging ideas" immediately, in particular not by those members who were well used to the traditional way of conducting business. Although professional organizations, such as the Institute of Electrical and Electronics Engineers (IEEE) and the International Council on Systems Engineering (INCOSE), provide slightly different definitions, the community of practitioners generally understands systems engineering as an interdisciplinary collaborative approach to derive, evolve, and verify a life cycle-balanced system solution. Systems engineering is about engineering systems that provide functionality to the user as required, when required, and where required.

Following the traditional approach, such as described by Buede and Miller (2016) or Dickerson and Mavris (2016), the traditional systems engineering process starts with the recognition of the necessity to provide a system to fill a capability gap and provide the functionality a user requires. Depending on the modeling language and development method chosen for the project, system engineers develop a concept of operations from which the system-level requirements flow. They engineer a physical and a functional architecture that define logical and physical structures of functions and components, which results in the overall design of the system, and bring them together into a system specification – defining information exchange between functions and interfaces between components – that can be verified and validated using the set of requirements used to specify the system's behavior originally. This sequential method is captured in the "Vee" model, as among many others described by Forsberg and Mooz (1992). Figure 5.1 shows the "Vee" model as depicted in MITRE's Systems Engineering Guide (2014).

In recent years, the use of Model-based Systems Engineering (MBSE), as introduced by Wymore (1993), has gained more popularity. MBSE replaces documents with a common system and architecture model that allows the definition of many viewpoints, highlighting different aspects of the system without challenging the consistency of the overall solution. All concepts, attributes, and relations are captured in a common model. MBSE facilitates the use of simulation in support of such processes because the information that makes up the system is captured in a consistent model.

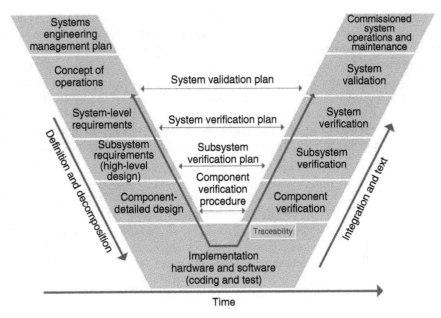

FIGURE 5.1 The "Vee" model of systems engineering.

The traditional approach was successfully applied to engineer not only simple systems but also to more complicated ones. Complicated systems that require the orchestration of many processes, like the Apollo project of the National Aeronautics and Space Administration (NASA), were not easy to design and develop, but by applying the principles of reductionism, the system can be broken down into manageable parts. Systems engineering helped successfully in solving such challenges.

As Ottino (2004) observes: "Engineering is about making things happen, about convergence, optimum design and consistency of operation. Engineering is about assembling pieces that work in specific ways – that is, designing complicated systems." However, the pure notion of system boundaries and behavior is challenged when it comes to complexity, which can be best described by the phrase usually attributed to Aristotle: "the whole is more than the sum of its parts." What systems expose characteristics that are not guided by reductionism? Can systems engineering still be applied? These questions will be addressed in the next sections.

System of Systems Engineering

Whenever systems are no longer providing their functionality to a well-defined user, but these functions are interwoven to provide a new set of capabilities, new system behavior of the coupled systems emerges as a result of the new interplay of systems.

The system build is referred to as a supersystem, a metasystem, or SoSs, which is made up of components that are large-scale systems themselves. This new structure is more than just another big system, as it is no longer monolithic. Maier (2015) observes that the participating systems are operationally and managerially independent, as they already fulfill a purpose on their own, requiring operational control and management for this original purpose. They are geographically distributed, and the systems evolve by themselves. However, as the functions are interconnected to provide a new set of capabilities, the systems are no longer independent. Coupling the systems punches stove pipes of their often-monolithic structure. Depending on the governance, the MITRE's Systems Engineering Guide (2014) distinguishes four categories of SoS, which are defined as a set or arrangement of systems that results when independent and useful systems are integrated into a larger system that delivers unique capabilities.

- *Directed system of systems*: objectives, management, funding, and authority are subsumed under one governance, resulting in the systems being subordinated. The operational and managerial independence is dissolved, and evolution is replaced by planned updates.
- *Acknowledged system of systems*: objectives, management, funding, and authority are aligned and orchestrated by the governance, but systems retain their own management, funding, and authority in parallel.
- *Collaborative system of systems*: No top–down objectives, management, authority, responsibility, or funding at the governance level. The systems voluntarily work together to address shared or common interest via enhanced communication.
- *Virtual system of systems*: No top–down objectives, management, authority, responsibility, or funding at the governance level. The systems may not even know about each other, but simply discover possible services that can be utilized.

The lack of common management will surely lead to unintended and often surprising consequences in all cases but with fully directed SoS. Garcia and Tolk (2015) show such counterintuitive effects for a new netcentric weapon system that is integrated into an existing portfolio. Although the new netcentric weapon system behave liked expected, lethalizing assigned targets effectively, the use of bandwidth to operate decreased the efficiency of the legacy systems more than the new system increased it. With the additional new system, the overall success rate went down. These insights were based on the use of massive simulation systems used to evaluate systems and missions under various constraints, already hinting at the need to relay on modeling and simulation methods to gain more insight into the dynamic supersystems.

As discussed by Keating (2008), supersystems can be interpreted in a metasystem that balances the need for self-organization and following a purposeful design, modulating stability versus change, and balancing autonomy of the systems versus their integration. The four categories of SoS can be interpreted as balance points within this metasystem. Keating also distinguishes between systems engineering processes and system-based methodology, pointing out the need for the latter. Although processes are specific, offering a set of steps that must be accomplished in a clearly prescriptive form, a methodology offers a general framework with sufficient detail to guide formulations of high-level approaches. Therefore, processes are problem-specific and often not repeatable for another SoS, but methodologies can contribute to a broader set of solutions. He finally observes the need to consider the role of the environment and context of the SoS way more than it is the case for traditional approaches, as even small interactions with the environment can result in significant changes within the SoS, as the effect may be amplified by the relations between the systems.

Complex Systems Engineering

Over the past decades, complexity has been identified as a new challenge. Complexity adds a new dimension to systems engineering. The borderline between complicated systems and complex systems can be fuzzy, as shown in an introduction paper by Rouse (2003). He observes that the degree by which a system is perceived as complicated or complex may vary with the level of education of the system engineer. What seems complicated to a novice may become simpler for an expert. He also observes that some simple systems may expose very complex behavior, in particular when the functions and interconnections are discontinuous and nonlinear. Such nonlinear and/or discontinuous nature of the elements of the system may lead to behaviors labeled as catastrophes, chaos, and so on. Kurtz and Snowden (2003) were among the first to group systems into four categories, as shown in Figure 5.2, that resulted in the today accepted groups of simple, complicated, and complex systems.

- *Simple systems* are straightforward to deal with using traditional systems engineering approaches, as they are within the KNOWN realm.
- *Complicated systems* are KNOWABLE, but their scale in components, interfaces, and relations requires extensions to the traditional approaches.
- *Complex systems* are characterized by high interdependence of their components that have no common management, but evolve and self-organize, often producing emergent behavior. COMPLEX addresses all forms identified by Wolf and Holvoet (2005).

Kurtz and Snowden (2003) included *chaotic or catastrophic systems* in their overview. Such systems often present weird problems and cannot be controlled. It needs to be pointed out that chaos can be observed in simple, complicated, and complex systems. Mathematically, whenever a nonlinear function is first stretched beyond system borders and then folded back to the allowed domain, chaos is likely to occur. The effect is that arbitrary close starting points will lead to significantly different trajectories over time, leading to a systemic unpredictability of the system behavior. Nonetheless, it may be possible to govern the system to avoid the realm and regions of chaos, if it is known which parameters influence the chaotic behavior, or at least which parameter constellation result in such behavior. Like a sailor avoids vortexes by staying in calmer portions of a river, so a system engineer can avoid chaotic behavior of a known system by observation and control.

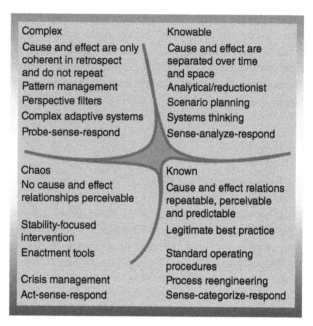

FIGURE 5.2   System categories as defined by Kurtz and Snowden (2003). Reprint Courtesy of International Business Machines Corporation, © 2003 International Business Machines Corporation.

Within this book, we are interested in complex systems that do not simply depart from our expectations of continuous, linear behavior, but that indeed produce emerging behavior and ask, to what degree such behavior can be discovered, understood, managed, or maybe even engineered into such complex systems. Keating (2008) observes that these emerging properties exist only on a macro-level, existing beyond the properties that can be dealt with by reductionism. They are not held by any of the constructing components, and they are irreducible. They are part of the whole, but not of any of its components. Sheard and Mostashari (2009) propose the following definition after a review of systems engineering relevant literature from complexity theory:

> Complex systems are systems that do not have a centralizing authority and are not designed from a known specification, but instead involve disparate stakeholders creating systems that are functional for other purposes and are only brought together in the complex system because the individual "agents" of the system see such cooperation as being beneficial for them.

This lack of coordination and shared responsibility, the absence of requirements, boundaries, and unforeseen behavior is diametric to the design philosophy of traditional systems engineering. This is true for systems of systems that do not fall into the directed category as well.

As Ottino (2004) observes: "The emergent properties of complex systems are far removed from the traditional preoccupation of engineers with design and purpose." How can we overcome this gap between intended design and emerging behavior in complex systems?

A first set of recommendations was proposed by Bar-Yam (2003). He observes that four principles for large engineering projects are still applied: (i) substantially new technology is used; (ii) the new technology is based on a clear understanding of the basic principles and equations that govern the system; (iii) the goal of the project and its more specific objectives and specifications are clearly understood; (iv) a design will be created essentially from scratch based on the equations and then will be implemented. Although the Manhattan project and the space program successfully applied these principles, many newer projects overrun their budget, take much longer than planned for, or are only partially fulfilling the requirements. Bar-Yam (2003) proposes an evolutionary approach that is targeting a much less set of objectives. Instead of collecting all requirements in the beginning, they grow with the progress of the system, similar to the agile development processes known from software development. Practitioners of many domains apply these ideas when following the moto "build a little, test a little."

Norman and Kuras (2006) use their experiences of applying systems engineering methods in the context of developing solutions for an Air Operation Center, which they identify as a complex system. They compare traditional and complex systems engineering observations in Table 5.1.

TABLE 5.1 Observations in Traditional and Complex Systems Engineering

Traditional Systems Engineering	Complex Systems Engineering
Products are reproducible	No two enterprises are alike
Products are realized to meet pre-conceived specifications	Enterprises continually evolve so as to increase their own complexity
Products have well-defined boundaries	Enterprises have ambiguous boundaries
Unwanted possibilities are removed during the realizations of products	New possibilities are constantly assessed for utility and feasibility in the evolution of an enterprise
External agents integrate products	Enterprises are self-integrating and re-integrating
Development always ends for each instance of product realization	Enterprise development never ends – enterprises evolve
Product development ends when unwanted possibilities are removed and sources of internal friction (competition for resources, differing interpretations of the same inputs, etc.) are removed	Enterprises depend on both internal cooperation and internal competition to stimulate their evolution

TABLE 5.2   Supporting Methods for Complex Systems Engineering

Analyze	Diagnose	Model	Synthesize
Data mining	Algorithmic complexity	Uncertainty modeling	Design structure matrix
Splines	Monte Carlo methods	Virtual immersive modeling	Architectural frameworks
Fuzzy logic	Thermodynamic depth	Functional/behavioral models	Simulated annealing
Neural networks	Fractal dimension	Feedback control models	Artificial immune system
Classification and regression trees	Information theory	Dissipative systems	Particle swarm optimization
Kernel machines	Statistical complexity	Game theory	Genetic algorithms
Nonlinear time series analysis	Graph theory	Cellular automata	Multi-agent systems
Markov chains	Functional information	System dynamics	Adaptive networks
Power law statistics	Multi-scale complexity	Dynamical systems	
Social network analysis		Network models	
Agent-based models		Multi-scale models	

Although all approaches agree that extensions and new methods are needed to support complex systems engineering, no common method comparable to the Vee model for complex systems has been established. Instead, methodologies are recommended that combine methods and approaches from a variety of disciplines that have to cope with complexity. The INCOSE Complex Primer for System Engineers (Sheard *et al.*, 2015) provides the following matrix in support of analyzing, diagnosing, modeling, and synthesizing complex systems, shown in Table 5.2.

A common denominator is the use of modeling and simulation-related methods, in particular the application of artificial intelligence. A recent National Science Foundation workshop on research challenges in modeling and simulation for engineering complex systems (Fujimoto *et al.*, 2017) contributed to consolidate and augment this methodology matrix, addressing the domains of conceptual modeling, computational challenges, uncertainty modeling, and reuse, composition, and adaptation.

Modeling and simulation provides the necessary means to detect, manage, and control emergence in the virtual realm before challenges produce risks in the real world. It helps understand how emergence can be contained or potentially even engineered by choosing the proper boundaries and constraints or allowing appropriate degrees of freedom when systems are composed or integrated into open architectures, such as the Internet of Things, Smart Cities, and other SoSs that evolutionary grow from independently developed components that by themselves are systems.

As a rule, complex systems are not derived from an overarching design, but they evolve from the sometimes loose-coupling of usually well-engineering systems or components for a new and often unforeseen purpose. Complex systems are therefore not ill-defined or insufficiently engineered, but they require a new form of governance to ensure that emergence is discovered, understood, managed, and utilized. In the next section, we will have a closer look at the state of the art of using complex systems engineering means to support the engineer with these tasks.

UNDERSTANDING AND MANAGING EMERGENCE

When looking at other chapters of this book, it becomes obvious that what we mean by emergence can be sometimes contradictory from one discipline to another and that we seem to accept that. Emergence has ontological, epistemological, and methodological perspectives, and they address different sets of challenges. Taxonomies of emergent behavior have been discussed in different contexts, some of them referred to in other chapters as well, and they help to specify the conceptualization of emergence used in that given context. Even within the same discipline, the focus may be on methodological aspects – how we solve challenges in the domain – or referential aspects – what challenges of the domain do we solve (Hofmann, 2013).

For this chapter, we focus on a taxonomy being applicable to complex systems engineering. The proposed taxonomy is derived from the descriptions in complex adaptive systems theory at the fundamental level, extended toward SoSs engineering, and eventually their manifestation using modeling and simulation. We identify four categories of emergence as already addressed in the introduction to this chapter, as they were introduced by Maier (2015):

- *Simple emergence*: the emergent property is readily predicted by simplified models of the system.
- *Weak emergence*: the emergent property is reproducible and consistently predicted with simulation.
- *Strong emergence*: the emergent property that is consistent with known properties but not reproducible in the simulation. It is unpredictable and inconsistent in simulation.
- *Spooky emergence*: the emergent property is inconsistent with the known properties of the system.

Of these four categories, the concept of weak and strong emergence is captured more often than the additional categories as introduced by Maier (2015). As a rule, weak emergent properties can be formally specified using mathematical principles and reproduced in a simulation environment. They are known a priori. In contrast, strong emergent properties are discovered by observing the system or the simulation thereof in the context of their environment (Bar-yam, 2004). Strong emergent behavior is bounded by the knowledge of the existing properties of the system.

Some of the fundamental work on complex systems engineering methods in support of meeting the challenge of emerging behavior is focused on these two categories. Fromm (2005) and Holland (2007) focus on the feedback types and causal relationships between weak and strong emergence and propose multiple sub-categories based on the applicability of the feedback and causation. Mittal (2013a,b) considers only weak and strong emergence and describes how they can be formally modeled and eventually simulated. Mittal (2014) also provides information about the cyclic relationship between strong and weak emergent behavior and provides methods to address these challenges.

The emerging types are not mutual exclusive. It is possible that the same system exposes different categories of emergence, depending on what aspects are evaluated in the spaciotemporal spectrum. We may be able to predict the system behavior in space, but not on time, and vice versa.

Emergence Categories and System Properties

Mapping emergence categories to system properties with the objectives first to exclude unwanted emergence and second to provide the ability to select the right system properties when certain emergence categories shall be observable for further research has been the focus of other recent work. If certain emergence categories can be excluded based on the system properties, then the wish of systems engineers to be in control of a system may be obtainable by keeping the system within a certain propertied region. On the other side, if certain properties connected with emergence are exposed by a system, the system engineers can be prepared to cope with emergence as soon as it is observed.

Mittal and Rainey (2015) related the four categories of emergence (simple, weak, strong, and spooky) with two broad classes of complex systems (predictive and not predictive). They propose that the behavior of a system can be categorized into predictable and unpredictable regions, respectively. As a rule, predictable systems can be modeled in detail and it is possible to navigate from one state to another, if the systems do not contain chaotic functions. The mathematics within control theory being applicable is modern control theory or digital control theory. Unpredictable systems do not only have a systemic element of uncertainty, it maybe that not even all states can be defined. It is therefore generally not possible to predict the transition from one state to another. The mathematics that would apply in this region is estimation theory.

In Complex Systems Engineering section, we introduced three system categories, namely simple, complicated, and complex systems. When revisiting the emergence categories simple, weak, strong, and spooky, a relationship between those categories can be established, when we assume that we use models of the systems – or even simulations of the systems – to gain insight into the dynamic behavior.

- Simple systems are known. There cause-and-effect relations are perceivable, repeatable, and predictable. As such, they are closely related to simple emergence, which is readily predicted.

- Complicated systems are knowable, and the cause-and-effect chains can be understood by reductionism and detailed analysis. Emergent behavior in this system category is reproducible and consistently predictable, hence the category is weak emergence.

- In complex systems, the cause-and-effect relations are only coherent in retrospect and usually do not repeat. Although the behavior is consistent and explainable within the system, it is not reproducible (otherwise the systems were complicated, not complex). Therefore, complex system exhibit strong emergence.

- Spooky emergence has no counterpart, as the definition places it outside of the system. Spooky behavior is inconsistent with the known properties of the predicting system – the model or the simulation – and points to the inappropriateness of the used prediction system. Although complex systems are not predictable and produce strong emergence due to their complexity, spooky emergence is not predictable due to the absence of a system model that can explain the observation.

This view is consistent with Ashby (1956), who proposed that the main source of emergent behavior is the lack of understanding of the system itself. Although we know today that emergence can also be systemic, the lack of understanding can still be an important human factor, as also observed by Rouse (2003). This view assumes that our system is not complex, but complicated, and that the reason for our perception of complexity and emergence is that we use an insufficient model of the system capturing our knowledge. If a system exposes several categories of emergence in the spaciotemporal spectrum, we may already understand some aspects of the system well enough to recognize all complicated relations and interoperations, whereas other aspects still seem to be complicated to us.

Computationally, simple and complicated systems can be understood by the engineer and reproduced with models and executed as simulations. It is furthermore consistent with recent research that concluded that M&S support to emergent behavior can fully incorporate the weak emergent behavior but can only augment the strong emergent behavior studies (Mittal, 2013a; Szabo and Teo, 2015), as we can only simulate appropriately what we understand sufficiently, leading to a simple or complicated simulation.

It is therefore a logical step to understand such a complicated simulation becoming the immersive environment to study complex systems, but not in the closed world of the simulation. Instead, simulations can be used to perpetually increase our knowledge about a system by incrementally adding knowledge to the simulation and continuing to compare the results with the observed behavior. The real system is our object of interest, its simulation the best possible representation of our knowledge.

Insufficient knowledge about the system can result from the lack of subject matter experts, in particular in domains that require experts from more than one discipline. As academic disciplines, as we know them today, often result from decades of reductionism and specialization, one of the first challenges to overcome in the context of multi-, inter-, and transdisciplinary research is the establishment of a common understanding to allow such experts to collaborate (Stock and Burton, 2011). Mittal and Cane (2016) provide a process supporting the gap analysis in cases where not enough expertise is present within a team to support building an accurate model. Using analysis methods accepted in the community, such as functional dependency networks as proposed by Garvey and Pinto (2009) or general dependency analysis proposed by Guariniello and Delaurentis (2013), common use cases or mission threads are evaluated in the light of underlying research questions. By developing actionable improvements, the computational model is improved gradually until all emergence lies within the boundaries of this system model, and hopefully the emergence can be reduced to simple and weak emergence, as observed above.

This approach depends on the availability of human subject matter experts, which can become easily a challenge. Therefore, other approaches are evaluated to utilize big data and deep learning efforts to autogenerate system descriptions by observation, as envisioned in Tolk (2015): big data provide the necessary correlation observations, and deep learning can help to establish causal relations between such observations. Mittal and Risco-Martín (2017) provide a framework for such data-driven modeling of complex adaptive systems. Their section on data-driven analytics for complex adaptive system model evolution shows how to apply machine-learning methods to develop and continuously adapt a simulation system using data derived from the real-world system observations.

Implications for Systems Behavior and Emergence

Summarizing the observations in the literature referenced in this chapter, we concluded the equivalency of system categories with observable emergence categories in the last section. We can build systems, but we can also better understand natural systems by building a simulation system. Figure 5.3 depicts these equivalencies.

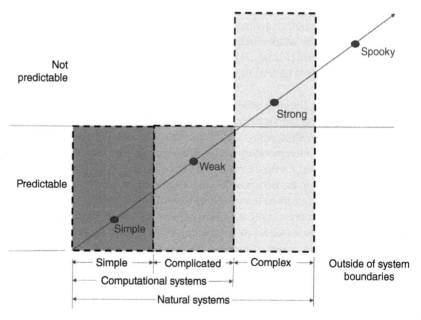

FIGURE 5.3   Systems and emergence.

Simple and complicated systems are predictable. They can be used to produce and explain simple and weak emergence. Complex systems have not predictable elements and can produce strong emergence. If an emergent behavior is completely out of the boundaries of a system, which means we have no system model, we can apply to understand it, it is spooky. We furthermore use computational systems, like simulation systems or computational science methods, to mimic and better understand natural systems.

Natural systems, as described in several of our other book chapters, can be complex, but a computational system by itself is not. As pointed out by Chaitin (1987), computer programs are simply executing algorithms implemented as computable functions to transform input parameters into output parameters. There is no creativity in this process. They can become very complex, but due to the finite solution space and fully specified transformation functions, they are not complex. As computer simulation systems are computer programs, they underlie the same constraints. This observation is valid for closed computational systems.

Ashby's (1956) claim that the source of strong emergent behavior is the lack of understanding of the system itself is often true. If we use a simulation to better understand the system, we reduce the complexity and – in case we can create a simulation that provides the identical observed behavior as the system, we reduce the complexity to being complicated. If we still observe strong emergence that we cannot predict with our simulation system, we have a truly complex system. Managerial and governance choices can increase or decrease complexity in SoSs.

The Elusiveness of Spooky Emergence

As mentioned before, spooky emergence is a *characteristic non-grata* in engineering-related discipline. Bar-yam (2004) introduces several types of strong emergence that include the environment of the system and shows how strong emergence can be observed due to the relationship of the system and its environment. It should be pointed out that in Maier's (2015) definition, this form is also recognized as strong emergence, as the environment and the system are modeled as a "super system" in which emergence is consistent with known properties of system and environment, but not reproducible in the simulation, unpredictable in nature.

Spooky emergence is more than that, and nearly touches the domain of metaphysics. Craver (2007, p. 217) observes in his work neuroscience the following.

Behaviors of mechanisms are sometimes emergent in this epistemic sense. However, one who insists that there is no explanation for a nonrelational property of the whole in terms of the properties of its component parts-plus-organization advocates a spooky form of emergence. Indeed, levels of mechanisms are levels of ontic mechanistic explanation. Advocates of the spooky emergence of higher-level properties must have in mind a different sense of "level" altogether. Advocates of spooky emergence cannot therefore appeal to levels of mechanisms to make their view seem familiar and unmysterious.

These critical observations on spooky emergence are shared by many experts, who feel uncomfortable with something that lies not only outside of the system boundaries – of the extended system understanding that includes the context of the system as well – but that may originate outside of the way of thinking. As long as emergence remains at least consistent with our knowledge, although unpredictable and irreproducible, we still have the comfort that we may gain more insights into the near future. At least in principle, we may regain control.

However, if emergence happens unpredictable and outside of our realm of knowledge – or even beyond our way of thinking – it is "spooky" and requires radically new ways to deal with it. As such, spooky emergence is a challenge of our current approaches, as it shows the limits of our approaches and methods at hand. Like Newtonian physicists were challenged by Einstein's relativity and Heisenberg's uncertainty, our view of system is challenged by emergence, and most of all by spooky emergence.

SUMMARY AND DISCUSSION

Within this chapter, we introduced three categories of systems: simple, complicated, and complex! As shown, the borderlines between them can be fluid. As a rule, simple systems are well understood, and all their interactions are known. The simple system behavior is fully traceable and, as such, predictable. With an increased scale of components, interfaces, and relations, complicated systems are no longer as easy to understand as simple ones, but the system behavior can still be traced, allowing to understand the cause-and-effect chains. Using the right tools, complicated systems are as manageable as simply systems. However, when systems become complex, this is no longer the case. Complex systems expose unpredictable behavior, and this element of unpredictability is inherent to the complexity of the system. No better management nor breakdown into less-complex subsystem can help to predict the emergence.

This categorization of systems was accompanied by categorizes of emergence: simple, weak, strong, and spooky! This emergence categorization, first introduced by Maier (2015), extends the traditional view of weak and strong emergence and maps well to our system categories. Simple emergences maps to simple system, weak emergence maps to complicated systems, and strong emergence maps to complex systems. Spooky emergence has a special place. Whenever spooky emergence occurs, our system approach was insufficient to understand it, as the spooky emergence is inconsistent with the system properties. Our systems approach reached a frontier with spooky emergence.

Systems engineering methods can not only help us to design and develop systems but they can also help us to design and develop artificial systems in order to better understand and study the real-world system of interest. We presented several ways how to use in particular simulation solutions to do this. In the context of this book, it is worth mentioning that all computational sciences are applying the ideas captured here as well: the build a model using the knowledge captured in their discipline and execute this model (which is a simulation), and then they use feedback from the comparison with the real-world system to gradually improve their models. As observed by Bedau (2008, p. 454):

Note that these patterns and regularities produced by computational systems are not mere simulations of emergent phenomena. Rather, they are computational embodiments of real emergent phenomena. That is, the computer produces something that is weakly emergent in its own right. If the computer happens to be simulating some natural system, that natural system might also exhibit its own emergent phenomena. Further, if the simulation is accurate in the relevant respects, it might explain why and how the natural system's phenomena is weakly emergent. But the computer simulation itself, considered as an object it its own right, is also exhibiting emergent behavior.

Although, it is agreed upon that simple and weak emergence can be reproduced by simulation and that spooky emergence for sure lies out of the scope of simulation by definition. The question whether strong emergence can generally be reproduced by simulation systems or not is not yet answered. If an observed system provides strong emergence, and we can produce a simulation system with the same behavior, we can argue that by producing the simulation system, we increased our knowledge enough to transform what looked like a complex system into a complicated system. If, however, the system is systemically complex, then we cannot reproduce the strong emergence. This bears the question if complexity may not be the final frontier of systems thinking and that a new approach may be needed to tackle these problems.

DISCLAIMER

The authors' affiliation with The MITRE Corporation is provided for identification purposes only and is not intended to convey or imply MITRE's concurrence with, or support for, the positions, opinions, or viewpoints expressed by the author. The publication has been approved for public release; distribution unlimited, case number 17-3081-1.

REFERENCES

Ashby, W. (1956) *An Introduction of Cybernetics*, Chapman and Hall.

Bar-Yam, Y. (2003) When systems engineering fails-toward complex systems engineering. Systems, Man and Cybernetics, 2003. IEEE International Conference on, vol. 2. IEEE, pp. 2021–2028.

Bar-Yam, Y. (2004) A mathematical theory of strong emergence using multiscale variety. *Complexity*, 9 (6), 15–24.

Bedau, M.A. (2008) Is weak emergence just in the mind? *Minds & Machines*, 18, 443–459.

Buede, D.M. and Miller, W.D. (2016) *The Engineering Design of Systems: Models and Methods*, John Wiley & Sons.

Chaitin, G.J. (1987) *Algorithmic Information Theory*, Cambridge Tracts in Theoretical Computer Science, Cambridge University Press.

Craver, C.F. (2007) *Explaining the Brain: Mechanisms and the Mosaic Unity of Neuroscience*, Oxford University Press.

Dickerson, C. and Mavris, D.N. (2016) *Architecture and Principles of Systems Engineering*, CRC Press.

Forsberg, K. and Mooz, H. (1992) The relationship of systems engineering to the project cycle. *Engineering Management Journal*, **4** (3), 36–43.

Fromm, J. (2005) Types and forms of emergence. *arXiv preprint nlin/0506028*, doi: 10.1002/9781119288091.ch12.

Fujimoto, R., Bock, C., Chen, W. *et al.* (2017) *Research Challenges in Modeling and Simulation for Engineering Complex Systems*, Springer International Publishing.

Garcia, J.J. and Tolk, A. (2015) Executable architectures in executable context enabling fit-for-purpose and portfolio assessment. *Journal of Defense Modeling and Simulation*, **12** (2), 91–107.

Garvey, P.R. and Pinto, C.A. (2009) *Introduction to functional dependency network analysis*. Second International Symposium on Engineering Systems, Cambridge, MA.

Guariniello, C. and Delaurentis, D. (2013) Dependency analysis of System-of-Systems operational and developmental networks. Conference on Systems Engineering Research, Atlanta, GA. 2013.

Hofmann, M. (2013). Ontologies in modeling and simulation: an epistemological perspective. *Ontology, Epistemology, and Teleology for Modeling and Simulation*, A. Tolk Springer, Berlin, Heidelberg 59–87, 10.1007/978-3-642-31140-6.

Holland, O.T. (2007) Taxonomy for the modeling and simulation of emergent behavior systems. *Proceedings of the 2007 spring simulation multiconference*, vol. 2, Society for Computer Simulation International, pp. 28–35.

Keating, C.B. (2008) Emergence in system of systems, in *System of Systems Engineering* (ed. M. Jamshidi), Wiley, Hoboken, NJ, pp. 169–190.

Kurtz, C.F. and Snowden, D.J. (2003) The new dynamics of strategy: Sense-making in a complex and complicated world. *IBM Systems Journal*, **42** (3), 462–483.

Maier, M. (1998) Architecting principles for system-of-systems. *Systems Engineering*, **1**, 267–284.

Maier, M. (2015) The role of modeling and simulation in system of systems development, in *Modeling and Simulation Support for System of Systems Engineering Applications* (eds L.B. Rainey and A. Tolk), John Wiley and Sons, Hoboken, NJ, pp. 11–44.

MITRE (2014) Systems engineering guide: collected wisdom from MITRE's systems engineering experts, MITRE Corporate Communications and Public Affairs, http://www.mitre.org/sites/default/files/publications/se-guide-book-interactive.pdf (accessed September 2017).

Mittal, S. (2013a) Emergence in stigmergic and complex adaptive systems: a formal discrete event systems perspective. *Cognitive Systems Research*, **21**, 22–39.

Mittal, S. (2013b) Netcentric complex adaptive systems, in *Netcentric Systems of Systems Engineering with DEVS Unified Process* (eds S. Mittal and J.L. Risco Martin), CRC Press, pp. 639–661.

Mittal, S. (2014) Model engineering for cyber complex adaptive systems. European Modeling and Simulation Symposium.

Mittal, S. and Cane, S.A. (2016) Contextualizing emergent behavior in system of systems engineering using gap analysis. *Proceedings of the Modeling and Simulation of Complexity in Intelligent, Adaptive and Autonomous Systems 2016 (MSCIAAS 2016)*, Society for Computer Simulation International.

Mittal, S. and Rainey, L. (2015) Harnessing emergence: the control and design of emergent behavior in system of systems engineering. *Proceedings of Summer Computer Simulation Conference*, Society for Computer Simulation International.

Mittal, S. and Risco-Martín, J.L. (2017) Simulation-based complex adaptive systems, in *Guide to Simulation-Based Disciplines*, Springer International Publishing, Cham, Germany, pp. 127–150.

Norman, D.O. and Kuras, M.L. (2006) Engineering complex systems, in *Complex Engineered Systems* (eds D. Braha, A.A. Minai, and Y. Bar-Yam), Springer, Berlin, Heidelberg, pp. 206–245.

Ottino, J.M. (2004) Engineering complex systems. *Nature*, **427** (6973), 399.

Rainey, L. and Tolk, A. (eds) (2015) *Modeling and Simulation Support for System of Systems Engineering Applications*, John Wiley and Sons, Hoboken, NJ, USA.

Rouse, W.B. (2003) Engineering complex systems: Implications for research in systems engineering. *IEEE Transactions on Systems, Man, and Cybernetics Part C: Applications and Reviews*, **33** (2), 154–156.

Sheard, S. and Mostashari, A. (2009) Principles of complex systems for systems engineering. *Systems Engineering*, **12** (4), 295–311.

Sheard, S., Cook, S., Honour, E., Hybertson, D., Krupa, J., McEver, J., McKinney, D., Ondrus, P., Ryan, A., Scheurer, R. and Singer, J. (2015). A complexity primer for systems engineers. INCOSE Complex Systems Working Group White Paper, http://www.incose.org/docs/default-source/ProductsPublications/a-complexity-primer-for-systems-engineers.pdf (accessed September 2017).

Stock, P. and Burton, R.J.F. (2011) Defining terms for integrated (multi-inter-trans-disciplinary) sustainability research. *Sustainability*, **3** (8), 1090–1113.

Szabo, C. and Teo, Y.M. (2015) Formalization of weak emergence in multiagent systems. *ACM Transactions on Modeling and Computer Simulation*, **26** (1), 6:1–6:25.

Tolk, A. (2015) The next generation of modeling & simulation: integrating big data and deep learning. *Proceedings of the Conference on Summer Computer Simulation*, Society for Computer Simulation International, San Diego, CA, USA, pp. 1–8.

Wolf, T. and Holvoet, T. (2005) Emergence versus self-organization: different concepts but promising when combined. *Lecture Notes in Artificial Intelligence*, **3464**, 1–15.

Wymore, A.W. (1993) *Model-Based Systems Engineering*, CRC Press.

6

EMERGENCE IN COMPLEX ENTERPRISES

William B. Rouse

Center for Complex Systems and Enterprises, Stevens Institute of Technology, 1 Castle Point Terrace, Hoboken, NJ 07030, USA

SUMMARY

Complex enterprises often behave in unexpected ways. This chapter addresses the nature of enterprises and emergent behavior, including the path-dependent aspects of emergent behaviors. The notion of complexity and complex systems are reviewed. Multi-level models are then discussed to provide a framework for consideration of emergence in societies, cities, institutions, and companies. Throughout, the roles of modeling and simulation as a means to understanding emergence are considered and illustrated.

INTRODUCTION

This chapter is concerned with the tendencies of complex enterprises to behave in unexpected ways. This phenomenon is often referred to as emergent behavior. It is important at the outset to define the terms "enterprise" and "emergence," as well as the notion of "complexity."

An enterprise is a goal-directed organization of resources – human, information, financial, and physical – and activities, usually of significant operational scope, complication, risk, and duration (Rouse, 2006). Enterprises can range from corporations to supply chains, to markets, to governments, to economies, where the latter may be seen as enterprise ecosystems.

Emergent Behavior in Complex Systems Engineering: A Modeling and Simulation Approach,
First Edition. Edited by Saurabh Mittal, Saikou Diallo, and Andreas Tolk.

Supply chains can be viewed as extended enterprises linking upstream and downstream providers and consumers of raw materials, components, products, services, and so on. Markets can be viewed as further extensions, often involving several supply chains. Governments provide networks of services ranging from trash collection to national defense. Economies include an ecosystem of entities that produce, distribute, and consume goods and services.

Emergence has been studied for millennia, first formalized by Lewes (1875–1879) and articulated in the context of complexity and complex systems by Corning (2002). Emergence is a process of becoming visible and meaningful. It is a process of the local becoming global. Two forms are of note. First are unexpected global patterns that, in retrospect, could have been predicted. Climate change is a good example. Second are unexpected global patterns that, due to interactions of agents pursuing local goals, could not have been predicted. Political movements and conflicts are often of this nature.

The second form of emergence often exhibits path dependence. The meaning and interpretation of the current state of an enterprise depends on how it arrived at that state. For example, the emergence of World War II depended on the circumstances of how World War I ended, followed by the global experience of the Great Depression. Path dependence often makes it difficult to predict the specific nature of emergence.

It is important at the outset that I explicate the nature of many of the models and simulations discussed in this chapter. In almost all instances, these computational constructions include humans in the loop, typically teams of humans and occasionally broader populations. In some cases, natural processes such as hurricanes and flooding are included in the loop. Much of the emergence observed arises from humans' reactions to computational projections. These projections cause them to behave in ways other than anticipated, often in ways they could not have anticipated. Further, they are typically the source of assessments of whether an outcome is good or bad.

Thus, the system is the entity being modeled. The model is the representation of the system, which includes the designed system and the natural system in which it operates. The simulation is the way the model is "solved" to yield projections of consequences of assumptions and actions taken. For many of the examples in this chapter, humans in the loop rather than software representations of humans take many of the actions of interest. In this way, the models and simulations serve as key elements of decision support systems.

In the next section, the notion of complexity and complex systems are reviewed. Multi-level models are then discussed to provide a framework for consideration of emergence in societies, cities, institutions, and companies. The roles of modeling and simulation as a means to understanding emergence are considered and illustrated throughout the chapter.

COMPLEX SYSTEMS

A system is a group or combination of interrelated, interdependent, or interacting elements that form a collective entity. Elements may include physical, behavioral, or symbolic entities. Elements may interact physically, computationally, and/or by exchange of information. Systems tend to have goals and purposes, although in some cases, the observer ascribes such purposes to the system from the outside, so to speak (Rouse, 2003, 2007a, 2015).

A complex system is one whose perceived complicated behaviors can be attributed to one or more of the following characteristics: large numbers of relationships among elements, nonlinear and discontinuous relationships, and uncertain characteristics of elements and relationships. From a functional perspective, the presence of complicated behaviors, independent of underlying structural features, may be sufficient to judge a system to be complex. Complexity is "perceived" because apparent complexity can decrease with learning (Rouse, 2003, 2007a, 2015).

Snowden and Boone (2007) have argued that there are important distinctions that go beyond those outlined above. Their framework includes simple, complicated, complex, and chaotic systems. Simple systems can be addressed with best practices. Complicated systems are the realm of experts. Complex systems represent the domain of emergence. Finally, chaotic systems, for example, extreme weather events, require rapid responses to stabilize potential negative consequences. The key distinction with regard to the types of contexts discussed in this chapter is complex versus complicated systems. There is a tendency, they contend, for experts in complicated systems to perceive that their expertise, methods, and tools are much more applicable to complex systems than is generally warranted.

Poli (2013) also elaborates the distinctions between complicated and complex systems. Complicated systems can be structurally decomposed. Relationships such as those listed above can be identified, either by decomposition or in some cases via blueprints. "Complicated systems can be, at least in principle, fully understood and modeled." Complex systems, in contrast, cannot be completely understood or definitively modeled. He argues that biology and all the human and social sciences address complex systems. In these domains, problems cannot be solved as they are in complicated systems; they can only be influenced so that unacceptable situations are at least partially ameliorated.

The distinctions articulated by these authors are well taken. Complicated systems have often been designed or engineered. There are plans and blueprints. There may be many humans in these systems, but they are typically playing prescribed roles. In contrast, complex systems, as they define them, typically emerge from years of practice and precedent. There are no plans and blueprints. Indeed, much research is often focused on figuring out how such systems work. As noted by Poli, a good example is human biology.

As an interesting aside, models and simulations tend to represent complex systems as complicated systems in that the models and simulations are engineered to emulate the complex system of interest. One of the primary reasons for including humans in the loop in our models and simulations is to avoid assuming away the complexity. Another reason is that the models and simulations are intended to provide decision support to these humans.

The nature of human and social phenomena within enterprise systems is a central consideration. Systems where such phenomena play substantial roles are often considered to belong to a class of complex systems termed *complex adaptive systems* (Rouse, 2000, 2008, 2015). Systems of this type have the following characteristics:

- They tend to be nonlinear, dynamic, and do not inherently reach fixed equilibrium points. The resulting system behaviors may appear to be random or chaotic.
- They are composed of independent agents whose behaviors can be described as based on physical, psychological, or social rules, rather than being completely dictated by the physical dynamics of the system.
- Agents' needs or desires, reflected in their rules, are not homogeneous and, therefore, their goals and behaviors are likely to differ or even conflict – these conflicts or competitions tend to lead agents to adapt to each other's behaviors.
- Agents are intelligent and learn as they experiment and gain experience, perhaps via "meta" rules, and consequently change behaviors. Thus, overall system properties inherently change over time.
- Adaptation and learning tends to result in self-organization and patterns of behavior that emerge rather than being designed into the system. The nature of such emergent behaviors may range from valuable innovations to unfortunate accidents.
- There is no single point(s) of control – system behaviors are often highly distributed, seemingly unpredictable and uncontrollable, and no one is "in charge." Consequently, the behaviors of complex adaptive systems usually can be influenced more than they can be controlled.

As might be expected, understanding and influencing enterprise systems having these characteristics create significant complications for modeling and simulation. For example, the simulation of such models often does not yield the same results each time. Random variation may lead to varying "tipping points" among stakeholders for different simulation runs. These models can be useful in the exploration of leading indicators of the different tipping points and in assessing potential mitigations for undesirable outcomes.

MULTI-LEVEL MODELS

We need to computationally model the functioning of the complex enterprise of interest to enable decision makers, as well as other significant stakeholders, to explore the possibilities and implications of changing their enterprises in various ways. The goal is to create organizational simulations that will serve as "policy flight simulators" for interactive exploration by teams of often disparate stakeholders who have inherent conflicts, but need and desire an agreed upon way forward (Rouse and Boff, 2005; Rouse, 2014a).

Consider the architecture of complex enterprises, defined broadly, shown in Figure 6.1 (Rouse, 2009; Rouse and Cortese, 2010; Grossman *et al.*, 2011).

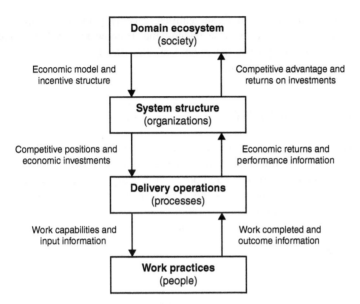

FIGURE 6.1   Architecture of complex enterprises.

The levels relate to levels of abstraction of the physical, human, economic, and social phenomena associated with the functioning of an enterprise. These levels are very useful for understanding and conceptually designing an enterprise model or simulation. The distinctions among these levels play less of a role computationally.

The efficiencies that can be gained at the lowest level (work practices) are limited by the nature of the next level (delivery operations). Work can only be accomplished within the capacities provided by available processes. Further, delivery organized around processes tends to result in much more efficient work practices than for functionally organized business operations.

However, the efficiencies that can be gained from improved operations are limited by the nature of the level above, that is, system structure. Functional operations are often driven by organizations structured around these functions, for example, marketing, engineering, manufacturing, and service. Each of these organizations may be a different business with independent economic objectives. This may significantly hinder process-oriented thinking.

And, of course, potential efficiencies in system structure are limited by the ecosystem in which these organizations operate. Market maturity, economic conditions, and government regulations will affect the capacities (processes) that businesses (organizations) are willing to invest into enable work practices (people), whether these people be employees, customers, or constituencies in general. Economic considerations play a major role at this level (Rouse, 2010a,b).

These organizational realities have long been recognized by researchers in sociotechnical systems (Emery and Trist, 1973), as well as work design and system ergonomics (Hendrick and Kleiner, 2001). Policy flight simulators enable computational explorations of these realities, especially by stakeholders without deep disciplinary expertise in these phenomena. Empowering decision makers to fly the future before they write the check can dramatically increase confidence and commitment to courses of action.

EMERGENCE IN SOCIETY

The evolution of society has been laced with emergent phenomena. In *Sapiens*, Harari provides a narrative of humanity's creation and evolution (Harari, 2015). He considers how homo sapiens prospered and prevailed despite their being several different species of "humans" 100 000 years ago. He discusses the role humans played in the global ecosystem in the process of creating, and also destroying, great empires. He concludes by elaborating the ways in which science and technology have, in recent decades, begun to understand and modify the processes that have governed life for billions of years.

Tainter addresses the collapse of complex societies (Tainter, 1988). His central argument is that as societies evolve, they add layers of complexity to address various issues and problems. With each new layer, additional resources are needed to manage the increase in complexity. Eventually, all resources are being used to manage the layers of complexity. When new issues or problems emerge, such as internal conflicts or external intruders, additional resources are required for the mitigation of these issues. However, at each turn, the value of investing in further complexity declines. This declining marginal returns result in insufficient investment resources to deal with new issues and decline ensues, eventually leading to collapse.

Diamond also addresses societal collapse (Diamond, 2005), with an emphasis on climate change, population explosion, and political discord. He discusses the demise of historical civilizations, linking the factors driving these collapses to contemporary instantiations of these factors. For both Tainter and Diamond, collapse emerged from societal decisions, and sometimes lack of decisions, that over time had unintended and often irreversible consequences. In general, it is almost impossible for societal enterprises to eliminate layers of complexity, even when they are no longer needed. Although the foci of Tainter and Diamond are rather different, the consequences are similar.

Andrews addresses contemporary conflicts across all regions and many countries in the world (Andrews, 2015). He addresses long-standing ethnic and religious rivalries, as well as resource-driven conflicts. He discusses the reasons why global conflict is ubiquitous. Country by country, he summarizes the causes, participants, impacts, and likely outcomes. His treatise is a contemporary guide to how and why war and terrorism persist. Conflict is an emergent phenomenon that seems to be inherent whenever differences in preferences, opinions, and resources are sufficiently disparate.

Casti addresses extreme events, which he terms X-events (Casti, 2012). Casti argues (in line with Tainter, whom he acknowledges) that our overly complex societies have grown highly vulnerable to extreme events that will ultimately lead to collapse. He is fascinated in particular by events for which there inherently are no historical data and, therefore, no empirical basis for prediction. He also questions society's abilities to recover from the range of global catastrophes he characterizes.

Approach to Modeling and Simulation

Societies emerge, driven by biological, social, and economic forces, all in the physical context where they happen to exist. The development of societies, and perhaps their collapse, is very difficult to predict. Yet, we feel compelled to address such the contingencies. Looking at the overall system that needs to be influenced can facilitate addressing the challenges of, for example, climate change and likely consequences.

As shown in Figure 6.2, the world can be considered as a collection of different phenomena operating on different time scales (Rouse, 2014b). Loosely speaking, there are four interconnected systems: environment, population, industry, and government. Some agents operate in multiple systems, usually from different perspectives and on different time scales. In this notional model, population consumes resources from the environment and creates by-products. Industry also consumes resources and creates by-products, but it also produces employment. The government collects taxes and produces rules and, of course, consumes some resources. The use of the environment is influenced by those rules.

Each system component has a different associated time constant. In the case of the environment, the time constant is decades to centuries. The population's time constant can be as short as a few days. Government's time constant may be a bit longer, thinking in terms of years. Industry is still longer, on the order of decades. These differences in time constants tend to differentiate roles of agents in multiple systems, for example, being a consumer as well as a government or industry decision maker.

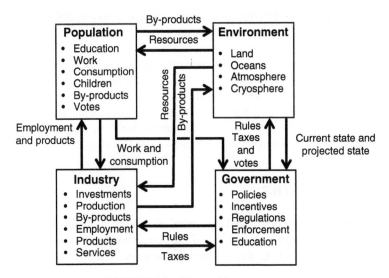

FIGURE 6.2   The world as a system.

These systems can be represented at different levels of abstraction and/or aggregation. A hierarchical representation does not capture the fact that this is a highly distributed system, all interconnected. It is difficult to solve one part of the problem, as it affects other pieces. By-products are related to population size, so one way to reduce by-products is to moderate population growth. Technology may help to ameliorate some of the by-products and their effects, but it is also possible that technology could exacerbate the effects. Clean technologies lower by-product rates but tend to increase overall use, for instance.

Sentient stakeholders include population, industry, and government. Gaining these stakeholders' support for such decisions will depend on the credibility of the predictions of behavior, at all levels in the system. Central to this support are "space value" and "time value" discount rates. The consequences that are closest in space and time to stakeholders matter the most and have lower discount rates; attributes more distributed in time and space are more highly discounted. These discount rates will differ across stakeholders.

People will also try to "game" any strategy to improve the system, seeking to gain a share of the resources being invested in executing the strategy. The way to deal with that is to make the system sufficiently transparent to understand the game being played. Sometimes, gaming the system will actually be an innovation; other times, prohibitions of the specific gaming tactics will be needed.

The following three strategies are likely to enable addressing the challenges of climate change and its consequences.

- *Share Information*: Broadly share credible information so all stakeholders understand the situation.

- *Create Incentives*: Develop long-term incentives to enable long-term environmental benefits while assuring short-term gains for stakeholders.

- *Create an Experiential Approach*: Develop an interactive visualization of these models to enable people to see the results.

Of course, these strategies may not overcome the skepticism evidenced by many who are inconvenienced by phenomena of climate changes.

An experiential approach can be embodied in a policy flight simulator that includes large interactive visualizations that enable stakeholders to take the controls, explore options, and see the sensitivity of results to various decisions (Rouse, 2014a). This notion of policy flight simulators is elaborated later in this chapter.

EMERGENCE IN CITIES

Society is increasingly dominated by cities, as people globally move to cities seeking prosperity. To provide a sense of the concentration of resources in cities, metropolitan New York City, defined broadly, generates roughly 10% of the GDP in the United States. Edward Glaeser, in his book *Triumph of the City*, provides an insightful treatise on the nature and roles of cities (Glaeser, 2011).

Cities have been seen as dirty, poor, unhealthy, and environmentally unfriendly. This was certainly true before modern urban infrastructure emerged replacing, for example, New York City's 100 000 horses and 1200 metric tons of horse manure daily. Glaeser argues that cities are now actually the healthiest, greenest, and – economically speaking – the richest places to live. He argues that cities are humanity's greatest creation and our best hope for the future.

Barber (2013) agrees and would let city mayors rule the world. He contrasts dysfunctional nations with rising cities. Mayors, he argues, have no choice but make sure that schools are open, traffic lights work, and the garbage is picked up. Otherwise, they will be turned out in the next election. Members of national legislative bodies, on the other hand, can manage to accomplish nothing and still be reelected.

Brook (2013) provides a fascinating look into developing world "instant cities" such as Dubai and Shenzhen. He anticipates development of such cities by referencing previous instant cities – St. Petersburg, Shanghai, and Bombay, now Mumbai. Tsar Peter the Great personally oversaw the construction of St. Petersburg in early eighteenth century. In the middle of the following century, the British facilitated Shanghai becoming the fastest growing city in the world, an English-speaking, Western-looking metropolis. During the same period, the British Raj invested in transforming Bombay into a cosmopolitan hub.

All the three cities were gleaming exemplars of modernity, all with ambiguous legacies. Impoverished populations surrounded these cities. These populations could daily see what they did not have. Eventually, they revolted in one manner or another, providing evidence that development can have both positive and negative consequences. The visionaries who planned these cities and supervised their construction had likely not imagined these emergent phenomena.

Approach to Modeling and Simulation

Consider how modeling and simulation can be employed to address urban resilience. Assuring urban resilience involves addressing three problems. First, there is the technical problem of getting things to work, keeping them working, and understanding impacts of weather threats, infrastructure outages, and other disruptions. The urban oceanography example discussed by Rouse (2015) provides a good illustration of addressing the technical problem of urban resilience.

Second, there is the behavioral and social problem of understanding human perceptions, expectations, and inclinations in the context of social networks, communications, and warnings. Behavioral and social phenomena are often emergent, precipitated by media to which people pay attention as well as the social relationships at hand.

Third is the contextual problem of understanding how norms, values, and beliefs affect people, including the sources of these norms, values, and beliefs.

The needs–beliefs–perceptions model (Rouse, 2015) is relevant for this problem, augmented by historical assessments of how the neighborhoods and communities of interest evolved in terms of risk attitudes, for example.

Addressing these three problems requires four levels of analysis:

- *Historical Narrative*: Development of the urban ecosystem in terms of economic and social change – What happened when and why?
- *Ecosystem Characteristics*: Norms, values, beliefs, and social resilience of urban ecosystem – What matters now and to whom?
- *People and Behaviors*: Changing perceptions, expectations, commitments, and decisions – What are people thinking and how do they intend to act?
- *Organizations and Processes*: Urban infrastructure networks and flows – water, energy, food, traffic, and so on – How do things work, fail, and interact?

Returning to the technical problem, the organization and process levels of analysis involve projecting and monitoring urban network flows and dynamics by season and time of day. It requires understanding impacts of weather scenarios such as hurricanes, nor'easters, and heat waves. Also important are impacts of outage scenarios including power loss, Internet loss, and transportation link loss (e.g., bridge or tunnel). Finally, unfortunately, there is a need to understand the impacts of terrorist scenarios, ranging from localized threats like 9/11 to pervasive threats such as might affect the water or food supply.

Issues at the people level, where individual behaviors of population members arise, can be characterized in terms of people's questions. At first, their questions include: What is happening? What is likely to happen? What do others think? Later, their questions become: Will we have power, transportation? Will we have food and water? What do others think? Further on, their questions become: Where should we go? How can we get there? What are others doing?

These questions bring us back to the behavioral and social problem. We need to be able to project and monitor answers to people's questions by scenario as time evolves. This involves understanding impacts of content, modes, and frequencies of communications. Integral to this understanding is how best to portray uncertainties associated with predictions of technical problems.

Finally, all of this understanding must occur in the context of this city's norms, values, and beliefs, as well as its historical narrative. Thus, we must keep in mind that cities are composed of communities and neighborhoods that are not homogenous. This leads us to the levels of analysis of ecosystem characteristics and historical narrative. These levels of analysis provide insights into who should communicate and the nature of messages that will be believed.

To become really good at urban resilience, we need to address the following research questions:

- How can cities best be understood as a collection of communities and neighborhoods, all served by common urban infrastructures?

- How does policy (e.g., zoning, codes, and taxes), development (e.g., real estate, business formation, and relocation), immigration, and so on affect the evolution of communities and neighborhoods within a city?

- When technical problems arise, what message is appropriate and who should deliver it to each significantly different community and neighborhood within the city?

- How can we project and monitor the responses of each community and neighborhood to the information communicated, especially as it is shared across social networks?

Consider the approaches needed to address these questions. Certainly, empirical studies enable historical narratives and case studies of past events. Historical and anthropological research efforts are central in this regard. They provide the basis for interpreting the context of the city of interest.

Mathematical and computational models of environmental threats, infrastructure functioning, and organizational performance, as well as human response, provide more formal, albeit abstract, explanations of urban events and evolution. Rouse (2015) elaborates a wealth of phenomena and approaches for addressing these ends. These computational components interact as shown in Figure 6.3.

Finally, virtual worlds with decision makers in the loop can enable "what if" explorations of sources of problems and solution opportunities. Such policy flight simulators can support decision makers and other key stakeholders to understand issues, share perspectives, and negotiate policies and solutions (Rouse, 2014a). We return to policy flight simulators later in this chapter.

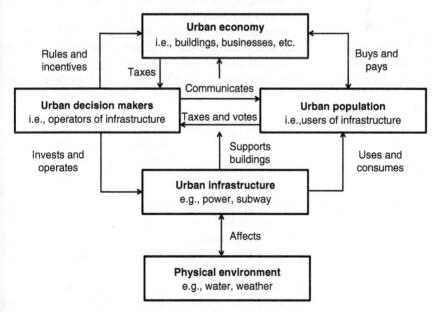

FIGURE 6.3 A city as a system.

EMERGENCE IN INSTITUTIONS

Institutional change can be very difficult, in part, due to the relative lack of forces of creative destruction (Schumpeter, 1942). This section addresses two institutions – healthcare delivery (Rouse and Cortese, 2010; Rouse and Serban, 2014) and higher education (Rouse, 2016). Healthcare delivery in the United States was the poster child for runaway costs; higher education now plays this role.

Healthcare Delivery

For several years, the National Academies has been engaged in a systemic study of the quality and cost of healthcare in the United States (IOM, 2000, 2001; National Academy of Engineering and Institute of Medicine, 2005). Clearly, substantial improvements in the delivery of healthcare are needed and, many have argued, achievable via value-based competition (Porter and Teisberg, 2006). Of course, it should be kept in mind that our healthcare system did not get the way it is overnight (Stevens et al., 2006).

There are several contemporary drivers of change in healthcare delivery. Insurance reform will surely be followed by healthcare reform. The ongoing health policy agenda is shifting the emphasis from simply covering more people under the Medicaid and Medicare programs to changing delivery practices. The range of changes included in future healthcare reform is highly uncertain. Providers and payers need to be able to consider very different hypothetical scenarios. All in all, the Affordable Care Act (ACA) has created both hope and apprehension (Goodnough, 2013). Anticipated changes to the ACA following the 2016 election have increased uncertainties.

Employers' economic burden for providing healthcare – due, in part, to cost shifting from Medicare and Medicaid patients – is unsustainable (Rouse, 2009, 2010a). This hidden tax leads to competitive disadvantages in the global marketplace (Meyer and Johnson, 2012). Employees' unhealthy lifestyles are increasing the incidence and cost of chronic diseases, leaving employers to absorb both increased healthcare costs and the costs of lost productivity (Burton et al., 1998).

Healthcare providers will have to adapt to new revenue models. For example, they may be paid for outcomes rather than procedures as currently attempted under some stipulations of ACA. Improved quality and lower costs will be central. Providers will have to differentiate good (profitable) and bad (not profitable) revenue, which means that they will need to understand and manage their costs at the level of each step of each process.

Many studies by the National Academies and others have concluded that a major problem with the healthcare system is that it is not really a system. Earlier in this chapter, I elaborated on the differences between traditional systems and complex adaptive systems. Healthcare delivery is a great example of complex adaptive system, as is higher education.

Before elaborating on these characteristics in the context of healthcare, it is useful to reflect on an overall implication for systems with these characteristics. With any conventional means, one cannot command or force such systems to comply

	Traditional system	Complex adaptive system
Roles	Management	Leadership
Methods	Command and control	Incentives and inhibitions
Measurement	Activities	Outcomes
Focus	Efficiency	Agility
Relationships	Contractual	Personal commitments
Network	Hierarchy	Heterarchy
Design	Organizational design	Self organization

FIGURE 6.4 Comparison of organizational behaviors.

with behavioral and performance dictates. Agents in complex adaptive systems are sufficiently intelligent to game the system, find "workarounds," and creatively identify ways to serve their own interests.

The best way to approach the management of complex adaptive systems is with organizational behaviors that differ from the usual behaviors, such as adopting a human-centered perspective that addresses the abilities, limitations, and inclinations of all stakeholders (Rouse, 2007b). Figure 6.4 contrasts the organizational behaviors needed in complex adaptive systems with those behaviors typical in traditional systems (Rouse, 2008).

Given that no one is in charge of a complex adaptive system, the management approach should emphasize leadership rather than traditional management techniques – influence rather than power. Because none, or very few, of the stakeholder groups in the healthcare system are employees, command and control have to be replaced with incentives and inhibitions. No one can require that stakeholders comply with organizational dictates. They must have incentives to behave appropriately.

Not only are most stakeholders independent agents but they are also beyond direct observation. Thus, one cannot manage their activities but can only assess the value of their outcomes. In a traditional system, one might attempt to optimize efficiency. However, the learning and adaptive characteristics of a complex adaptive system should be leveraged to encourage agility rather than throttled by optimization focused on out-of-date requirements.

Of course, there are contractual commitments in complex systems, but because of the nature of these systems, stakeholders can easily change allegiances, at least at the end of their current contracts. Personal commitments, which can greatly diminish the risks of such behaviors, imply close social relationships rather than weak social relationships among stakeholder groups and transparent organizational policies, practices, and outcomes.

Work is done by heterarchies, whereas permissions are granted and resources provided by hierarchies. To the extent that the heterarchy has to stop and ask the hierarchy for permission or resources, the efficiency and effectiveness of the system is undermined. Decision-making authority and resources should be delegated to the heterarchy with, of course, the right incentives and inhibitions.

Finally, because complex adaptive systems self-organize, no one can impose an organizational design. Even if a design was imposed, stakeholders would inevitably morph it as they learn and adapt to changing conditions. In that case, the organization that the management of such a system would think it is running would not really exist. To the extent that everyone agrees to pretend that it still exists, or ever existed, value will be undermined.

Higher Education

As noted earlier, higher education has become the poster child for out-of-control costs, replacing healthcare, which now seems more or less being controlled, albeit much too high for many segments of the population. Tuition increases have far out-paced increases of the overall cost of living. This is due to the relative decline of public support for higher education, whereas administrative costs have been steadily growing much faster than the costs of teaching and research. A primary enabler and consequence of this cost growth has been the availability of student loans that have led to debt levels that exceed the total credit card debt in the United States (Rouse, 2016).

We need to get a grip on the economics of higher education, with a goal of transforming the system to improve the overall value proposition. Academia provides a wide variety of offerings that serve a diverse mix of constituencies. Delivery processes for these offerings can be quite creative but are often burdened with inefficiencies. This is complicated by academic governance processes, which can be overwhelming, more so when the university is also a public sector agency. There is much to build on, but nevertheless much to overcome.

Figure 6.5 depicts a multi-level architecture of academic enterprises (Rouse, 2016). The practices of education, research, and service occur in the context of processes, structure, and ecosystem. Understanding the relationships among practices, processes, structure, and ecosystem provides the basis for transforming academia, leveraging its strengths, and overcoming its limitations. I explicitly address these relationships in terms of both conceptual and computational models of academic enterprises in a recent book (Rouse, 2016).

The architecture in Figure 6.5 helps us to understand how various elements of the enterprise system either enable or hinder other elements of the system, all of which are embedded in a complex behavioral and social ecosystem. Practices are much more efficient and effective when enabled by well-articulated and supported processes for delivering capabilities and associated information, as well as capturing and disseminating outcomes.

Processes exist to the extent that organizations (i.e., campuses, colleges, schools, and departments) invest in them. These investments are influenced by economic

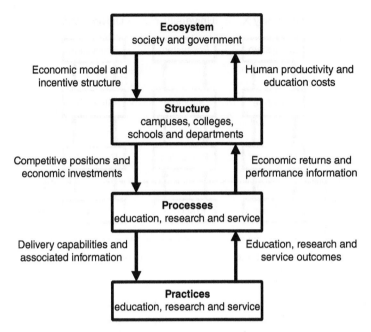

FIGURE 6.5   Architecture of academic enterprises.

models and incentive structures and are made in pursuit of competitive positions and economic returns. These forces hopefully coalesce to create an educated and productive population, at an acceptable cost.

When we employ Figure 6.5 to understand relationships among universities, the interesting phenomenon in Figure 6.6 emerges. The hierarchical structure of Figure 6.5 dovetails with the heterarchical nature of academic disciplines. The dotted rectangle in Figure 6.6 represents how faculty disciplines both compete and define standards across universities.

The disciplines define the agenda for "normal" science and technology, including valued sponsors of this agenda and valued outlets for research results. Members of faculty disciplines at other universities have an enormous impact on promotion and tenure processes at any particular university. Such professional affiliations also affect other types of enterprises (healthcare, for example). However, universities seem to be the only enterprise that allows external parties to largely determine who gets promoted and tenured internally. This has substantial impacts on understanding and modeling the performance of any particular university.

More specifically, the standards set at the discipline level determine:

- Agenda for "normal" science and technology;
- Valued sponsors of this agenda; and
- Valued outlets for research results.

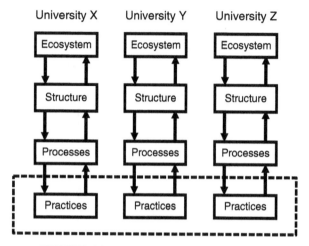

FIGURE 6.6   Hybrid architecture of academia.

Consequently, almost everyone chases the same sponsors and journals, leading to decreasing probabilities of success with either. In addition, each faculty member produces another faculty member every year or so, swelling the ranks of the competitors. Recently, retirements are being delayed to refill individuals' retirement coffers, which decrease numbers of open faculty slots.

As probabilities of success decrease, faculty members

- write an increasing number of proposals,
- submit an increasing number of journal articles,
- resulting in constantly increasing costs of success,
- congested pipelines, which foster constantly increasing times until success, and
- bottom line is less success, greater costs, and longer delays.

The models discussed in Rouse (2016) enable exploration of these phenomena.

Universities can hold off these consequences by hiring fewer tenure-track faculty members, that is, using teaching faculty and adjuncts. But this will retard their march up the rankings and hence slow the acquisition of talented students, who will succeed in life and later reward the institution with gifts and endowments. The tradeoff between controlling cost and enhancing brand value is central.

Alternatively, universities can pursue "niche dominance" and only hire tenure-track faculty in areas where they can leapfrog to excellence. This will, unfortunately, result in two classes of faculty – those on the fast track to excellence and those destined to teach a lot. The first class will be paid a lot more because of the great risks of their being attracted away to enhance other universities' brands.

As indicated earlier, higher education is also a good example of a complex adaptive system. The right column in Figure 6.4 will be immediately recognizable to anyone who has served on the faculty of a research university. This does not make change intractable. It just requires different methods and tools.

Approach to Modeling and Simulation

Four levels of the overall healthcare delivery enterprise are shown in Figure 6.7 (Rouse and Cortese, 2010; Rouse and Serban, 2014). This framework provides a conceptual model for understanding relationships among the various elements of this enterprise, ranging from patient–clinician interactions at the bottom to policy and budget decisions by Congress and Centers for Medicare and Medicaid Services at the top. In the middle, providers and payers make investment decisions that balance the patients' needs from below and the "rules of the game" from above.

Note that "medicine" is delivered on the bottom level of this figure. Additional functions associated with healthcare delivery occur on the other three levels. To foster high-value health for everyone, learning has to occur at all of the levels.

As discussed earlier, this enterprise is being challenged by needs for fundamental change. The ACA, new payment schemes, and market and technology opportunities and threats have resulted in an enormous number of initiatives at all levels of the enterprise. Orchestrating these initiatives is very much complicated by the complex adaptive nature of the healthcare system. Put simply, there are millions of independent

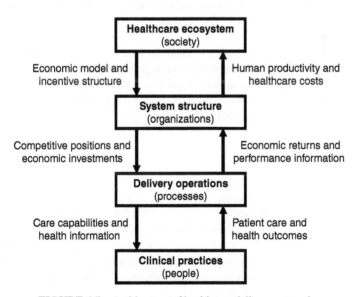

FIGURE 6.7   Architecture of healthcare delivery enterprise.

agents reacting to the forces driving change. All of these agents are rapidly learning in the process, but the enterprise as a whole is learning much more slowly.

The framework in Figure 6.7 has been used to undertake the following modeling and simulation projects:

- *Emory Health*: Prevention and wellness for type 2 diabetes and heart disease (Park *et al.*, 2012).
- *Vanderbilt Health*: Chronic disease management for hypertension, type 2 diabetes, and heart disease.
- *Indiana Health*: Chronic disease management for dementia and Alzheimer's disease (Boustany *et al.*, 2016).
- *Penn Medicine*: Transition care for high-risk elderly patients (Rouse *et al.*, 2017b).
- *New York City Health Ecosystem*: Mergers and acquisitions, partially in response to ACA (Yu *et al.*, 2016).
- *Population Health*: Addressing healthcare and social determinants of health in an integrated manner (Rouse *et al.*, 2017a).

At the lower levels of Figure 6.7, these models and simulations involved agent-based and discrete event simulations. At the higher levers were microeconomic and macroeconomic models, sometimes instantiated as system dynamics models. All models were parameterized using data from sponsors and public sources. Substantial interactive visualizations were created to enable decision makers to explore relevant policy options. The importance of such visualizations is further discussed later in this chapter.

It is important to note that emergent phenomena are often precipitated by interactions among the models across the multi-level representation. For example, entangled states across levels can require simultaneous solutions of the multiple models, which can be difficult. One type of emergent understanding is that phenomena that had been assumed to be independent of each other may interact in subtle ways.

EMERGENCE IN COMPANIES

Unlike institutions that can ignore market forces, companies must address needs to change. Consider the evolution of the Fortune 500 (Schumpeter, 2009). In the period 1956–1981, an average of 24 firms dropped out of the Fortune 500 list every year, that is, 120% turnover in 25 years. In 1982–2006, an average of 40 firms dropped out of the Fortune 500 list every year, yielding 200% turnover in 25 years. Fundamental change is very difficult, becoming more difficult, and the failure rate due to "creative destruction" is very high (Schumpeter, 1942).

Consider Eastman Kodak. During most of the twentieth century, Kodak held a dominant position producing photographic film. The company's ubiquity was such that its "Kodak moment" tagline entered the common lexicon to describe a personal event that demanded to be recorded for posterity. Kodak began to struggle financially in the late 1990s, as a result of the decline in sales of photographic film and its slowness to accept the need to transition to digital photography. As a part of a turnaround strategy, Kodak attempted to focus on digital photography and digital printing and to generate revenues through aggressive patent litigation, but they were too late. They filed for bankruptcy in 2012 and emerged in 2013 a much smaller company.

A more recent example is Nokia, who was, for a period, the largest vendor of mobile phones in the world. Nokia's dominance also extended into the smartphone industry through its Symbian platform, but was eventually overshadowed by competitors Apple and Samsung, the latter employing Google's Android operating system. Microsoft eventually bought Nokia's mobile phone business for $7.17 billion. The emergence of the Apple's iPhone enabled the creative destruction of Nokia.

Enterprise Transformation

How do companies transform when facing such challenges. Our earlier studies (Rouse, 2005, 2006) have led us to formulate a qualitative theory, "Enterprise transformation is driven by experienced and/or anticipated value deficiencies that result in significantly redesigned and/or new work processes as determined by management's decision-making abilities, limitations, and inclinations, all in the context of the social networks of management in particular and the enterprise in general."

There is a wide range of ways to pursue transformation. Figure 6.8 summarizes conclusions drawn from a large number of case studies. The ends of transformation can range from greater cost efficiencies to enhanced market perceptions, to new product and service offerings, to fundamental changes of markets. The means can range from upgrading people's skills to redesigning business practices, to significant infusions of technology, to fundamental changes of strategy. The scope of transformation can range from work activities to business functions, to overall organizations, to the enterprise as a whole.

The framework in Figure 6.8 has provided a useful categorization of a broad range of case studies of enterprise transformation. Considering transformation of markets, Amazon leveraged IT to redefine book buying, w Wal-Mart leveraged IT to redefine the retail industry. In these two instances at least, it can be argued that Amazon and Wal-Mart just grew; they did not transform. Nevertheless, their markets were transformed.

Illustrations of transformation of offerings include UPS moving from being a package delivery company to a global supply chain management provider, IBM's transition from manufacturing to services, Motorola moving from battery eliminators to radios to cell phones, and CNN redefining news delivery. Examples of transformation of perceptions include Dell repositioning computer buying, Starbucks repositioning coffee purchases, and Victoria's Secret repositioning lingerie buying. The many instances of transforming business operations include Lockheed Martin merging three aircraft companies, Newell Rubbermaid resuscitating numerous home products companies, and Interface adopting green business practices.

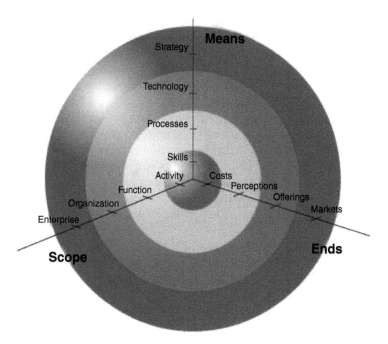

FIGURE 6.8   Transformation framework.

The costs and risks of transformation increase as the endeavor moves farther from the center in Figure 6.8. Initiatives focused on the center will typically involve well-known and mature methods and tools from industrial engineering and operations management. In contrast, initiatives toward the perimeter will often require substantial changes of products, services, channels, and so on, as well as associated large investments.

It is important to note that successful transformations in the outer band of Figure 6.8 are likely to require significant investments in the inner bands also. In general, any level of transformation requires consideration of all subordinate levels. Thus, for example, successfully changing the market's perceptions of an enterprise's offerings is likely to also require enhanced operational excellence to underpin the new image being sought. As another illustration, significant changes of strategies often require new processes for decision making, for example, for R&D investments.

The transformation framework can be applied to thinking through a range of scenarios. The inner circle in Figure 6.8 focuses on enterprise efficiency by, for example, focusing on particular activities, the skills needed for these activities, and the costs of these activities. In contrast, the outer circle of Figure 6.8 might focus on totally new value propositions, addressing the whole enterprise, rethinking strategy, and fundamentally changing the marketplace.

Changes in the outer circle will very likely require changes in the adjacent circle. New offerings in a range of organizations will be enabled by new technologies. Success of these offerings is likely to involve changes of perceptions in the next circle at the functional level, enabled by new processes. Thus, we can see that embracing a totally new value proposition will require reconsideration of everything the enterprise does.

This does not imply that everything will change. Instead, it means that every-thing needs to be considered in terms of how things consistently fit together, function smoothly, and provide high-value outcomes. This may be daunting, but is entirely fea-sible. The key point is that one cannot consider transforming the marketplace without considering how the enterprise itself should be transformed.

We hasten to note that, at this point, we are only addressing what is likely to have to change, not how the changes can be accomplished. In particular, we are not considering how to gain the support of stakeholders, manage their perceptions and expectations, and sustain fundamental change (Rouse, 2001, 2006, 2007b).

Scenarios such as outlined above can be explored computationally (Rouse and Boff, 2005). Thus, we are proposing that computational transformation should pre-cede physical transformation. This enables exploration of a wide range of scenarios and, in particular, helps to get rid of bad ideas quickly. The good ideas that remain can be carefully refined for empirical investigation.

Approach to Modeling and Simulation

We have been involved with a large number of companies facing challenges of change. Numerous case studies are summarized in Rouse (2006, 2007b). The most common scenario we faced was the need to create new product offerings and invest in the technologies needed to enable these new offerings. We developed a suite of four software tools for addressing this scenario.

Figure 6.9 depicts key components and relationships of two of these software tools. The *Product Planning Advisor* helped decision makers explore new product offerings. The *Technology Investment Advisor* was used to "backcast" the investments needed

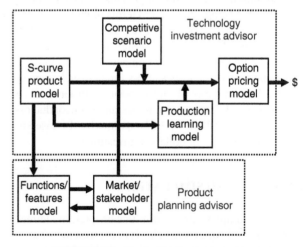

FIGURE 6.9   Product/technology strategy.

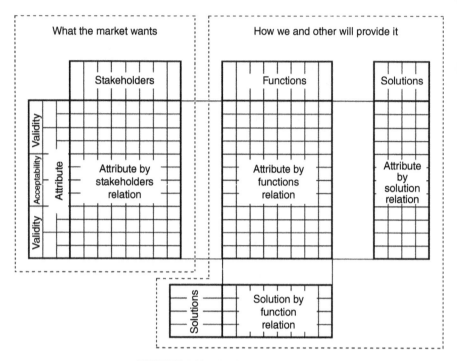

FIGURE 6.10   Product/market model.

years earlier to subsequently enable the new offerings. The projections emanating from the right of this figure are the firm's projected Income Statements over the time horizon of interest, typically 10 years.

The product planning process was quite detailed. It involved filling in the computational model summarized in Figure 6.10. This involved identifying the major stakeholders in the market being addressed in terms of their attributes of interest and the relative importance of these attributes. The process required identifying how the firm and its competitors would strive to please stakeholders. Finally, alternative offerings (solutions), from both the firm and its competitors, were functionally defined.

This approach to helping companies address fundamental competitive challenges was employed for a large number of engagements with large technology-oriented companies in aerospace and defense, electronics and semiconductors, pharmaceuticals, and beverages. Rouse and Boff (2004) summarize 14 case studies using this approach that resulted in billions of dollars of investments.

ROLE OF MODELING AND SIMULATION

Throughout the many examples discussed in this chapter, modeling and simulation was used to address "what if" questions. This is in contrast to "what is" questions, the realm of data analytics. To address "what if" questions, we need to make structural and parametric assumptions about the processes that will generate the future. This often requires further assumptions about independence, conservation, and continuity. We also have to assure that the collective set of assumptions is consistent across the multiple levels of abstraction employed to represent the phenomena of interest. This tends to be easier to do when the component models are sets of equations rather than legacy software codes (Tolk, 2014).

The types of models and simulations discussed in this chapter present validation challenges (Rouse, 2015). One is typically predicting the future of an enterprise, often laced with significant changes, several or many years into the future. In other words, one is predicting the future of a system that does not yet exist. Empirical validation is seldom possible.

A weak form of validation involves predicting the current state of the enterprise from, say, a period of 5 years ago. If the predictions are off, there may be something wrong with the model. Alternatively, you may be predicting a future that likely could have happened but did not. Nevertheless, if the predictions are accurate, one can feel more comfortable, but the assumptions about enterprise changes in the future still have not been validated. Note that some refer to this process as model calibration rather than weak validation.

An alternative position on this issue is that one is looking for insights rather than predictions. Interactive visualizations as shown in Figure 6.11 can be used to help decision makers explore the complexity of their enterprise, in this case the healthcare delivery ecosystem of New York City (Yu *et al.*, 2016). We have found that decision makers use such environments to develop shared mental models, get rid of bad ideas quickly, and refine good ideas for empirical evaluation.

We were asked, during a healthcare-related workshop, what one calls the dynamic interactive visualizations as shown in Figure 6.11, we first provided a phrase laced

FIGURE 6.11   Simulation of NYC Health Ecosystem in *Immersion Lab*.

with engineering jargon. The participants rejected this. Then we said, "It's a policy flight simulator where you can fly the future before you commit to decisions." This name stuck.

We have studied how such environments affect decision makers across hundreds of engagements over many years and in several domains (Rouse, 1998, 2014a). We have found that decision makers are much more focused on exploration than answers. The insights they seek come from group discussions and debates in the context of the models and simulations. The creativity comes from the group as they explore what affects what and how possible changes affect outcomes. After one workshop in the semiconductor industry, using the processes in Figures 6.9 and 6.10, an electrical engineer commented, "Now I know what the key financial metrics are and how they are calculated. Before this workshop, I didn't even know such metrics existed."

Decision Making

Enterprises have differing abilities to predict their futures, as well as differing abilities to respond to these futures. What strategies might enterprise decision makers adopt to address the alternative futures they have explored using modeling and simulation? As shown in Figure 6.12, we have found that there are four basic strategies that decision makers can use: optimize, adapt, hedge, and accept (Pennock and Rouse, 2016).

If the phenomena of interest are highly predictable, then there is little chance that the enterprise will be pushed into unanticipated territory. Consequently, it is in the best interest of the enterprise to optimize its interventions to be as efficient as possible. In other words, if the unexpected cannot happen, then there is no reason to expend resources beyond process refinement and improvement.

If the phenomena of interest are not highly predictable, but interventions can be appropriately adapted when necessary, it may be in the best interest for the enterprise

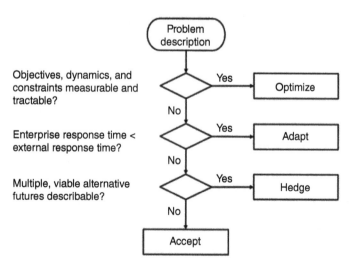

FIGURE 6.12   Strategy framework for enterprise decision makers.

to plan to adapt. For example, agile capacities can be designed to enable their use in multiple ways, to adapt to changing demands. In this case, some efficiency has been traded for the ability to adapt.

For this approach to work, the enterprise must be able to identify and respond to potential issues faster than the ecosystem changes. For example, consider increased patient flows that tax capacities beyond their designed limits. Design and building of new or expanded facilities can take considerable time. On the other hand, reconfiguration of agile capacities should be much faster.

If the phenomena of interest are not very predictable and the enterprise has a limited ability to respond, it may be in the best interest of the enterprise to hedge its position. In this case, it can explore scenarios where the enterprise may not be able to handle sudden changes without prior investment. For example, an enterprise concerned about potential obsolescence of some lines of service may choose to invest in multiple, potential new services. Such investments might be pilot projects that enable learning how to deliver services differently or perhaps deliver different services.

Over time, it will become clear which of these options makes most sense and the enterprise can exercise the best option by scaling up this line of service based on what they have learned during the pilot projects. In contrast, if the enterprise were to take a wait and see approach, it might not be able to respond quickly enough, and it might lose out to its competitors.

If the phenomena of interest are totally unpredictable and there is no viable way to respond, then the enterprise has no choice but to accept the risk. Accept is not so much a strategy as a default condition. If one is attempting to address a strategic challenge where there is little ability to optimize the efficacy of offerings, limited ability to adapt offerings, and no viable hedges against the uncertainties associated with these offerings, the enterprise must accept the conditions that emerge. Learning is still possible, however.

We have found it useful to think about enterprise learning at two levels – single-loop and double-loop (Argyris, 1991; Rouse et al., 2017a). Single-loop learning is akin to business process improvement. One tunes a process to continually make it better. Double-loop learning is embodied in enterprise transformation. One comes to realize that a process is fundamentally flawed and has to change. This realization may enable moving from accept in Figure 6.12 back to hedge and perhaps even adapt.

Thus, the relevant strategy is very situation dependent. The situation may change over time, in part, due to external forces, but also, in part, due to learning that results in new perspectives, prompting new ideas for how to proceed.

CONCLUSIONS

This chapter has focused on the use of modeling and simulation to address emergence in complex enterprises. The examples elaborated in this chapter focused on understanding and addressing complex problems and not necessarily advancing the discipline of modeling and simulation. Nevertheless, the use of modeling and simulation was key to uncovering emergent phenomena, either provided by the computations or the humans in the loop. These humans were typically quite impressed with what modeling and simulation enabled them to do.

The wide range of examples presented shows the breadth of applicability of modeling and simulation. The emphasis on insights rather than predictions, all in the context of group discussion and debate, allows for consideration of behavioral and social phenomena that may be difficult to incorporate in the models and simulations. For example, reluctance to buy into change is likely to be much better understood by the group than by the models.

It is not reasonable to expect that everything can be modeled and simulated. Beyond the simple intractability of such aspirations, the reality is that models can inform decisions but cannot make them, at least not in complex enterprises. The necessary competencies for decision makers when leading fundamental change include leadership, vision, strategy, communication, and collaboration (Rouse, 2011). Respondents to this survey also valued technical skills, albeit not so highly. However, change initiatives seldom fail due to lack of technical skills; they fail because leadership, vision, strategy, communication, and/or collaboration are lacking.

In fact, we have found that our models and simulations are most successful when supporting decision-making teams with strong leadership skills and commitment to moving the enterprise forward. In such situations, access to subject matter experts becomes easy, necessary data are quickly available, and group meetings are readily convened. Such groups often attribute much value to the interactive models and simulations. We know that much of the success is due to the supportive context they provided, as well as their creative interpretations of their discoveries when using a policy flight simulator to explore alternative futures.

REFERENCES

Andrews, J. (2015) *The World in Conflict: Understanding the World's Troublespots*, Public Affairs, New York.

Argyris, C. (1991) Teaching smart people how to learn. *Harvard Business Review*, **69** (3), 99–109.

Barber, B.R. (2013) *If Mayors Ruled the World: Dysfunctional Nations, Rising Cities*, Yale University Press, New Haven, CT.

Boustany, K., Pennock, M.J., Bell, T. *et al.* (2016) *Leveraging Computer Simulation Models in Healthcare Delivery Redesign*, Indiana University Health, Aging Brains Care Center, Indianapolis, IN.

Brook, D. (2013) *A History of Future Cities*, Norton, New York.

Burton, W.N., Chen, C.-Y., Schultz, A.B., and Edington, D.W. (1998) The economic costs associated with body mass index in a workplace. *Journal of Occupational & Environmental Medicine*, **40** (9), 786–792.

Casti, J. (2012) *X-Events: The Collapse of Everything*, Morrow, New York.

Corning, P.A. (2002) The re-emergence of "emergence:" a venerable concept in search of a theory. *Complexity*, **7** (6), 18–30.

Diamond, J. (2005) *Collapse: How Societies Choose to Fail or Succeed*, Viking, New York.

Emery, F. and Trist, E. (1973) *Toward a Social Ecology*, Plenum Press, London.

Glaeser, E.L. (2011) *Triumph of the City: How Our Greatest Invention Makes Us Richer, Smarter, Greener, Healthier, and Happier*, Penguin, New York.

Goodnough, A. (2013) A Louisville clinic races to adapt to the health care overhaul. *New York Times*(June 23), p. 1.

Grossman, C., Goolsby, W.A., Olsen, L., and McGinnis, J.M. (2011) *Engineering the Learning Healthcare System*, National Academy Press, Washington, D.C.

Harari, Y.N. (2015) *Sapiens: A Brief History of Humankind*, Harper, New York.

Hendrick, H.W. and Kleiner, B.M. (2001) *Macroergonomics: An Introduction to Work System Design*, Human Factors and Ergonomics Society, Santa Monica, CA.

Institute of Medicine (2000) *To Err is Human: Building a Safer Health System*, National Academies Press, Washington, D.C.

Institute of Medicine (2001) *Crossing the Quality Chasm: A New Health Systems for the 21st Century*, National Academies Press, Washington, D.C.

Lewes, G.H. (1875–1879) *Problems of Life and Mind*. 5 Volumes, Trübner & Co., London.

Meyer and Johnson (2012) Cost shifting in healthcare: an economic analysis. *Health Affairs*, **2**, 20–35.

National Academy of Engineering and Institute of Medicine (2005) *Building A Better Delivery System: A New Engineering/Health Care Partnership*, National Academies Press, Washington, D.C.

Park, H., Clear, T., Rouse, W.B. *et al.* (2012) Multi-level simulations of health delivery systems. A prospective tool for policy, strategy, planning and management. *Journal of Service Science*, **4** (3), 253–268.

Pennock, M.J. and Rouse, W.B. (2016) The epistemology of enterprises. *Systems Engineering*, **19** (1), 24–43.

Poli, R. (2013) A note on the difference between complicated and complex social systems. *Cadmus*, **2** (1), 142–147.

Porter, M.E. and Teisberg, E.O. (2006) *Redefining Health Care: Creating Value-Based Competition on Results*, Harvard Business School Press, Boston, MA.

Rouse, W.B. (1998) Computer support of collaborative planning. *Journal of the American Society for Information Science*, **49** (9), 832–839.

Rouse, W.B. (2000) Managing complexity: disease control as a complex adaptive system. *Information, Knowledge, Systems Management*, **2** (2), 143–165.

Rouse, W.B. (2001) *Essential Challenges of Strategic Management*, Wiley, New York.

Rouse, W.B. (2003) Engineering complex systems: implications for research in systems engineering. *IEEE Transactions on Systems, Man, and Cybernetics – Part C*, **33** (2), 154–156.

Rouse, W.B. (2005) A theory of enterprise transformation. *Journal of Systems Engineering*, **8** (4), 279–295.

Rouse, W.B. (ed.) (2006) *Enterprise Transformation: Understanding and Enabling Fundamental Change*, John Wiley, Hoboken, NJ.

Rouse, W.B. (2007a) Complex engineered, organizational & natural systems: issues underlying the complexity of systems and fundamental research needed to address these issues. *Systems Engineering*, **10** (3), 260–271.

Rouse, W.B. (2007b) *People and Organizations: Explorations of Human-Centered Design*, Wiley, New York.

Rouse, W.B. (2008) Healthcare as a complex adaptive system: implications for design and management. *The Bridge*, **38** (1), 17–25.

Rouse, W.B. (2009) Engineering perspectives on healthcare delivery: can we afford technological innovation in healthcare? *Journal of Systems Research and Behavioral Science*, **26**, 1–10.

Rouse, W.B. (2010a) Impacts of healthcare price controls: potential unintended consequences of firms' responses to price policies. *IEEE Systems Journal*, **4** (1), 34–38.

Rouse, W.B. (ed.) (2010b) *The Economics of Human Systems Integration: Valuation of Investments in People's Training and Education, Safety and Health, and Work Productivity*, Wiley, New York.

Rouse, W.B. (2011) Necessary competencies for transforming an enterprise. *Journal of Enterprise Transformation*, **1** (1), 71–92.

Rouse, W.B. (2014a) Human interaction with policy flight simulators. *Journal of Applied Ergonomics*, **45** (1), 72–77.

Rouse, W.B. (2014b) Earth as a system, in *Can Earth's and Society's Systems Meet the Needs of 10 Billion People?* (ed. M. Mellody), National Academies Press, Washington, D.C., pp. 20–23.

Rouse, W.B. (2015) *Modeling and Visualization of Complex Systems and Enterprises: Explorations of Physical, Human, Economic, and Social Phenomena*, Wiley, Hoboken, NJ.

Rouse, W.B. (2016) *Universities as Complex Enterprises: How Academia Works, Why it Works These Ways, and Where the University Enterprise is Headed*, Wiley, Hoboken, NJ.

Rouse, W.B. and Boff, K.R. (2004) Value-centered R&D organizations: ten principles for characterizing, assessing & managing value. *Systems Engineering*, **7** (2), 167–185.

Rouse, W.B. and Boff, K.R. (eds) (2005) *Organizational Simulation: From Modeling and Simulation to Games and Entertainment*, Wiley, New York.

Rouse, W.B. and Cortese, D.A. (eds) (2010) *Engineering the System of Healthcare Delivery*, IOS Press, Amsterdam.

Rouse, W.B. and Serban, N. (2014) *Understanding and Managing the Complexity of Healthcare*, MIT Press, Cambridge, MA.

Rouse, W.B., Johns, M.M.E., and Pepe, K. (2017a) Learning in the healthcare enterprise. *Journal of Learning Health Systems*, **1** (4).

Rouse, W.B., Naylor, M., Pennock, M.J. *et al.* (2017b) *The Use of Policy Simulation to Implement the Transitional Care Model*, Center for Complex Systems and Enterprises, Stevens Institute of Technology, Hoboken, NJ.

Schumpeter, J. (1942) *Capitalism, Socialism, and Democracy*, Harper, New York.

Schumpeter (2009) Taking flight. *The Economist* (September 19), p. 78.

Snowden, D.J. and Boone, M.E. (2007) A leader's framework for decision making. *Harvard Business Review* (November), pp. 69–76.

Stevens, R.A., Rosenberg, C.E., and Burns, L.R. (eds) (2006) *History and Health Policy in the United States*, Rutgers University Press, New Brunswick, NJ.

Tainter, J.A. (1988) *The Collapse of Complex Societies*, Cambridge University Press, Cambridge, UK.

Tolk, A. (2014) *Ontology, Epistemology, and Teleology for Modeling and Simulation: Philosophical Foundations for Intelligent M&S Applications*, Springer, Berlin.

Yu, Z., Rouse, W.B., Serban, S., and Veral, E. (2016) A data-rich agent-based decision support model for hospital consolidation. *Journal of Enterprise Transformation*, **6** (3/4), 136–161.

7

EMERGENCE IN INFORMATION ECONOMIES: AN AGENT-BASED MODELING PERSPECTIVE

Erika Frydenlund[1] and David C. Earnest[2]

[1]Virginia Modeling, Analysis and Simulation Center, Old Dominion University, Suffolk, VA 23435, USA
[2]Department of Political Science, University of South Dakota, Vermillion, SD 57069, USA

INTRODUCTION

We see instances of technological path-dependency all around us. Fossil fuel-based cars have dominated the automobile industry even though electric cars existed in the mid-nineteenth century. Operating systems for personal computers have converged on two choices even though alternative operating systems exist – including free ones. A similar convergence has occurred in software for mobile devices, with the software of two companies constituting 99.6% of the global market for tablets and smart phones operating systems (Moscaritolo, 2017). Even the spelling of words eventually standardized while alternative spellings were dropped. What drives the market to center around one or two main inventions when there are so many possibilities at the outset? How does this convergence happen? Preferential attachment, where certain technologies gain momentum and people are drawn to adopt technologies that already have a number of users, informs a number of insights into network structure. In this chapter, we use the idea of technological lock-ins to demonstrate the value of focusing on the idea of *emergence* as a primary concept in computational social sciences. Here, we model the emergence of scaling in communication networks to show how simple individual rules can lead to technology adoption patterns that are robust to

Emergent Behavior in Complex Systems Engineering: A Modeling and Simulation Approach,
First Edition. Edited by Saurabh Mittal, Saikou Diallo, and Andreas Tolk.
© 2018 John Wiley & Sons, Inc. Published 2018 by John Wiley & Sons, Inc.

starting communication network structures. Rather than align technological choices with the most popular adoptions (preferential attachment), our agent decision-making relies on dropping connections with agents who are most dissimilar to them, randomly adding new connections, and adopting slightly different technologies until they are satisfied with their network and technology choice. Unlike traditional social sciences where experimentation is impractical or impossible, we use an agent-based model to test differing technological complexities and network structures to show how populations end up locked-in to a small set of technologies. The discussion begins with an overview of how *emergence* as a concept features in computational social sciences and proceeds to describe the model, experiment, and results.

EMERGENCE IN THE SOCIAL SCIENCES

Observations of emergence in social systems predate discussions of agent-based modeling by nearly a century, but computer simulation and the formalized study of complex adaptive systems have brought the concept of emergence under direct investigation (Gilbert and Troitzsch, 2005). Epstein (2006) argues that the classical notions of emergence are not the same as what is described in agent-based modeling and simulation contexts. Classical emergentism sees macro-level phenomena that arise from micro-level decisions and interactions as irreducible. Computational social science, on the other hand, seeks to understand the relationship between the micro and macro in order to discern patterns and the root of the emergent phenomena (Epstein, 2006).

   Over three decades ago, Nobel Laureate and a founding father of agent-based modeling, Schelling, noted that social phenomena intrinsically lead to emergence through simple rules. He did not use the word "emergence," but instead described that what we see in social settings develops from individual behaviors and choices that form "system[s] of interaction between individuals and their environment" (Schelling, 1978, p. 14). To illustrate this, Schelling observes how a group of people might fill an auditorium, noting that their decisions and motivations, understood at the individual level, may lead to surprising aggregate results. For instance, an audience may fill the seats from back to front without any instruction to do so. Self-organizing behaviors such as this often give rise to observable macroscopic phenomena.

   Emergence is not a concept of interest to social scientists alone; complexity scholars from a wide variety of academic disciplines investigate the self-organization and emergent properties of entities in biological, social, and engineered complex systems. In fact, even some engineered systems such as infrastructure may be based on emergent network structures that originate from human behavioral choices. Where people choose to live and work can dictate the creation of infrastructure such as transportation and utilities networks that connect them. The structure of these networks could be seen as an emergent property of human agency. Humans in social systems interact with not only their built environment, but also the socially constructed environment created by other people in the system. This is particularly apparent in rural communities on the outskirts of large cities. Outside of Washington, D.C., for instance, people moved to surrounding rural or small towns such that became "exurbs" where housing development facilitated real estate booms outside the expensive main city (Berube *et al.*, 2006). Soon after, commuter transit options developed and businesses and services moved into support the population increase and demographic shift. Infrastructure networks, then, had to evolve to service the dramatic growth in population. If one looks at one of these exurbs today, the distribution of economic and residential growth areas, as well as the road layout and infrastructure grid, derives from individual-level interactions and decisions: people choosing to move and build houses in a rural county, businesses and government officials negotiating zoning plans, and utilities being installed to accommodate these choices.

This is a simple example for which we can untangle the rationale for the emergent infrastructure networks. Macro-level phenomena derived from micro-level interactions and decisions characterize many social systems and can be much more difficult to reduce to simple rules. For instance, how mobs form from mostly ambivalent groups of people (Epstein, 2013); how actors form cooperative relationships with little information about the others' motives (Axelrod, 1984); or even how decentralized transnational networks organize in our increasingly globalized world (Earnest, 2015). The plethora of examples related to emergence is one reason why the topic is so prominent in computational social science conversations. Epstein (2006) goes so far as to argue that the only way to understand macro-level phenomena is to "grow" the society from the bottom up – what he calls "generative social science."

But, although we can agree that emergent phenomena are everywhere, what does "emergence" really mean and why is it important to social scientists? The answer is not so simple. In fact, entire volumes and articles have been written about the idea of emergence, the difficulty of narrowing down an agreed-upon definition, and the debate about the specificity of existing uses and definitions (Baker, 2010; Holland, 1999; Olaya, 2007). In its most generic form, *emergence* is individual-level behaviors, decisions, and interactions that produce unexpected, novel, and irreducible macro-level phenomena (Huneman, 2008). These aggregate results that are often difficult to predict even given known starting conditions and behavioral rules represent "emergent" behavior observed at the macroscopic level (Mitchell, 2009, p. 13). Miller and Page (2007) add to that definition the principle that small perturbations in localized behavior do not greatly affect the observed aggregate phenomena. For some, however, the very use of the term *emergent* is problematic in that in encompasses contradictory properties and expectations, requiring narrower and more specific limitations to what the term should mean (Epstein, 2006; Huneman, 2008; Olaya, 2007). Wilson (2010) proposed a narrow definition in an attempt to quantify "weak emergence," while Gilbert (2002) and Cederman (2005) each offer a taxonomy of "emergence" in relation to modeling social behavior and norms.

Why is emergence so important that complexity scientists engage in extensive debate over defining, refining, and quantifying the term *emergence*? In an increasingly interconnected world, where seemingly inconsequential events have dramatic ripple effects through global economic or environmental systems, we look to complexity theories to help describe the phenomena and events we are witnessing. Thinking through social phenomena using a complexity theory lens helps us to break away from attempting to explain real-world observations by relying on central limit theorems and bending the assumptions of statistical measures. We can move away from framing social phenomena in terms of equilibrium and begin to capture the immense entanglement of feedback loops that include micro- and macro-level processes (Miller and Page, 2007). Complexity theory increasingly motivates research in a number of social science fields, including political science and international relations (Cindea, 2006; Kavalski, 2015; Macy and Willer, 2002). The rise of complexity theory has given way to more modeling and simulation approaches to social science research, in turn focusing our attention on understanding and theorizing about the potential for micro-level interactions to generate macro-level phenomena. One application of the complexity theory and exploration of emergence is in information economies where we can theorize on the mechanisms that drive adoption of some technologies over others.

INFORMATION ECONOMIES AND COMPUTATIONAL SOCIAL SCIENCE

Information economies exhibit network effects, a type of increasing return. In juxtaposition to the attenuating effects of decreasing returns, systems with increasing returns tend to grow by a factor greater than one for each unit of input (Arthur, 1994). Network effects produce such exponential growth because each adopter's choice of a technology increases the value of subsequent adopters' choices. To use a dated example, a fax machine is useless without others who also use fax machines; the device's utility derives fundamentally from the number of and growth in the community of adopters. This process of adoption tends to produce a nonlinear s-shaped growth curve as the technology becomes more valuable to both users and to those who have yet to adopt. However, network effects and increasing returns tend to produce several inefficiencies in markets. For one, the choices of early adopters strongly condition the incentives for subsequent adopters, creating a path-dependent process that often leads to "lock-in" (Arthur, 1989). That is, consumers may converge on a single product choice that may be technically inferior to alternative products (Cowan, 1990; Cowan and Hultén, 1996; Foray, 1997). Network effects also tend to produce markets characterized by monopolistic or oligopolistic competition (Arthur, 1990).

In communication networks, increasing returns tend to produce growth by "preferential attachment." That is, the probability of a node receiving new connections grows as the number of existing connections (i.e., its degree) increases. Preferential attachment is how early adopters of a technology attract new adopters, producing exponential growth that tends to lock in a single technology. The resulting network among adopters exhibits a scale-free structure, one in which the degree follows a power law distribution (Barabási and Albert, 1999). Such networks tend to have a large number of nodes with just a few connections but a small number of "hubs," or nodes with a large number of connections. Previous research has found that the scale-free structure characterizes a variety of communication networks including email communications (Ebel *et al.*, 2002); links among pages on the worldwide web (Barabási *et al.*, 2000), social media (Mislove *et al.*, 2007), Wikipedia (Spek *et al.*, 2012), mobile phones

(Eagle *et al.*, 2009), transportation networks (Guimera *et al.*, 2005), interbank lending (van Lelyveld, 2014), and even semantics (i Cancho and Solé, 2001).

Complex systems theory conceives of these attributes of information economies – scale freeness, technological lock-in, oligopolistic competition, and others – as emergent properties of the system. As discussed elsewhere in this volume, emergence refers to attributes or behaviors of a system as a whole that differ qualitatively from its constituent parts. It consists of "novel and coherent structures, patterns, and properties [that arise] during the process of self-organization in complex systems" (Goldstein, 1999). Emergent properties are macro-level ones that are distinct from, but arise from, interactions among agents at the micro-level. Barabási and Albert have identified scale freeness as one such emergent property of networks that grow through preferential attachment (Barabási and Albert, 1999). Likewise, Arthur notes that technological lock-in and monopolistic competition are emergent features of markets characterized by increasing returns (Arthur, 1990, 1994).

Our study departs from this previous scholarship by investigating whether scaling emerges in information economies through mechanisms other than preferential attachment. Rather than model the probability of attachment directly as a function of node degree, our study examines how agents choose information technologies. By focusing on agents' search among choices of communication technologies, the study examines the relationship between technological lock-in (i.e., convergence of a market on a single technology) and agent choices under conditions of imperfect knowledge of market-level technological trends. Our simulation approach moves beyond explicitly modeling preferential attachment; instead, to the degree preferential attachment occurs in our model, the process itself is an emergent phenomenon arising from agents' imperfect choices of communication technologies.

WHY AGENT-BASED MODELING?

Our focus on agents' choices to "grow" preferential attachment lends itself well to an agent-based modeling approach. Agent-based modeling provides a mechanism to endow individual agents with heterogeneous attributes and allow interaction between agents according to explicit rules for behavior. This approach of "growing" phenomena from micro-level factors and interactions also allows the scientist to look across levels of analysis and investigate the effects of adaptive agent behaviors and the emergence of macro-level phenomena (Epstein, 2006; Epstein and Axtell, 1996; Gilbert, 2008). The ABM approach is well suited to social phenomena where behaviors are adaptive, nonlinear, path-dependent, and heterogeneous in ways that lead to network effects (Bonabeau, 2002). In the case of information economies, all of these features of ABM apply.

Although much of modeling and simulation in the engineering-related disciplines focuses on prediction, this is not the prime motivating factor for social science modeling. Observations of agent interactions and emergence of norms or system-level phenomena are useful for thinking through theory, explaining phenomena, investigating potential factors that contribute to emergent phenomena, and even guiding future studies among other non-predictive reasons (Epstein, 2008). In this instance, we use ABM to guide an investigation of how individual behaviors might explain scaling in communication networks. The goal is not to predict the shape of scaling economies, but rather to explore how individual behaviors may shape network structures under certain conditions. Additionally, experimentation as performed in biological and physical sciences is often unethical or infeasible in social contexts. When looking to the emergent phenomena of information economies, the scale of the problem makes experimentation an impractical choice. ABM here offers the flexibility to change the scale of the simulation and explore the subsequent effects on emergent phenomena (Bonabeau, 2002). ABMs have received criticism for the complexity of their computational designs and the difficulty reproducing the model or results. We describe our model using the ODD Protocol as original proposed by Grimm *et al.* (2006) to follow a standard documentation procedure that conveys the mechanisms of the underlying agent-based model.

MODEL OVERVIEW

Purpose

The purpose of this model is to understand "network effects," or how users' adoption of a technology produces scaling in communication networks. In particular, it examines how communication "hubs" emerge over time as users choose technologies based on the adoption choices of others. It also examines how a market-dominant technology emerges from the interdependent choices of technology users.

State Variables and Scales

The model comprises two levels: adopters and the communication network. The model uses three variables to characterize adopters: a technology choice, a set of initial network neighbors, and a true/false parameter that characterizes the adopter's satisfaction with its choice of technology. Adopters in the model are functionally identical. The communication network has one state variable: the structure of the network at initialization. The network is the highest hierarchical level in the model.

Process Overview and Scheduling

The model proceeds in time steps equivalent to adopters' decision to keep or change a technology. Within each step, adopters execute four phases. First, adopters measure their satisfaction by comparing their technology choice to the choices of their network neighbors. Second, each dissatisfied adopter drops its connection to the network neighbor whose technology choice is the most dissimilar from the adopter's choice. Third, dissatisfied adopters create a network connection to another randomly chosen adopter. Finally, a dissatisfied adopter changes one randomly chosen attribute of its technology. Adopters who are satisfied maintain both their current network neighbors and technology choice. Within each time step, the simulation processes adopters in a (pseudo) random order.

Design Concepts

Emergence: The structure of the communication network emerges as adopters drop dissimilar neighbors from their network neighborhood and randomly choose new neighbors. A dominant technology also emerges as adopters search for neighbors with similar technologies. By "dominant," we mean a technology that is the choice of a large plurality or majority of adopters. Such "lock-in" of a technology is similar to monopolistic or oligopolistic markets in information technologies.

Adaptation: We explicitly model adaptation. Adopters improve their satisfaction by dropping neighbors who have dissimilar choices of technology and by randomly searching for new network neighbors whose technology choices are more similar.

Fitness: Fitness seeking is modeled explicitly. Adopters calculate satisfaction with their technology choice as the similarity between its technology and the technologies of their network neighbors.

Sensing: Adopters know the technology choices of their network neighbors. They do not know the (emergent) global dominant technology.

Stochasticity: At initialization, adopters receive random "technologies" encoded as a bit string generated using a binomial probability distribution. Adopters also receive initial network neighbors that the model generates using one of three network generators: a scale-free, small-world, or random network.

Collectives: Through learning about their neighbors' technologies choices, adopters change the structure of the network of technology users.

Observation: For experimentation, we record adopter- and network-level variables. Adopter-level variables include the fitness of the adopter's technology choice, the number of network neighbors it has, and its satisfaction with its technology choice. At the network level, we record the emergent network structure using conventional measures including the degree distribution, average path length, density, clustering coefficient, and maximum cluster size. These measures of the structure of the communication network come from basic network theory (Wasserman and Faust, 1994; Watts, 1999). The degree of an adopter is simply the number of ties it has to other adopters; the degree distribution describes this parameter for the population of adopters. The average path length is simply the mean of the shortest paths through the network between every pair of adopters. The clustering coefficient measures the probability that any two adopters that share a tie are also connected to a randomly chosen third adopter. The density of the network is the ratio of actual ties among adopters to the maximum possible number of such ties. Together, these measures of the network's properties allow us to discern whether the simulated adopters are reorganizing the communication network in a manner that replicates the structure of real-world communication networks.

DETAILS

Initialization

At the simulation's initialization, the model generates a population of 250 adopters. All adopters are functionally identical. Adopters then receive a random initial technology T modeled as a bit string, or a set that is a random sequence of zeros and ones, for example, $T = \{\ 0\ 1\ 1\ 0\ 1\ \}$. Although this is an abstract representation of a product, we consider it a general representation of a technology's "genome." Each position in the bit string represents an attribute of a technology. For example, if one considers a common personal communication technology – a smart mobile phone – we can differentiate among mobile phone types by their screen size, camera resolution, "apps" or software applications, and other attributes. If one thinks about these attributes of smart phones as binary attributes (e.g., a low-resolution vs a high-resolution camera), we can represent the variety of mobile phones in a market by varying the length of the bit string (Figure 7.1). A five-position bit string implies a population of mobile phone "phenotype" (or observable traits) that is 2^5 or 32 possible phone types. This implementation allows the model to vary the technology's complexity by varying the length of the bit string, with larger sets representing more complex technologies. The model uses a pseudo random number generator to populate each position on the technology list T with a uniform probability $P(Y = 1)$ of 0.5.

At initialization, the model generates an initial network among adopters. Ties among adopters are undirected, that is, if adopter A has adopter B as a neighbor, then B by definition also has A as a neighbor. To test for the simulation's sensitivity to initial network structure, we use three different network generation algorithms. The "random" network generator uses the Erdos–Renyi model (Erdos and Rényi, 1960). The "small world" generator relies upon the Kleinberg Model (Kleinberg, 2000). Finally, the "preferential attachment" generator uses the Barabasi–Albert algorithm (Albert and Barabási, 2002). The three generators produce initial networks

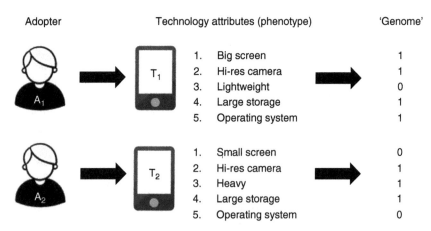

Adopter	Technology attributes (phenotype)	'Genome'
A_1 → T_1	1. Big screen 2. Hi-res camera 3. Lightweight 4. Large storage 5. Operating system	1 1 0 1 1
A_2 → T_2	1. Small screen 2. Hi-res camera 3. Heavy 4. Large storage 5. Operating system	0 1 1 1 0

FIGURE 7.1   Technology phenotype examples.

with different attributes, including notably very different degree distributions. The histograms in Figure 7.1 illustrate the initial degree distributions for the three network structures.

Because interactions among adopters occurs in networked relations, the model's spatial representation of physical space (e.g., where the visualization places agents) is immaterial to the model's dynamics. This assumption is consistent with modern communication networks whose costs do not scale with physical distance. Nonetheless, to aid with visual comprehension of the model, we use the Tutte layout to represent the adopters in a circle (Tutte, 1947). To visualize the emergence of "hubs" in the communication network, the simulation dynamically scales the size of adopters according to their degree.

Input

The model requires two inputs: the "complexity" of the technology and the structure of the initial communication network.

The complexity of the technology is a function of the length of the technology bit string T. Because the simulation allows us to vary the length of T, we can examine the effect on the communication network of growing technology complexity. Each additional bit to the technology's genome increases its complexity by a factor of 2. The experiment simulates the complexity of technology using $5 \leq T \leq 10$. This implies a minimum of $2^5 = 32$ and a maximum of $2^{10} = 1024$ possible technologies. Increases in the length of bit string T thus complicate each adopter's search for a technology that satisfies it.

Submodels

The simulation uses two submodels for the adopters' fitness and learning.

Fitness: Each adopter i evaluates its choice of technology T_i by comparing it to the choices T_j of its network neighbors j. Our representation of a technology choice as a bit string allows us to use basic information theory to compare T_i and T_j. The "Hamming distance" D_H between two bit strings A and B of length L is simply the number of positions in which the bits disagree:

$$D_H = \sum_{i=1}^{l} |A_i - B_i|$$

For example, the distance between [1 0 0 0] and [1 0 0 1] is 1 because the lists disagree on the last bit. Likewise, the Hamming distance between [1 0 0 0] and [1 1 0 0] is 1. Thus, an adopter whose technology has a high Hamming distance is one that poorly matches the technologies its network neighbors have chosen.

We use a simple algorithm to have adopters evaluate the fitness of their technology choice. Each adopter measures its Hamming distance from each of its neighbors. It then calculates the average of its Hamming distances. Finally, it compares its average Hamming distance to the average Hamming distances of its neighbors. An "unfit"

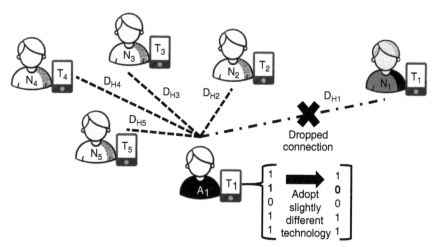

FIGURE 7.2   Agents assess Hamming distances and, if unhappy, eliminate the "farthest" neighbor and adopting a new technology.

adopter is one whose average Hamming distance is greater than the mean of the average Hamming distances of its neighbors (Figure 7.2).

Learning: Each adopter evaluates its fitness by comparing its mean Hamming distance to the mean Hamming distances of its network neighbors. This evaluation allows adopters to improve their fitness in two ways. First, if one or more of an adopter's neighbors has a higher (less fit) mean Hamming distance, the adopter drops the least fit neighbor from its network. This reproduces the principle of homophily or the tendency in social networks of agents to associate with similar agents (or, in our implementation, to dissociate from dissimilar agents). The adopter then creates a new tie with another randomly chosen adopter. Second, if an adopter is less fit on average than its neighbors (i.e., its mean Hamming distance is greater than the average of the mean Hamming distances of its network neighbors), the adopter changes its technology. It does so by randomly selecting a position of T and changing the bit, either from 0 to 1 or 1 to 0. Rather than changing technologies entirely, then, adopters gradually change their technology T in a stochastic fashion (Figure 7.2). This procedure allows adopters to explore new technologies in response to the adoption choices of their network neighbors.

EXPERIMENT AND FINDINGS

We ran the simulation 180 times. For each of the three initial network structures (random, small world, and scale-free), we varied the technology bit string T from 5–10 positions, representing increasingly complex products. We repeated the simulation 10 times for each of six values of T and three initial network structures, for a total of 180 runs. Each simulation ran for 500 steps, a time point by which all simulations had reached a steady state in terms of (i) the lock-in to a particular technology genotype and (ii) the equilibration of network attributes including density and clustering coefficient. Although we collected aggregate measures of the network structure and adopter fitness for each step of each run, in the following discussion, we focus on network structure and adopter fitness at the beginning of each simulation ($t = 0$; see Figure 7.3) and the end ($t = 500$).

The Emergence of Scaling

In network theory, a "scale-free" network is one for which the degree distribution follows a power law. Although there is some debate about how best to estimate the scaling exponent of a degree distribution (Clauset et al., 2009), a visual inspection of the degree distribution can suggest the presence of scaling. Figure 7.3 illustrates the initial ($t = 0$) degree distribution of the runs for each of the three initial network configurations; Figure 7.4 illustrates the distributions at the end of each run. A comparison of the histograms in the top half of Figure 7.4 shows some evidence of scaling. Irrespective of the structure of the network at initialization, the simulations produce degree distributions with a large number of low-degree nodes – that is, adopters that have fewer than 10 network neighbors – as well as a small number of high-degree adopters. This pattern characterizes the emergence of "hubs" in the communication network in which the value of an adopter's choice of technology grows as others adopt the same technology.

Although the degree distribution illustrates the emergence of "hubs" in the simulated communications network, the degree distribution does not follow a classic power law distribution. Figure 7.5 superimposes the natural log of the degree frequencies on the histograms of the final degree distributions for each of the three initial network structures. For a classic scale-free network, the natural log of the degree frequencies would illustrate an approximate straight line. The s-shaped curves in all three plots of Figure 7.5 indicate an overrepresentation of high-degree hubs, that is, a relatively large number of adopters with a large network neighborhood.

Technological "Lock In"

It is possible that, although hubs emerge in the simulated communication networks, they may not necessarily represent neighborhoods or cliques of adopters of the same technology. To examine this possibility, the scatterplots in Figure 7.4 (bottom row) plot each adopter's mean Hamming distance from its neighbors against its degree. The patterns in the scatterplot illustrate that adopters within the same network neighborhood are converging on similar technologies. High-degree adopters tend to have relatively low Hamming distances, whereas low-degree adopters tend to have higher Hamming distance.

This pattern suggests that a "core" of high-degree adopters lock into a similar or identical technology, whereas a "periphery" of low-degree adopters adopt different technologies. To determine whether adopters are locking into a single or small number of technologies, one can simply convert each adopter's technology from a bit string to a decimal format number, and plot adoption choices as a histogram. Figure 7.6 illustrates the final distribution of technology adoptions for a set of nine of

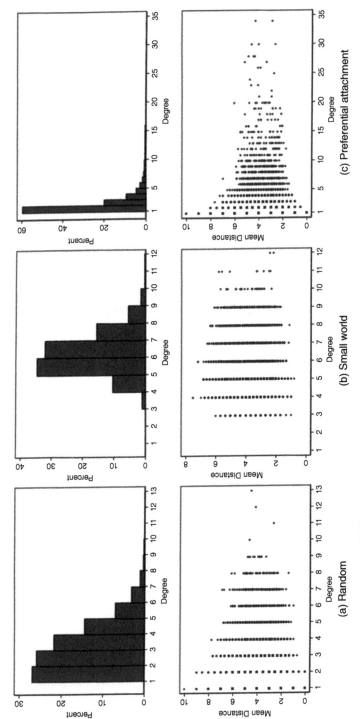

FIGURE 7.3 Network structures and Hamming distances at initialization.

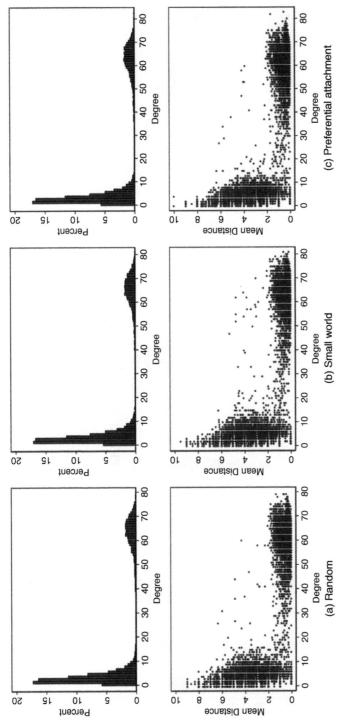

FIGURE 7.4 Network structures and Hamming distances at $t = 500$.

141

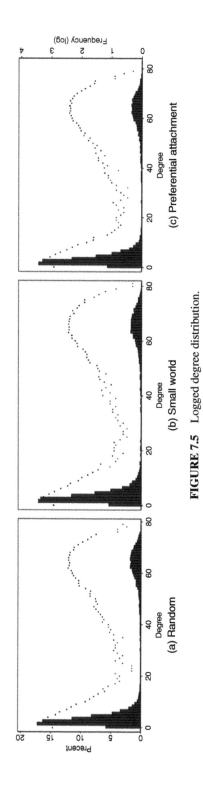

FIGURE 7.5 Logged degree distribution.

142

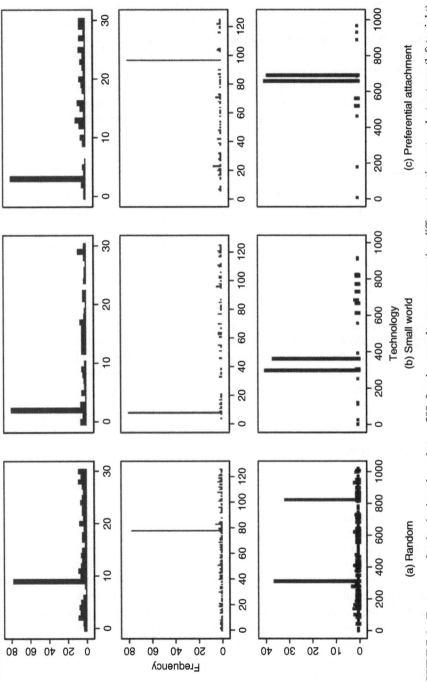

FIGURE 7.6 Frequency of technologies adopted at $t = 500$, for nine sample runs representing different starting network structures (left to right) and different technology complexities (5, 7, and 10 bit string lengths top to bottom).

the runs of the simulation: one each for the three initial network distributions and the three lengths of the technology bit string (5, 7, and 10 positions, from top to bottom). The figure shows that adopters consistently lock into one or two technologies, irrespective of the structure of the initial network. For technologies of five- and seven bit positions, the adopters lock into a single dominant technology. For all three simulations of a 10-position technology, the adopters lock into two competing technologies. This finding is consistent with real-world products characterized by network effects. Because the value of a technology increases as the number of other adopters grows, markets tend to converge on a single dominant technology that results in oligopolistic competition.

Equifinality

The simulated communications network exhibits the condition of equifinality – that is, the network converges on a similar structure and technological lock-in, irrespective of its starting conditions. As Figure 7.3 illustrates, the experiment tested three different initial network structures by using algorithms for random, small-world, and preferential attachment networks. The three network structures presented both different degree distributions (top row of Figure 7.3) and patterns of Hamming distance by degree (bottom row). Yet by $t = 500$, the simulation converged on similar degree distributions, as illustrated in the top row of Figure 7.4. Likewise, irrespective of initial conditions, the simulated adopters produced networks of a small number of high-degree adopters with very similar technologies and a large number of low-degree adopters. From all three initial conditions, the simulated adopters tended to lock into one or two technologies (Figure 7.6). That a relatively simple algorithm produces such similar end states attests to the power of network effects to produce oligopolistic competition and technological lock-in.

CONCLUSIONS

This chapter presented an example of emergence in social phenomena. Heterogeneous agents following simple behavioral rules interacted repeatedly in the agent-based model to produce emergent macro-level effects, namely technological lock-in. We diverge from the prevailing notion of preferential attachment to simulate technological lock-in by "growing" network structures through agent-level decisions. Rather than adopt the most popular technology, agents evolve network structures composed of others who share similar technology choices while simultaneously adopting a slightly different technology. Networks arising from preferential attachment are often scale-free. Our networks, regardless of starting network structure, showed evidence of scale-free qualities, namely hubs; however, unlike scale-free networks, the resulting network structures did not follow a power law. Instead, we witness the emergence of a larger than expected number of technology adopters with large network neighborhoods. Looking further into this phenomenon, we found that regardless of starting network structure, we witness technological lock-in of one or two technologies, in keeping with dominant technologies in environments of oligopolistic competition. From the simple algorithms driving agents to make decisions given imperfect knowledge of the larger market, we generate technological lock-in that reflects phenomena in the real-world.

Rather than focus on prediction, we used simulations to explore the robustness of varying starting conditions to "grow" technological lock-ins as observed in real-world social systems. Technological lock-ins emerge from real-world social systems, and existing theory can guide our understanding of that process. The power of ABMs to dynamically simulate simple agent rules that lead to emergent technological adoption patterns allows us to test the robustness of network effects in ways that traditional theories and statistics cannot. Complexity theory and inquiry into emergent phenomena add experimental capabilities to social sciences that might not be possible otherwise.

REFERENCES

Albert, R. and Barabási, A.-L. (2002) Statistical mechanics of complex networks. *Reviews of Modern Physics*, **74** (1), 47.

Arthur, W.B. (1989) Competing technologies, increasing returns, and lock-in by historical events. *The Economic Journal*, **99** (394), 116–131.

Arthur, W.B. (1990) 'Silicon Valley' locational clusters: when do increasing returns imply monopoly? *Mathematical Social Sciences*, **19** (3), 235–251.

Arthur, W.B. (1994) *Increasing Returns and Path Dependence in the Economy*, University of Michigan Press.

Axelrod, R. (1984) *The Evolution of Cooperation*, Basic Books, New York.

Baker, A. (2010) Simulation-based definitions of emergence. *Journal of Artificial Societies and Social Simulation*, **13** (1). doi: 10.18564/jasss.1531

Barabási, A.-L. and Albert, R. (1999) Emergence of scaling in random networks. *Science*, **286** (5439), 509–512.

Barabási, A.-L., Albert, R., and Jeong, H. (2000) Scale-free characteristics of random networks: the topology of the world-wide web. *Physica A: Statistical Mechanics and its Applications*, **281** (1), 69–77.

Berube, A., Singer, A., Wilson, J.H. and Frey, W.H. (2006) Finding Exurbia: America's fast-growing communities at the metropolitan fringe. Retrieved from Washington, D.C., https://www.brookings.edu/wp-content/uploads/2016/06/20061017_exurbia.pdf

Bonabeau, E. (2002) Agent-based modeling: methods and techniques for simulating human systems. *Proceedings of the National Academy of Sciences*, **99** (suppl 3), 7280–7287. doi: 10.1073/pnas.082080899

i Cancho, R.F. and Solé, R.V. (2001) The small world of human language. *Proceedings of the Royal Society of London B: Biological Sciences*, **268** (1482), 2261–2265.

Cederman, L.-E. (2005) Computational models of social forms: advancing generative process theory. *American Journal of Sociology*, **110** (4), 864–893. doi: 10.1086/426412

Cindea, I. (2006) Complex systems: new conceptual tools for international relations. *Perspectives*, **26**, 46–68.

Clauset, A., Shalizi, C.R., and Newman, M.E. (2009) Power-law distributions in empirical data. *SIAM Review*, **51** (4), 661–703.

Cowan, R. (1990) Nuclear power reactors: a study in technological lock-in. *The Journal of Economic History*, **50** (3), 541–567.

Cowan, R. and Hultén, S. (1996) Escaping lock-in: the case of the electric vehicle. *Technological Forecasting and Social Change*, **53** (1), 61–79.

Eagle, N., Pentland, A.S., and Lazer, D. (2009) Inferring friendship network structure by using mobile phone data. *Proceedings of the National Academy of Sciences*, **106** (36), 15274–15278.

Earnest, D.C. (2015) *Massively Parallel Globalization: Explorations in Self-Organization and World Politics*, State University of New York Press, Albany, NY.

Ebel, H., Mielsch, L.-I., and Bornholdt, S. (2002) Scale-free topology of e-mail networks. *Physical Review E*, **66** (3), 035103.

Epstein, J.M. (2006) *Generative Social Science: Studies in Agent-Based Computational Modeling*, Princeton University Press, Princeton.

Epstein, J.M. (2008) Why model? *Journal of Artificial Societies and Social Simulation*, **11** (4), 12.

Epstein, J.M. (2013) *Agent Zero: Toward Neurocognitive Foundations for Generative Social Science*, Princeton University Press, Princeton and Oxford.

Epstein, J.M. and Axtell, R. (1996) *Growing Artificial Societies: Social Science from the Bottom Up*, Brookings Institution Press, Washington, D.C.

Erdos, P. and Rényi, A. (1960) On the evolution of random graphs. *Publications of the Mathematical Institute of the Hungarian Academy of Sciences*, **5** (1), 17–60.

Foray, D. (1997) The dynamic implications of increasing returns: technological change and path dependent inefficiency. *International Journal of Industrial Organization*, **15** (6), 733–752.

Gilbert, G.N. (2002, February 2003) Varieties of Emergence. Paper presented at the Workshop on Agent 2002 Social Agents: Ecology, Exchange, and Evolution Conference, University of Chicago.

Gilbert, G.N. (2008) *Agent-Based Models*, Sage Publications, Los Angeles.

Gilbert, G.N. and Troitzsch, K.G. (2005) *Simulation for the Social Scientist*, Open University Press, Berkshire, England.

Goldstein, J. (1999) Emergence as a construct: history and issues. *Emergence*, **1** (1), 49–72.

Grimm, V., Berger, U., Bastiansen, F. *et al.* (2006) A standard protocol for describing individual-based and agent-based models. *Ecological u*, **198** (1), 115–126. doi: 10.1016/j.ecolmodel.2006.04.023

Guimera, R., Mossa, S., Turtschi, A., and Amaral, L.N. (2005) The worldwide air transportation network: anomalous centrality, community structure, and cities' global roles. *Proceedings of the National Academy of Sciences*, **102** (22), 7794–7799.

Holland, J.H. (1999) *Emergence: From Chaos to Order*, Perseus Books, Cambridge, MA.

Huneman, P. (2008) Emergence made ontological? computational versus combinatorial approaches. *Philosophy of Science*, **75** (5), 595–607. doi: 10.1086/596777

Kavalski, E. (ed.) (2015) *World Politics at the Edge of Chaos: Reflections on Complexity and Global Life*, State University of New York University Press, Albany, NY.

Kleinberg, J. (2000) The small-world phenomenon: an algorithmic perspective. Paper presented at the Proceedings of the thirty-second annual ACM symposium on Theory of computing.

van Lelyveld, I. (2014) Finding the core: network structure in interbank markets. *Journal of Banking & Finance*, **49**, 27–40.

Macy, M.W. and Willer, R. (2002) From factors to actors: computational sociology and agent-based modeling. *Annual Review of Sociology*, **28**, 143–166.

Miller, J.H. and Page, S.E. (2007) *Complex Adaptive Systems: An Introduction to Computational Models of Social Life*, Princeton University Press, Princeton, NJ.

Mislove, A., Marcon, M., Gummadi, K.P., Druschel, P. and Bhattacharjee, B. (2007) Measurement and analysis of online social networks. Paper presented at the Proceedings of the 7th ACM SIGCOMM Conference on Internet Measurement.

Mitchell, M. (2009) *Complexity: A Guided Tour*, Oxford University Press, New York.

Moscaritolo, A. (2017) With 99.6 percent of the market, can anyone topple iOS, Android? *PC Magazine*.

Olaya, C. (2007) Can the whole be more than the computation of the parts? A reflection on emergence, in *Worldviews, Science And Us* (eds C. Gershenson, D. Aerts, and B. Edmonds), World Scientific Publishing Company, River Edge, USA, pp. 81–98.

Schelling, T.C. (1978) *Micromotives and Macrobehavior*, W. W. Norton & Company, New York, NY.

Spek, S., Postma, E. and Van Den Herik, J. (2012) Wikipedia: organisation from a bottom-up approach.

Tutte, W.T. (1947) A ring in graph theory. Paper presented at the Mathematical Proceedings of the Cambridge Philosophical Society.

Wasserman, S. and Faust, K. (1994) *Social Network Analysis: Methods and Applications*, Cambridge University Press, Cambridge.

Watts, D.J. (1999) *Small Worlds: The Dynamics of Networks between Order and Randomness*, Princeton University Press, Princeton, NJ.

Wilson, R. (2010). The third way of agent-based social simulation and a computational account of emergence *Journal of Artificial Societies and Social Simulation*, **13**(8), http://jasss.soc .surrey.ac.uk/13/3/8.html.

8

MODELING EMERGENCE IN SYSTEMS OF SYSTEMS USING THERMODYNAMIC CONCEPTS

John J. Johnson IV[1], Jose J. Padilla[2], and Andres Sousa-Poza[3]

[1]*Systems Thinking & Solutions, Ashburn, VA 20148, USA*
[2]*Virginia Modeling Analysis and Simulation Center, Old Dominion University, Suffolk, VA, USA*
[3]*Engineering Management & Systems Engineering, Old Dominion University, Norfolk, VA 23529, USA*

SUMMARY

Emergence, in its earliest discussions, is exemplified by thermodynamic transitions in chemical systems that produce unexplainable and unpredictable effects. We posit that characterizing thermodynamic transitions provides insight into modeling emergence in engineered systems, including those that are system of systems (SOS). In this case, we map factors in chemical systems that affect the occurrence of chemical reactions (like temperature and molecular freedom) to factors that affect the occurrence of emergence in SOS (like interoperability and component degrees of freedom). Understanding factor mappings, and the underlying interactions that connect them, contributes to our ability to characterize and model emergence in SOS and engineered systems in general. We provide an initial conceptualization of emergence using causal loops diagrams and discuss a path forward.

Emergent Behavior in Complex Systems Engineering: A Modeling and Simulation Approach,
First Edition. Edited by Saurabh Mittal, Saikou Diallo, and Andreas Tolk.
© 2018 John Wiley & Sons, Inc. Published 2018 by John Wiley & Sons, Inc.

INTRODUCTION

One aspect of the emergence concept is the apparent absence of traceability between the nature of parts in a system and the system effects (Lewes, 1875). This absence presents a potential problem for engineers and stakeholders in the design and operation of systems, as we cannot identify intentional consequences from those emerging from the system. Further, "classical" engineering design seeks to eliminate unexpected and unintended effects (Mina *et al.*, 2006). The problem of unintentional system effects is recognized in government and private sectors as having broad ranging financial and security consequences (Guckenheimer and Ottino, 2008; Schroeder and Gibson, 2010; ASME, 2011; Willman, 2014; NIST, 2016). However, some new concepts of engineering systems seek to harness the potential benefits of emergence and to encourage it as the intentional product of engineering efforts (Valerdi *et al.*, 2008; Tolk and Rainey, 2015; Mittal and Rainey, 2015). Whether defending against emergence or encouraging it; understanding the causal factors of emergence is an important research effort for engineering systems.

The systems of interest in this text are SOS, which are a type of engineered system. Engineered systems are ensembles of parts designed to produce effects that fulfill a purpose that cannot be satisfied by the individual parts of the system (Ackoff, 1971; Checkland, 1999; Blanchard and Fabrycky, 2006). The current body of knowledge on emergence in systems abounds with theories about its nature, yet there are persistent gaps that support the need to further our understanding. Silberstein and McGeever (1999), Corning (2002), Campbell (2015), and Sartenaer (2016) are among those that discuss persistent gaps in the emergence body of knowledge. One of the identified gaps is the absence of a characterization that explains the causal factors of emergence. This need is amplified by proposed research agendas from commercial industries, defense companies, and academia. Mogul (2006) focuses on problems created by emergent behavior ("misbehavior") in complex software systems. He proposes several research agenda to support his objective including the development of ... "a taxonomy of frequent causes of emergent misbehavior." Valerdi *et al.* (2008) discuss the risks to the resilience of a system that is posed by emergent behavior. They recommend research that will lead to architecture strategies for guided emergence to " ... steer emergent behavior in desired directions." Tolk and Rainey (2015) question the acceptance of positive emergence as " ... a welcomed coincident." Alternatively, they challenge the engineering community to conduct research that leads to emergence as the intentional " ... product of engineering efforts."

The common thread in these agenda items is the call for research that leads to " ... [gaining] a deeper understanding ... " of emergence and particularly how it occurs in systems that are engineered. The fundamental objectives in these agendas are summarized by Checkland's (1999) plan for a systems movement: " ... to search for conditions governing emergent properties and a spelling out of the relations between such properties and the wholes which exhibit them." Fulfilling this agenda has the potential to change how systems are designed and managed. If causal factors of systems that contribute to emergent effects are known, (i) the risk that unexplainable effects will occur could be assessed; (ii) design alternatives with fewer causal factors could be selected; (iii) if the causal factors are actually capable of being adjusted (i.e., they are mechanisms), then the likelihood of unexplainable effects could be controlled; and (iv) to the extent emergent effects are positive, their occurrence could be encouraged (i.e., engineered). These are the motivations for this chapter.

We use analogical reasoning to identify candidates for causal factors of SOS emergence. We depart from thermodynamics (specifically thermochemistry) and how thermodynamics explains emergence and transition to how we could explain emergence in SOS by its causal factors. Thermochemistry is the point of origin for the original concepts of emergence (Mill, 1846; Lewes, 1875). Lewes (1875), for instance, uses many examples of chemical reactions to explain the nature of emergent effects where some combinations produce properties that are different from the properties of their parts. One example he cites is the orange color produced from the combination of colorless oxygen and colorless nitrogen. The orange color is novel (i.e., new) and not traceable to the properties of its oxygen and nitrogen components or the process of combining them.

We build a conceptual model of emergence in SOS with the described analogies.

THERMODYNAMICS AND ITS EXPLANATORY ADVANTAGES

Our reasons for selecting thermochemistry as a medium of study goes beyond the general similarities of the chemical reaction transition phase and the change that occurs in SOS to producing emergent effects. There is broad applicability of thermodynamics as a source of analogies to explain concepts in general. There is also a well-established precedence in explanations specific to emergence and a general correspondence with characteristics of SOS.

Klein and Nellis (1991) and Ott and Boerio-Goates (2000) are among those who posit that the simplicity of its basic postulates makes thermodynamics applicable to " ... any discipline technology, application, or process." Consequently, thermodynamics has proved to be a rich source for analogies that are used for their explanatory power in a variety of ways. Table 8.1 lists a sample of analogies based on thermodynamic concepts.

The analogies in Table 8.1 are examples of how the consistency of the thermodynamic behavior and the broad acceptance of its governing principals are used to explain non-thermodynamic phenomena.

In addition to its broad variety of applications, there is a well-established precedence (over 150 years) for using principles of thermodynamics and more specifically thermochemistry, to explain the concepts of emergence in systems. Mill (1846) uses the thermochemical combination of substances (i.e., chemical reactions) as a contrast to his Composition of Cause principle. Chemical reactions produce "special and exceptional" cases where Composition of Cause does not apply because the joint effect of the combination is not the same as the sum of the separate effects of the substances. Although not specifically called emergence, this is considered one of the early expressions of the emergence concept.

TABLE 8.1 Thermodynamic Analogies

References	Application	Thermodynamic Concept
Sawada and Caley (1985)	Education systems and the process of learning	Entropy and thermodynamic equilibrium
Dyer (1996)	Effective scholarly conversations	Exothermic/endothermic reactions
Kotov (2002)	Dynamics of human culture	Biogeochemistry and dissipative structures
Chassin et al. (2004)	Control of complex adaptive systems	Carnot cycle
Sergeev (2006)	Economic equilibrium in financial markets	Entropy and thermodynamic equilibrium
Kauffman and Clayton (2006)	Emergence of order in biological systems	Chemical reactions
Dyer (2007)	Emergence of individual learning	Change in enthalpy
Bratianu and Andriessen (2008)	Knowledge as energy	Mechanical and thermal energy
Chew and Choo (2008)	Resistance to changing a banking system	Changing states of matter
Pati (2009)	Stress management and innovation in business systems	Enthalpy and the conservation of energy
Ortega and Braun (2013)	Rational decision making and maximum utility	Free energy and entropy
Kovacic (2013)	Effect of shared awareness within multiple cognitive representations of reality	Percolation theory (i.e., statistical mechanics)

Given its broad applicability as a source of analogies to explain concepts and a well-established precedence in explanations of emergence, we use thermodynamics (and thermochemistry in particular) as a suitable medium of study to develop explanations and characterizations of emergence in SOS.

EMERGENCE AND CHEMICAL REACTIONS

A chemical system is a hierarchical structure of chemical substances (i.e., components). The structure has macro-level properties and characteristics (i.e., macrostates) that are determined by combinations of individual substances (i.e., microstates) of the system. A chemical reaction is a thermochemical phenomenon characterized by changes in the macro and microstates of a chemical system. The changes are caused by interactions of chemical system components with the environment and other chemical systems. We establish a basic analogy to chemical systems if we consider that engineered systems are also structures with macro-level properties and characteristics that are determined by the configuration of their components. Interactions between components, the environment, and with other systems, can change the properties and behaviors of the engineered system.

Engineered systems and SOS are in the physical domain; therefore, we approach emergence as a phenomena that is theoretically derivable, even if doing so is inherently difficult (Holland, 1998; Bedau, 1997, 2008; Maier, 2015). Given its theoretical derivability, the identification of factors that cause emergence must also be possible. We begin the search for these factors by studying the chemical reaction phenomena to understand its factors and the underlying interactions that connect them.

Chemical Reactions

Components in a chemical system are held together in a certain configurations by forces (bonds). Each configuration is a microstate of the chemical system that determines the system's macrostates (its properties and behaviors). In order for a chemical reaction to take place, there must be a sufficient change in the systems internal energy (U) to break the bonds that are holding the molecules together. Chemical systems interact by exchanging energy with the environment (which can include other chemical systems). Endothermic reactions are an examples of chemical reactions where new microstates produce a new macrostate of the chemical system that has more energy than its initial state. The properties, behaviors, and the configuration of the molecules in the chemical system are a function of its internal energy. Therefore, a chemical reaction results in a change in the configuration, behavior, and properties. An example of a chemical reaction is depicted in Figure 8.1 (Brown *et al.*, 2014).

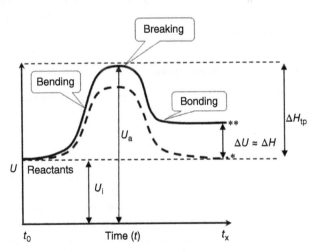

** Endothermic reaction → Emergent product

* No reaction → Original reactants

Brown *et al.* (2014)

FIGURE 8.1   Endothermic chemical reaction.

Consider the dashed line (*) in Figure 8.1. Chemical systems are initially in a state of equilibrium where its initial internal energy (U_i), the configuration of its components, and the associated properties and behaviors are constant. Although interacting with the environment or other chemical systems, the internal energy increases and causes the bonds that maintain the configuration to bend. If the change in internal energy ($U - U_i$), also known as change in enthalpy (ΔH), does not exceed the difference between the activation threshold (U_a) and U_i; eventually, the additional energy dissipates to the environment or another chemical system; and the system returns to its original state of equilibrium. U_a is the minimum amount of energy required to break the bonds that maintain the configurations of the system. $U_a - U_i$ is the enthalpy tipping point (ΔH_{tp}) that must be exceed before a chemical reaction can occur. In the case represented by the dotted line, the internal energy did not exceed ΔH_{tp}. Because the exchange of energy was insufficient to exceed the required threshold, final ΔH is zero, and the configuration, behavior, and properties of the chemical system return to their original state.

Consider the solid line (**) in Figure 8.1. During the interaction with the environment or other chemical systems, the internal energy increases and exceeds the activation threshold (U_a) and the enthalpy tipping point (ΔH_{tp}). The bonds that maintain the configuration break; some of the additional energy is absorbed into the system and some dissipates back to the environment or another chemical systems; new bonds are formed; and the system settles at a new state of equilibrium, and at a higher level of internal energy. In this case, the change in enthalpy (ΔH) is greater than zero and a chemical reaction has occurred. The configuration, behavior, and properties of the chemical system change and are not apparently derivable from the components and initial states of the chemical system; they are in essence emergents. Understanding the factors that cause the transition provides insight to cause–effect relationships for emergence in engineered systems.

Factors That Affect Chemical Reactions

There are generally five factors in chemical systems that affect the rate and likelihood that a chemical reaction will occur. The factors are derived from the law for the rate of a chemical reaction and the Arrhenius Equation (Brown *et al.*, 2014; Daintith, 2008; Linder, 2011): energy, temperature, molecular freedom, catalyst, and concentrations.

1. *Energy*: Energy is the capacity to do work or cause change in a system. In order for a chemical reaction to take place, there must be a sufficient change in the systems internal energy (U) to break the bonds that are holding its molecules together. Chemical systems and their substances interact by exchanging energy with the environment (which can include other chemical systems). Changes in interactions cause changes in the system's energy, which change the system's state and it properties.

2. *Temperature*: The kinetic energy in a chemical system is determined in part by the speed that the molecular components in the system are moving. The faster the components move, the greater the frequency and intensity (force) of

their collisions with each other. Molecules move faster at higher temperatures causing the kinetic energy and consequently the internal energy in the system to increase, which increase the possibility of new microstates. Thus, increases in temperature increase entropy (i.e., the number of microstates after the reaction relative to the initial number of microstates before the reaction). The rate of a chemical reaction grows exponentially as a function of temperature (T) where the change in the rate constant $= e^{-1/T}$.

3. *Molecular freedom*: There are a certain number particles in each molecule of the substances in a chemical system. Each particle has a certain charge (positive, negative, or neutral). Substances in a chemical system will only react if the number and charges of their particles are in the required alignment (i.e., orientation) when they collide with each other. The freedom of movement among molecules in a substance varies according to the state of the substance: solids have the least freedom, whereas plasma has the most. The variations in states affect the molecular movement and the chances that the particles will collide and be in the correct orientation to cause a chemical reaction. The chemical reaction rate is increased as a linear function of molecular freedom of the substances in the chemical system.

4. *Catalysts*: Increasing the volume of substances or energy from an external source causes the internal energy (U) of a chemical system to increase. New chemical products are formed when the change in the system's enthalpy (ΔH) plus its initial internal energy (U) exceed the system's activation energy (U_a). Some substances possess the ability to be added to a chemical system without changing the chemical products of the reaction. These substances (catalyst) maintain their structure while lowering the activation energy (U_a) and/or improving particle orientation during a collision. The rate of a chemical reaction grows exponentially as a function of reductions in U_a where the change in rate constant $= e^{-U_a}$.

5. *Concentrations*: The energy in a chemical system causes the molecule in its substance to move. In order for a chemical reaction to occur, the molecules in the chemical substances must collide with one another. The greater the number of molecules there are in the chemical system, the higher the probability that there will be a collision. The number of molecules is increased by increasing the concentration of one or more substances relative to the total volume. Increasing the concentration increases the frequency of collisions between molecules of the substances in the chemical system, which increases the rate of the reaction according to the chemical reaction rate law. The change in rate is exponential if the order of the substance is greater than 1. For example, increasing the volume of a substance by 50% (or a multiple of 1.5) with a reaction order of 2 would increase the reaction rate by a multiple of 1.5^2 or 2.25.

The factors for chemical reactions explain the cause and effect relationships that lead to emergence in chemical systems. We posit that there are also factors in engineered SOS that can support the development of a conceptual model for emergence in SOS.

DEVELOPING A CONCEPTUAL MODEL OF EMERGENCE IN SOS

We searched the literature for SOS factors that are conceptually similar to, or that can be reduced to, the chemical reaction factors mentioned above. Our interest is in the occurrence of emergence, rather than the nature of any particular emergent event. Understanding factors that determine whether or not emergence will occur and how fast the occurrence might take place will improve our ability to manage, mitigate, and/or encourage emergence. With this in mind, we looked for factors that may affect the likelihood of changes in system effects and that could impact the rate at which those changes might occur. Based on our review, four candidates were selected: information, interoperability, component degrees of freedom, and variety of regulators.

1. *Information*: Beer (1979) defines information as the actionable interpretation of data that causes a change in the systems state, where data are a statement of fact about a person, place, or thing. Information and energy are alike in that they both cause change in the system. The causal nature of information is also expressed by Deacon (2007) and Sunik (2011). Deacon (2007) defines information in context of regularly occurring processes and exchanges between entities. In this context, a signal (i.e., data) becomes information when it causes by its presence or absence, a change in the regularity (i.e., equilibrium) of the relationships between entities. Sunik (2011) defines information as the value of a variable in an algorithm that determines the "changes and movements" (i.e., the configuration) of objects. The change determined by the variables in the algorithm are actions that causes the object to have form, content, direction, nature, and so on that it would otherwise not have if the object was left undisturbed.

2. *Interoperability*: A successful exchange of messages (i.e., information) that results in the interaction of the system components requires that there must be meaning in the message. Observers of the message need "powers of discrimination" in order for the message to be fully understood and lead to action (Ashby, 1961). Tolk and Muguira (2003) and Tolk *et al.* (2007) discuss interoperability as a quality that goes beyond mere exchanging quantity of bits to a mutual understanding of what meant by the exchange. The greater the interoperability, the greater the information flow within and between systems. Tolk *et al.* (2012) adds that increases in interoperability cause increases in entropy (i.e., uncertainty about the future state of the system and disorder of its components). Increases in interoperability enables the system to form new configurations and ultimately increases entropy in the system.

3. *Component degrees of freedom*: There are various types of components in SOS including but are not limited to hardware, software, people, equipment, and processes. Each component is defined by the variables or dimensions over which they change. For example, hardware variables could be processor type, memory capacity, and operating system; people may vary be age, education, and experience; the proximity of the system components may vary over temporal and special dimensions. The degrees of freedom (or distinct number of possible

states) for the system is a function of the number of component types, and the number of variables or dimensions that define each component (Ashby, 1961; Mittal and Rainey, 2015). The greater the degrees of freedom, the more likely a component will be able to communicate (i.e., interact) with another component.

4. *Variety of regulators*: Some components act as regulators in the system. They control the outputs of other components by limiting the undesired outputs and allowing the desired outputs. In accordance with the Law of Requisite Variety, a system is regulated if the variety (or distinguishable elements) of its regulators are equal to or greater than the variety of the components being regulated (Ashby, 1961). Regulators suppress interaction and the production of new states while catalyst encourages it. If the variety of the regulators in the system is less than that of the other components, components will interact in unexpected ways and the system will become emergent. As the variety of regulators in a system falls below the requisite level, the threshold to cause emergent behavior is reduced.

Increasing or decreasing any of these factors will have a direct impact on the quantity and variety of effects that are produced by the system. This is a good starting point for building analogies to chemical reaction factors that determine the effects for chemical systems.

Analogies Between Chemical Systems and Engineered Systems

We extract conceptual similarities from the discussion on chemical reaction and SOS factors. The similarities are used to establish analogies that support the premise for transferring knowledge about the behaviors and relationships of chemical reaction variables to a set of variables in engineered systems. The analogous relationships are (i) energy versus information, (ii) temperature versus interoperability, (iii) molecular freedom versus component degrees of freedom, (iv) volume of catalyst versus variety of regulators, and (v) concentration of chemical substances versus concentration of system components. The analogies are defined by conceptual descriptions of the elements in the source domain (causal factors for endothermic chemical reactions) and target domain (engineered systems).

Energy Versus Information

Source Domain: Properties of a chemical systems are a function of the configuration of its molecules (i.e., its parts). Chemical systems exchange energy when they interact with other systems. Energy is the capacity to make a change in an entity's spatial position relative to other entities (i.e., configuration) and to change its capacity to transfer heat.

Target Domain: Information is the actionable interpretation of data that causes change in the configuration of system components. Changes in system configurations cause the system to have form, content, direction, a specific nature, and so on, that it would otherwise not have if the object were left undisturbed.

Temperature Versus Interoperability

Source Domain: Energy in a chemical system is heat. Heat is the energy that causes a change in the temperature (T) of the system. A positive change in temperature causes increases in the frequency and force at which molecules in the chemical system collide, which causes bonds between molecules in the system to break and eventually reconfigure. The end result of temperature increases is an increase in entropy and the rate of the chemical reaction that produces new system microstates (i.e., system properties).

Target Domain: Interoperability is "the ability of two or more systems or elements to exchange information and to use the information that has been exchanged" (IEEE, 2000, as cited in Morris *et al.*, 2004; Tolk, 2004). Interoperability is accomplished by establishing a common understanding of the information used by the participants in the exchange. Greater interoperability leads to more interactions and information exchange. The consequence of higher levels of interoperability and the successful exchange of information is an increase in entropy and the potential for the acquisition of new system functions/ capabilities (i.e., new system properties).

Molecular Freedom Versus Component Degrees of Freedom

Source Domain: Substances in a chemical system can be characterized the freedom of movement among their molecules. The greater the molecular freedom of movement, the more likely the molecules will collide in the required orientation to facilitate the exchange of energy, the breaking of bonds, and the forming of new configurations (i.e., a chemical reaction will occur).

Target Domain: Systems exchange information (i.e., communicate) by sending and receiving messages. The successful exchange of information exist to the extent the system or component receiving the message has degrees of freedom that are equal to or greater than the degrees of freedom for the message being sent. The greater the exchange of information between systems, the greater the potential for changes in system configurations resulting in new functions, capabilities, and properties.

Volume of Catalyst Versus Variety of Regulators

Source Domain: Increasing the volume of certain substances (catalyst) lowers the activation energy (Au) threshold required to cause a chemical reactions. As the threshold is approached, there are more interactions between the molecules in the chemical system and between the system and its environment (or other systems). This increased interaction causes energy to be exchanged and the rate (i.e., speed) of chemical reaction to grow exponentially. As a function of reductions in U_a, catalyst causes the likelihood of a chemical reaction to increase and new system configurations and properties to form.

Target Domain: Some components in engineered systems act as regulators that limit the results (i.e., states) of system and component interactions by blocking the flow of information. Ashby's (1961) law of Requisite Variety basically states that " ... only variety can destroy variety." Variety is the number of distinct possibilities. The variety of outcomes is limited (i.e., regulated) to the extent that the variety of the regulator (V_r) is greater than the variety of the inputs. In other words, the greater (V_r) is relative to the variety of inputs, the lower the variety or number of possible system states, and the lower (V_r) is relative to the variety of inputs, the greater the number of possible system states.

Concentration of Chemical Substances Versus Concentration of System Components

Source Domain: Concentration is the amount of one or more substances relative to the total number of substances in the system. Greater concentration results in great number of contacts or interactions for that substance. Increasing the concentration increases the frequency of collisions between molecules of the substances in the chemical system that increases the rate of the reaction.

Target Domain: The concentration concept is the same for components in an engineered system and substances in a chemical system. Increasing the quantity of a component relative to other components will increase the rate of interactions of that component in the system. To the extent that the component affects the system's state, it is assumed that changes in the component's interactions will cause changes in the state of the system.

Although some of the analogies are not as strong as others, they represent the general concept that chemical reaction factors are conceptually similar to some factors in engineered systems. The analogies are now used to define chemical reaction variables in terms of potential variables for engineered systems. As the systems of interest in this chapter are SOS, we define the variables as such (see Table 8.2).

At this point, we have defined a chemical reaction as a demonstration of emergence and identified factors in chemical systems that affect the rate and likelihood that a chemical reaction will occur. Analogies were used to convert the variables for chemical systems to variable for SOS. The next step is to use the converted variables to build a conceptual model of emergence in SOS.

A Conceptual Model of Emergence in System of Systems

Causal loop diagrams (CLDs) are a system dynamics modeling concept that represents the causal relationships between system variables and graphically depicts the behavior of the system (Sterman, 2000). CLDs are especially relevant in this chapter given that the essence of our objective concerns cause and effect relationships. The CLD conceptual model is developed by first defining the variables for the chemical reaction and then using analogies to convert those variables to conceptual variables for a SOS. A CLD is then constructed using the converted variables (see Figure 8.2).

The CLD is best understood by examining each causal loops in isolation. A causal loop is a continuous sequence of variables connected by arrows. Loops that reinforce

TABLE 8.2 Variable Conversions: Chemical System to System of Systems

Chemical System Variables	SOS Variables
Activation Energy (U_a): Minimum energy required to cause a chemical reaction	*Activation Information Threshold (I_a)*: Minimum information required to cause emergence
Catalytic Volume (V_c): The amount of substance in a chemical system that lowers the activation energy but does not react with the other substances	*Variety of Regulators (V_r)*: The degrees of freedom for components that regulate outputs/states/behaviors of other components
Concentration (C): The additional volume of a substance relative to total volume of substances	*Component Concentration (C_c)*: The quantity of a component relative to total number of components
Energy Differential (Q_e): Available heat to transfer from the environment to the system	*Information Differential (I_d)*: Available information for transfer from the environment to the system
Energy Transferred (Qx): Heat energy transferred from the system to the environment	*Information Transferred (I_x)*: Information transmitted from the system to the environment as the system returns to steady state
Enthalpy Change (ΔU): The difference between the external energy (U_e) and the initial internal energy of the system (U_i)	*Information Change (ΔI)*: The difference between the external information (I_e) and the initial internal information (I_i)
Enthalpy Ratio (H_r): The fraction of the tipping point that the system has reached for chemical reaction to occur	*Information Ratio of Emergence (IRE)*: The fraction of the tipping point that the system has reached for emergence occurs
Enthalpy Tipping Point (ΔH_{tp}): The difference between the activation energy (U_a) and initial internal energy (U_i)	*Information Tipping Point (ΔItp)*: The difference between the activation information threshold (I_a) and initial internal information (I_i)
External Energy (U_e): Energy that exist outside of the system	*External Information (I_e)*: Information that exist outside of the system and is not accounted for in the design
Initial Internal Energy (U_i): Internal energy of the system at $t = t_0$	*Initial Internal Information (I_i)*: Internal system information at $t = t_0$
Internal Energy (U): Total system energy at $t > t_0$	*Internal Information (I)*: Total system information at $t > t_0$
Molecular Freedom (F): Ability to move and change orientation	*Degrees of Freedom (Df)*: Ability of components to move and change orientation
Reception Rate (R_r): The amount of energy flowing into the system from the environment per unit of time	*Reception Rate (R_r)*: The amount of information flowing into the system from the environment per unit of time
Temperature (T): Average heat energy in the system	*Interoperability (Int)*: The degree that the system can exchange information
Xfer (Transfer) Rate (X_r): The amount of energy flowing into the system from the environment per unit of time	*X-Mission Rate (X_r)*: The amount of information flowing from the system to the environment per unit of time

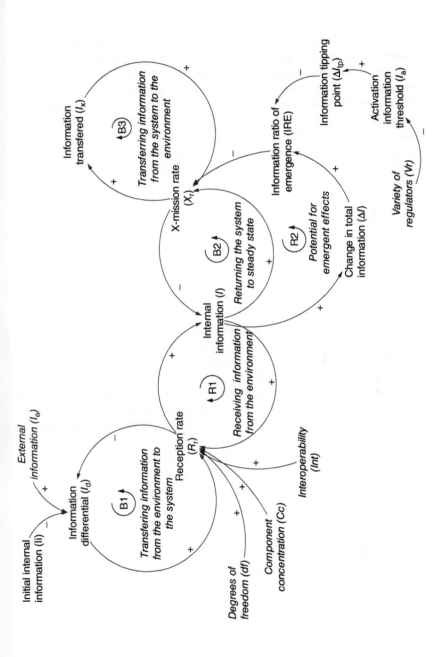

FIGURE 8.2 Conceptual model of emergence in system of systems.

161

behavior are indicated with the letter R in a circular arrow. Loops that provide balance and prevent the continuous growth of a behavior are indicated with the letter B in a circular arrow.

- *B1, Transferring Information from the Environment*: The interaction of the SOS with its environment begins with an input from an external information source (I_e). The information differential (I_d) between the system's initial internal information (I_i) and the external information source (I_e) is the amount of information that is available to be transferred to the system. The greater the differential, the more information there is for the system to receive information per unit of time, that is, reception rate (R_r). Rate of reception (R_r) can be increased by increasing component concentration (C_c), degrees of freedom (df), and system interoperability (Int). Given enough time and without the continuous addition of more information from an external source, the system will receive all available information, and the information differential will reduced to zero.

- *R1, Receiving Information from the Environment*: The system is designed to produce certain effects based on the quantity and types of information that it receives. When the information exchange begins at time (t) = 0, the information differential (i.e., the potential to receive additional information that was not considered in the design) is at its maximum; change in the system's internal information (ΔI) is at its minimum; and the system is in an initial steady state (information received = information transferred, and $\Delta I = 0$). Over the course of the interaction, the system receives external information (I_e) causing increases in the system's internal information (I) such that the system is no longer in steady state (information received ≠ information transferred from the system to the environment (I_x), and $\Delta I \neq 0$). The greater the relative amount of external information (I_e) to the required activation threshold for emergence (I_a), the greater the change in system internal information (ΔI). Internal information continues to grow and grow faster until all of the available information (I_d) has been received.

- *R2, Potential for Emergent Effects*: As the information differential continues to be received by the system, the change in the total internal information (ΔI) approaches the tipping point required for emergent effects to occur. ΔI approaching the information tipping point (ΔI_{tp}) indicates that the current configuration of the system is changing. The information ratio of emergence (IRE) = $\Delta I / \Delta I_{tp}$. If $\Delta I > \Delta I_{tp}$, that is, the IRE > 1, emergent effects will occur. If the internal information (I) does not reach the information activation threshold (I_a), ΔI will be $< \Delta I_{tp}$, IRE will be < 1, and emergent effects will not occur. This is analogous to the enthalpy tipping point (ΔH_{tp}) beyond which a chemical reaction will occur (see the discussion on Figure 8.1. Endothermic Chemical Reaction). Lowering the variety of regulators (V_r) can lower the information activation threshold (I_a) and increase the likelihood that IRE will be >1 between t_1 and t_x.

- *B2, Returning the System to Steady State*: SOS receive and transfer information. Initially, the R1 loop is dominant and the system is receiving more information than it transfers (i.e., ΔI is increasing). A tipping point will occur where the system will begin to transfer more information than it receives. At that point, dominance will shift to the B2 loop where ΔI and the system internal information will decline until the system returns to steady state. For each fraction of information, a certain amount of time to transfer to the environment is required to complete the interaction and return the system to steady state. The rate of information that transfers out of the system as it returns to steady state is the transfer rate (X_r).

- B3, *Transferring Information Back to the Environment*: SOS receive information as they interact with their environment and other systems. If the information tipping point (ΔI_{tp}) is not reached, the IRE will be <1. At this point, the system will continue to transfer information back to the environment until the gap between information transferred and available information to be received is zero, the interaction ends, the system returns to steady state at the original level of internal information (I), and emergent effects do not occur. However, if the peak of ΔI is >ΔI_{tp}, IRE will be >1. Under this condition, some of the available information will be absorbed by the system, the balanced will be transferred back to the environment, the system will reach a steady state at a new level of internal information, and emergent effects will occur.

From the conceptual model, we can characterize and explain emergence with a simple formula: if IRE $= \Delta I/\Delta I_{tp} \geq 1$, emergent effects will occur. The relationship between the IRE and the factors that influence whether or not the system becomes emergent could be initially defined as a tuple of the form IRE $[df, C_c, \text{Int}, V_r]$, where

df = degrees of freedom for system components

C_c = concentration of system components

Int = level of interoperability

V_r = variety of the system components that act as regulators

The factors in the IRE tuple impact the occurrence of emergence by increasing or decreasing the change in systems internal information (ΔI) and the information tipping point (ΔI_{tp}) of the system. Increases in df, C_c, or Int will increase ΔI, whereas decreasing V_r decreases ΔI_{tp}. As ΔI approaches ΔI_{tp}, IRE will approach one and the SOS will undergo a transition such that it will produce emergent effects. With these relationships in mind, a tree diagram is constructed in Figure 8.3 that captures causal factors of emergent effects in engineered systems.

FIGURE 8.3   Causal tree diagram: factors for emergent effects in engineered system.

A formula for emergence incorporating the factors is not offered. However, the IRE tuple and the causal relationships of the variables provide insights and a way forward for modeling emergence in engineered systems.

POTENTIAL IMPLICATION

The potential implications of the conceptual model and the proposed characterization of emergence are that engineers and managers may be able to use the identified factors and relationships to defend against or exploit the occurrence of emergence. The primary objective for engineers and managers of systems is to design, acquire, and operate systems that satisfy stakeholder requirements. The selection and configuration of components directly affects system's performance. Operational decisions in response to the system's environment can determine whether or not the system continues to meet its requirements. Addressing emergence can help engineers and managers accomplish these objectives. By identifying the causal factors of systems that contribute to emergent effects, engineers and managers will be able to (i) assess the risk that unexplainable effects will occur, (ii) select design alternatives with fewer causal factors, (iii) adjust the causal factors and control the likelihood of unexplainable effects, and (iv) encouraged (i.e., engineer) the occurrence of positive emergent effects.

The design and management implication can be grouped into four categories: risk assessment, analysis of alternatives (AoA), and system control.

Risk Assessment: After a system has been designed and deployed, it may be operationally important to know whether or not emergent effects are likely to occur. This can be determined by assigning a numerical value on the scale for each of the factors in the IRE and entering the values in one of the regression equations for IRE. The assignment can be made using a Delphi technique or other expert/group decision-making process. The IRE regression model provides a prediction of the value of IRE. The closer the value is to 1, the more likely an emergent effect will occur and that the effect will occur sooner rather than later.

Analysis of Alternatives (AoA): AoA is a required process in the acquisition of systems for the Department of Defense (DOD, 2015). AoA is an assessment of the various materiel solution alternatives that are being considered to satisfy the need for a SoS. Engineers and stakeholders can include the risk of emergent effects as a design consideration. Sensitivity analysis for each alternative can be performed by varying the assumptions used to determine the values for the IRE factors. As the value for IRE approaches 1, the potential design solution approaches the undesignable state space (where emergent effects will occur). Design solution alternatives can be considered on the basis of likelihood for emergent effects to occur and their sensitivity to the IRE assumptions.

System Control: The results of the design of experiment showed that the IRE factors go beyond correlation and actually have a causal relationship with the occurrence of emergents as defined by IRE $\geq$ 1. The challenge with using the IRE factors to control whether or not emergents will occur is in knowing how to control each of the IRE factors. Assuming that this knowledge exists, making adjustments to the design to increase or decrease the factors opens the door for controlling (or at least influencing) the occurrence of emergent effect. Note that the factors are not determinants of the nature of the actual emergent effect. The IRE factors only impact how fast and whether or not emergents effects will occur.

The potential implications are theoretical concepts and require additional research to make them ready for operational use.

CONCLUSIONS

Emergence in engineered systems is essentially a phenomenon that brings about unintended system effects, specifically, effects that have no apparent explanation in terms of the system's parts and their relationships. We approach emergence in this chapter as a phenomena that is theoretically derivable and therefore possible to identify factors that determine its occurrence. We posit that characterizing thermodynamic transitions in the form of chemical reactions provides insight into modeling emergence in engineered systems, including those that are SOSs. By mapping factors in chemical systems that affect the occurrence of chemical reactions to factors in engineered systems, we opened the door to transferring knowledge about the behaviors of a well-known phenomenon (chemical reaction) to help explain the lesser known phenomenon of emergence in engineered systems. Applying this knowledge, we propose the casual relationships between a set of variables for engineered systems. We provide an initial conceptualization of an emergence-dependent variable (IRE) and the causal relationships of its independent variables (component degrees of freedom, component concentration, interoperability, and variety of regulators). The identification of the engineered system variables and the conceptualization of their relationship to emergence provide a path forward for further research and contributes to modeling efforts of emergence in SOS.

RECOMMENDATIONS AND FUTURE WORK

A conceptual model and the characterization of emergence have been proposed. However, there are limitations that provide opportunities for improvement:

1. Analogies were used to convert thermodynamic factors and relationships in SOS. Validating knowledge transfer via analogical reasoning requires rigorous structural mapping between the domains beyond what has been presented in this chapter (Gentner, 1983; Lee and Holyoak, 2008).

2. The intent of Variable Conversions in Table 8.2 is to provide conceptual definitions of systems variables. Implementing the variables in a simulation model will require more formal definitions.

3. Conceptual models provide a coherent set of claims, assumptions, and constraints to reduce the concept's ambiguity, but they do not provide insight into how the system behaves over time. Converting the CLD to a stock and flow simulation model would enable the dynamic observation of the proposed factors and SOS relationships.

4. A set of factors has been proposed that affect the occurrence of emergence, but we have not addressed how to measure the factors, or how to control them.

5. The proposed characterization has not been applied to a real systems and its practical implications have not been empirically studied.

6. The chapter presents candidates for factors of emergence in engineered systems. A more exhaustive search of other potential factors may yield additional or even better candidates that should be considered.

These observations form the basis of a research agenda for future work. This agenda has the potential to further advance the body of knowledge on emergence and improve our ability to address emergence in engineering and managing SOSs.

REFERENCES

Ackoff, R.L. (1971) Towards a system of systems concepts. *Management Science*, **17** (11), 661–671.

Ashby, W.R. (1961) *An Introduction to Cybernetics*, Chapman & Hall Ltd.

ASME (2011) *Initiative to Address Complex Systems Failure: Prevention and Mitigation of Consequences*, American Society of Mechanical Engineers, Washington, D.C.

Bedau, M.A. (1997) Weak emergence. *Noûs*, **31** (s11), 375–399.

Bedau, M.A. (2008) Is weak emergence just in the mind? *Minds and Machines*, **18** (4), 443–459.

Beer, S. (1979) *The Heart of Enterprise*, vol. **2**, John Wiley & Sons.

Blanchard, B.S. and Fabrycky, W.J. (2006) Chapter 2, *Systems Engineering and Analysis* (eds B.S. Blanchard and W.J. Fabrycky), Pearson/Prentice Hall, Upper Saddle River, NJ, pp. 23–24.

Bratianu, C. and Andriessen, D. (2008). Knowledge as energy: a metaphorical analysis. Proceedings of the 9th European Conference on Knowledge Management, pp. 75–82.

Brown, T.E., LeMay, H.E.H., Bursten, B.E., and Murphy, C. (2014) *Chemistry the Central Science*, 13th edn, Prentice Hall.

Campbell, R. (2015) The concept of emergence, in *The Metaphysics of Emergence*, Palgrave Macmillan UK, pp. 192–231.

Chassin, D.P., Malard, J. and Posse, C. (2004) Managing complexity. International Conference on Computing, Communication and Control (CCCT)-04 invited session, Austin, TX, 16 August 2004. arXiv preprint nlin/0408051.

Checkland, P. (1999) *Systems Thinking, Systems Practice: Includes a 30-Year Retrospective,* John Wiley & Sons Ltd.

Chew, Y.T. and Choo, S.M. (2008) A study of the change management and challenges in a bank. *Research and Practice in Human Resource Management,* **16** (2), 100–118.

Corning, P.A. (2002) The re-emergence of "emergence": a venerable concept in search of a theory. *Complexity,* **7** (6), 18–30.

Daintith, J. (ed.) (2008) *A Dictionary of Chemistry,* Oxford University Press.

Deacon, T.W. (2007) Shannon–Boltzmann–Darwin: redefining information (Part I). *Cognitive Semiotics,* **2007** (15), 123–148.

DoD (2015) DoD Instruction 5000. 02. http://www.esd.whs.mil/Portals/54/Documents/DD/ issuances/dodi/500002_dodi_2015.pdf?ver=2017-08-11-170656-430

Dyer, G. (1996) Enthalpy as metaphor for the chemistry of conversations. *Systems Research,* **13** (2), 145–157.

Dyer, G. (2007) Enthalpy change: firing enthusiasm for learning. *Journal of Business Chemistry,* **4** (3), 116–126.

Gentner, D. (1983) Structure-mapping: a theoretical framework for analogy. *Cognitive Science,* **7** (2), 155–170.

Guckenheimer, J. and Ottino, J. (2008) Foundations for Complex Systems Research in the Physical Sciences and Engineering Report, National Science Foundation (NSF). www.siam .org/about/pdf/nsf_complex_systems.pdf.

Holland, J.H. (1998) *Emergence: From Chaos to Order,* Addison-Wesley, Reading, MA.

Kauffman, S. and Clayton, P. (2006) On emergence, agency, and organization. *Biology and Philosophy,* **21** (4), 501–521.

Klein, S. and Nellis, G. (1991) *Thermodynamics, Ch1 Basic Concepts,* Cambridge University Press, 978-0-521-19570-6.

Kotov, K. (2002) Semiosphere: a chemistry of being. *Sign Systems Studies,* **1**, 41–55.

Kovacic, S.F. (2013) Micro to macro dynamics of shared awareness emergence in situations theory: towards a general theory of shared awareness. (Order No. 3575224, Old Dominion University). ProQuest Dissertations and Theses, 162. https://search-proquest-com.proxy .lib.odu.edu/docview/304677003?accountid=12967

Lee, H.S. and Holyoak, K.J. (2008) The role of causal models in analogical inference. *Journal of Experimental Psychology: Learning, Memory, and Cognition,* **34** (5), 1111.

Lewes, G.H. (1875) *Problems of Life and Mind, First Series, The Foundations of a Creed,* vol. **II**, The Riverside Press, Cambridge, MA.

Linder, B. (2011) *Elementary Physical Chemistry,* World Scientific Publishing Company, Singapore.

Maier, M. (2015) Chapter 2, the role of modeling and simulation in systems-of-systems development, in *Modeling and Simulation Support for System of Systems Engineering Applications* (eds L.B. Rainey and A. Tolk), John Wiley & Sons, pp. 21–24.

Mill, J. S. (1846). *A System of Logic, Ratiocinative and Inductive [Electronic Resource]: Being a Connected View of the Principles of Evidence and the Methods of Scientific Investigation.* New York : Harper, 1846.

Minai, A.A., Braha, D., and Bar-Yam, Y. (2006) Complex engineered systems: a new paradigm, in *Complex Engineered Systems*, Springer, Berlin Heidelberg, pp. 1–21.

Mittal, S. and Rainey, L. (2015) Harnessing emergence: the control and design of emergent behavior in system of systems engineering. Proceedings of the Conference on Summer Computer Simulation, Society for Computer Simulation International, pp. 1–10.

Mogul, J.C. (2006) Emergent (mis) behavior vs. complex software systems. *ACM SIGOPS Operating Systems Review*, **40** (4), 293–304.

Morris, E., Levine, L., Meyers, C., Place, P. and Plakosh, D. (2004) System of systems interoperability (SOSI): final report (No. CMU/SEI-2004-TR-004). Carnegie-Mellon University Pittsburgh PA Software Engineering Institute.

NIST (2016) Measurement Science for Complex Information Systems. 2016. National Institute of Standards and Technology (NIST).

Ortega, P.A. and Braun, D.A. (2013). Thermodynamics as a theory of decision-making with information-processing costs. Proceedings of the Royal Society of London A: Mathematical, Physical and Engineering Sciences, vol. 469, no. 2153, p. 20120683, The Royal Society.

Ott, J.B. and Boerio-Goates, J. (2000) *Chemical Thermodynamics: Principles and Applications*, Academic Press, London, UK.

Pati, S.P. (2009) Stress management and innovation: a thermodynamic view. *Journal of Human Thermodynamics*, **5**, 22–32.

Sartenaer, O. (2016) Sixteen years later: making sense of emergence (again). *Journal for General Philosophy of Science*, **47** (1), 79–103.

Sawada, D. and Caley, M.T. (1985) Dissipative structures: new metaphors for becoming in education. *Educational Researcher*, **14** (3), 13–19.

Schroeder, B. and Gibson, G. (2010) A large-scale study of failures in high-performance computing systems. *IEEE Transactions on Dependable and Secure Computing*, **7** (4), 337–350.

Sergeev, V.M. (2006) Rationality, property rights, and thermodynamic approach to market equilibrium. (Author abstract). *Journal of Mathematical Sciences*, **4**, 1524.

Silberstein, M. and McGeever, J. (1999) The search for ontological emergence. *The Philosophical Quarterly*, **49** (195), 201–214.

Sterman, J. (2000) *Business Dynamics, Systems Thinking for a Complex World*, McGraw-Hill.

Sunik, B. (2011) Definition of Information. *BRAIN. Broad Research in Artificial Intelligence and Neuroscience*, **2** (4), 14–19.

Tolk, A. (2004) Moving towards a Lingua Franca for M&S and C3I – developments concerning the C2IEDM. Proceedings of the European Simulation Interoperability Workshop, Edinburgh, Scotland.

Tolk, A. and Muguira, J.A. (2003) The levels of conceptual interoperability model. Proceedings of the 2003 Fall Simulation Interoperability Workshop, September 2003, vol. 7, pp. 1–11.

Tolk, A. and Rainey, L.B. (2015) Chapter 22, towards a research agenda for M&S support of system of systems engineering, in *Modeling and Simulation Support for System of Systems Engineering Applications* (eds L.B. Rainey and A. Tolk), John Wiley & Sons.

Tolk, A., Diallo, S.Y. and Turnitsa, C.D. (2007, June) Data, Models, Federations, Common Reference Models, and Model Theory. European Simulation Interoperability Workshop, Genoa, Italy.

Tolk, A., Diallo, S.Y. and Padilla, J.J. (2012) Semiotics, entropy, and interoperability of simulation systems: mathematical foundations of M&S standardization. Proceedings of the Winter Simulation Conference, Winter Simulation Conference, pp. 2751-2762.

Valerdi, R., Axelband, E., Baehren, T. *et al.* (2008) A research agenda for systems of systems architecting. *International Journal of System of Systems Engineering*, **1** (1–2), 171–188.

Willman, D. (2014). $40-billion missile defense system proves unreliable. Los Angeles Times, 15. http://www.latimes.com/nation/la-na-missile-defense-20140615-story.html

9

INDUCED EMERGENCE IN COMPUTATIONAL SOCIAL SYSTEMS ENGINEERING: MULTIMODELS AND DYNAMIC COUPLINGS AS METHODOLOGICAL BASIS

Tuncer Ören[1], Saurabh Mittal[2], and Umut Durak[3]

[1]School of Electrical Engineering and Computer Science, University of Ottawa, Ottawa, Canada
[2]MITRE Corporation, McLean, VA, USA
[3]German Aerospace Center, Cologne, Germany

SUMMARY

Induced emergence is presented as a consequence of goal-directed steering of social systems. Multimodels offer a rich paradigm to model complex systems including complex social systems. Thirty types of multimodels are presented in an ontology-based dictionary where their definitions are given with their taxonomy. Formal modeling incorporating model structure and model coupling and especially dynamic coupling is emphasized with their relevance to multimodeling of social systems. Evaluation of the state-of-the-art tools that can implement the concept of induced emergence is presented.

Emergent Behavior in Complex Systems Engineering: A Modeling and Simulation Approach,
First Edition. Edited by Saurabh Mittal, Saikou Diallo, and Andreas Tolk.
© 2018 John Wiley & Sons, Inc. Published 2018 by John Wiley & Sons, Inc.

INTRODUCTION

The term *emergence* represents very important concepts in systems theories including complex systems theories, complex adaptive systems (CASs) theories, and systems of systems studies. However, in this chapter, the term *emergence* is deliberately used to denote a concept different from its usage in systems theories. For this reason, a brief clarification of the two usages of the term is in order, namely: (i) historical usage of the term *emergence* and (ii) emergence in systems theories.

Historical usage of emergence: The definitions of the terms *emergence* and *emerge* are found in the dictionary of etymology as follows:

"Emergence (n.): 1640s, "unforeseen occurrence," from French *émergence*, from *emerger*, from Latin *emergere* "rise up" (see *emerge*). Meaning "an emerging, process of coming forth" is from 1704.

emerge (v.): 1560s, from Middle French *émerger* and directly from Latin *emergere* "bring forth, bring to light," intransitively "arise out or up, come forth, come up, come out, rise," from assimilated form of *ex* "out" (see *ex-*) + *mergere* "to dip, sink" (see *merge*). The notion is of rising from a liquid by virtue of buoyancy. Related: *Emerged*; *emerging*." (ED-Emergence, 2000).

Emergence in systems theories: Most chapters of this book clarify the use of the term *emergence* in complex systems and systems of systems. As also quoted by Mittal and Rainey (2015), De Wolf and Holvoet (2005, p. 3) postulated that "There are actually four central schools of research that each influences the way emergence in complex systems is studied. They enumerate them (with additional clarifications and references) as: (i) CASs theory, (ii) nonlinear dynamical systems theory and Chaos theory, (iii) the synergetics (i.e., empirical study of systems in transition) school, and (iv) far-from-equilibrium thermodynamics." They succinctly recapitulate emergence as: "In short, the uses of the concept of emergence refer to two important characteristics: a global behaviour that arises from the interactions of the local parts, and that global behaviour cannot be traced back to the individual parts." They also clarify as well as exemplify differences and relationships of emergence and self-organization, namely self-organization without emergence, emergence without self-organization, as well as combining emergence and self-organization.

In this chapter, we focus on the effects of some external and often deliberate interventions on social systems. For this reason, (i) we adopt the historic definition of emergence, namely on "unforeseen occurrence" of behavior, properties, or structures in social systems; and (ii) we elaborate on goal-directed steering of social events to cause emergence or unforeseen occurrences in social systems.

Goal-directed steering of social events is like the proverbial double-edged sword. It can be done to advance a society, or at the hand of a political leader with dark triad personality, it can be detrimental to the society. A brief overview of dark triad personality is given by Furnham *et al.* (2013) and involves narcissism, Machiavellianism, and psychopathy. In this study, we explore feasibility of multimodels and dynamic couplings as methodological basis for the following: (i) to understand the mechanisms for goal-directed steering of social events. This part of the study entails how to prepare conditions so that predetermined state(s) may emerge, and based on perceived, deliberated, or anticipated characteristics of a system, how to anticipate and detect emergence of "desirable" as well as "unwanted" states. (ii) To anticipate undesirable situations to protect the system from adverse activities, before it becomes too late.

Simulation, by providing experimentation and experience possibilities, is becoming a very important infrastructure for many disciplines. Social systems, as CASs, are among such challenging and important systems. Studies on computational and especially simulation-based social systems are increasing and simulation-based techniques are being developed to tackle complex social issues (Mittal *et al.*, 2017; Ören *et al.*, 2017).

The offered methodological basis may be useful for modeling and simulation-based evaluation of alternatives in complex social systems. Appropriate formulation (i.e., modeling) of physical and social phenomena may improve our conception, perception, and understanding of reality and the way solutions may be developed. Based on this assumption, we elaborate on multimodeling and dynamic model coupling formalisms, which may be useful as methodological basis for induced emergence in social system engineering.

Agent-based modeling and multi-agent systems are increasingly used to conduct computational social experiments. We will discuss the current capabilities of many of the tools and will contrast with the needed capabilities to study induced emergence.

This chapter is organized as follows. In Computational Social Systems Engineering and Induced Emergence section, we review emergence and induced emergence and elaborate on social systems engineering. In Multimodels section, first, multimodeling is introduced in an intuitive way; afterwards, a systematic view of 30 types of multimodels is given in ontology-based dictionaries where definitions of concepts and their relationships are given. In Model Coupling section, model coupling (i.e., input/output relationships of models) is introduced. In Variable-Structure Models section, variable structure models including dynamic couplings, extensible models, as well as restrainable models are presented. In Induced Emergence in Computational Social Systems Engineering: Roles of Multimodels section, roles of multimodels for induced emergence in computational (and especially simulation-based) social systems engineering are elaborated on. In Beyond Agent-Based for Multimodeling and Simulating Emergence section, role of agents for simulation with multimodels is clarified and then follows the conclusion and future studies.

COMPUTATIONAL SOCIAL SYSTEMS ENGINEERING AND INDUCED EMERGENCE

Computational Social Science (CSS) is an emerging field that leverages the capacity to collect and analyze data with an unprecedented breadth, depth, and scale (Lazer *et al.*, 2009). As per the definition of Wikipedia (2015), it is a multidisciplinary and integrated approach to social survey focusing on information processing by means of advanced information technology. The computational tasks include the analysis of social networks and social geographic systems. Another perspective defines it as an interdisciplinary investigation of the social universe of many scales, ranging from individual actors to largest groupings, through the medium of computation (Cioffi-Revilla, 2014). Therefore, Computational Social Systems Engineering (CSSE) can be defined as follows:

> The use of computational and especially simulation-based technologies for performing social science toward engineering an improved social system.

CSSE incorporates input from multiple disciplines. As the discipline is still being defined, the outcome (as a contribution of these disciplines) is also in a fuzzy state. Although, broadly speaking, a better social system is the high-level desired outcome. Table 9.1 summarizes the contribution of a discipline and its impact on the broad outcome.

More specifically, the primary objective of CSS is to develop greater comprehension of the complexity of the interconnected global society and the ability to apply any gained insights into policy decisions for a better society (Conte, 2012). Based on this, CSSE then applies CSS for engineering better social systems.

Society is a CAS. It manifests all the properties of CAS such as emergent behavior, large number of interacting components/agents/systems engaged in goal-directed activities, and a dynamic environment that results in adaptation at the agent level. The notion of agent here symbolizes any entity that is capable of having a structure (in both software and hardware) and intrinsic behavior that allows it to interact with the environment. The complexity within the structure and behavior is ultimately reflected in the range of global macro-behaviors that the system eventually manifests. A parallel can be easily made between CSS and CAS on the basis that the path of both CSS and CSSE is not clearly defined. There are a lot of unknowns: how to use various contributing factors toward specification of a better social system. Wherever there are unknowns, there is potential of emergent behavior.

Emergence in Computational Social Systems Engineering

As per Ashby (1956), the source of emergent behavior is lack of understanding of the system itself. So, in defining the discipline of CSS and CSSE, the source of emergent behavior is threefold:

1. Emergent behavior resulting from incomplete understanding of the existing social system.
2. Emergent behavior resulting from specification of a "would-be" better system based on the results of multidisciplinary factors that are used to understand the existing social system.
3. Emergent behavior resulting from the computational implementation of such a multidisciplinary model. Such a model is difficult to validate and verify (Mittal, 2014a; Mittal and Zeigler, 2017; Zeigler, 2016; Vinerbi et al., 2010, Foo and Zeigler, 1985).

TABLE 9.1   Multidisciplinary Nature of Computational Social Science (CSS) and its Impact on Computational Social Systems Engineering (CSSE)

S. No.	Discipline	Contributing Factors	Impact on CSSE Outcome
1.	Data engineering	Information gathering, model building	Model validation, system validation
2.	Network science	Multi-level relationships	Association graphs, influence graphs, community formation
3.	Statistics	Law of large numbers, high-level patterns from low-level data	Development of indicators and triggers for guiding the system's behavior
4.	Computer science	Model building, model engineering, data visualization and representation	Enterprise-scalable systems accessible in a multimodel way (e.g., mobile and desktop)
5.	Computational sociology	Computational representation of surveys and other methods	Integration of legacy and novel methods, models, and data in the next-generation social systems
6.	Computational economics	Computational representation of economic activity	Economic indicators and triggers for an engineered social system
7.	Complexity science	Nonlinear behavior, attractor basins	Resilience and fault-tolerant social systems engineering
8.	Systems engineering	Modularity, interface design, community involvement	Development of cross-linkages, policy, and interrelationships between various elements of the social system
9.	Modeling and simulation	Abstraction, dynamic behavior, computational representation of structure and behavior	Experience, experimentation, design, test, and evaluation in a live, virtual, and constructive test bed for improved systems engineering
10.	Cybernetics	Communication and control	Policy making, law enforcement, information control using multiple media

If the complete knowledge about a system (its structure and behavior) would be represented as a circle, each pie will amount to a percentage of that knowledge. In CSS and CAS, in general, defining the circle is *the* hard problem. Consequently, each pie that is discovered because of contributing factors adds to the knowledge base and continues to add to the "current" understanding of circle. The newly discovered properties expand the circle's area and perimeter, thereby adding to the knowledge base. This was illustrated by the cyclical relationship between the *weak* and *strong* emergent behavior by Mittal (2014b). Weak emergent behavior is defined as the macro-behavior that is known and reproducible in a computational environment. Strong emergent behavior is defined as the holistic behavior that is unknown and inconsistently reproduced in a computational environment (Mittal and Rainey, 2015).

In the enumerated second case, this type of emergent behavior study stems from an incomplete team; that is, not enough expertise is present in the CSS team. A multidisciplinary team is a de facto requirement for any CSS endeavor, failure of which will lead to knowledge gaps and naive conclusions. Simply speaking, the whole pie is never looked at. Gap analysis then becomes an inclusive activity to understand the emergent behavior (Mittal and Cane, 2016).

In the enumerated third case, this type of emergent behavior stems from the deficiencies of either or both modeling and simulation paradigm (Mittal and Martin, 2017). A coarse model will not represent the reality to the extent needed and a high-fidelity model will require a lot of resources to represent something close to reality. A CAS model is a hybrid model that has both discrete event and continuous time elements (Mittal and Zeigler, 2017). A hybrid model, grounded in mathematics, which involves both the discrete and continuous complex dynamic systems, is the preferred option (Mittal, 2013). The model is then executed for dynamic behavior in a simulation infrastructure. The architecture of simulation system and the computational resources that it utilizes affect the model execution. Emergent behavior arising from model and simulation implementation must be known a priori and must be eliminated for meaningful CSS or CAS engineering (Mittal and Martin, 2017).

Induced Emergence in Simulation-Based Computational Social Systems Engineering

Contrary to Ashby's view (1956), which posits that the source of emergent behavior is lack of understanding of the system itself, the concept of induced emergence is based on goal-directed steering of social events. Thus, emergent social behavior, which would not have arisen without the intervention of a goal-directed steering of social events, arises due to this steering. Furthermore, the goal of the steering may be benevolent or not for the society. However, study of induced emergence is important for goals both benevolent and vicious. In the first case, alternatives can be tested by simulation to introduce desirable induced behavior in social systems. In the case of vicious goal-directed steering of social systems, for example, by a dark triad personality leader, the study of induced emergence may be useful for the citizens to apprehend some wrong doings in the society and to take precautions before the ill doings do not accumulate.

Goal-directed steering of social events can be direct or indirect. In the first case, that is, goal-directed direct steering of social events, the intervention may cause some direct induced behavior. For example, lowering the quality of education in a society may induce in the long run, a different social system. The desirability and undesirability of the emerging behavior depend on the observers and their value systems.

In the case of goal-directed indirect steering of social events, the intervention may cause some indirect induced behavior. Still in the same domain, not educating the girls in a society may lower their quality of life (direct influence) and the family education they can provide for next-generation men (indirect influence) may not be the same caliber as those which receive proper quality of education.

Sustainability of Civilizations, Induced Emergence, and Role of Simulation-Based Computational Social Systems Engineering

"Civilization is social order promoting cultural creation. Four elements constitute it: economic provision, political organization, moral traditions, and the pursuit of knowledge and arts. It begins where chaos and insecurity end. For when fear is overcome, curiosity and constructiveness are free, and man passes by natural impulse towards the understanding and embellishment of life" (Durant, 1935, p. 1). After clarification of conditions for civilization, Durant posits that:

> The disappearance of these conditions sometimes of even one of them may destroy a civilization. A geological cataclysm or a profound climatic change; an uncontrolled epidemic like that which wiped out half the population of the Roman Empire under the Antonines, or the Black Death that helped to end the Feudal Age; the exhaustion of the land, or the ruin of agriculture through the exploitation of the country by the town, resulting in a precarious dependence upon foreign food supplies; the failure of natural resources, either of fuels or of raw materials; a change in trade routes, leaving a nation off the main line of the world's commerce; mental or moral decay from the strains, stimuli and contacts of urban life, from the breakdown of traditional sources of social discipline and the inability to replace them; the weakening of the stock by a disorderly sexual life, or by an epicurean, pessimist, or quietist philosophy; the decay of leadership through the infertility of the able, and the relative smallness of the families that might bequeath most fully the cultural inheritance of the race; a pathological concentration of wealth, leading to class wars, disruptive revolutions, and financial exhaustion: these are some of the ways in which a civilization may die. For civilization is not something inborn or imperishable; it must be acquired anew by every generation, and any serious interruption in its financing or its transmission may bring it to an end. Man differs from the beast only by education, which may be defined as the technique of transmitting civilization. (Durant, 1935, p. 3).

Hence, for the sustainability and advancement of civilizations, several categories of activities are needed:

1. Proper education of the young generations, especially to pass the values of the civilization.
2. To prepare the citizens to preserve and to create new conditions to support and attract new talents to be interested to the maintenance and advancement of the civilization.
3. To protect the civilization from natural disasters, from external as well as internal dangers (such as from malevolent leaders).

The quality of education, as "the technique of transmitting civilization," thus becomes vital. An important computational approach to education, that is, simulation-based education, including sociopolitical simulation games, can offer several advantages to test alternatives about social situations and gaining experience for training on several aspects of social alternatives (Ören et al., 2017).

MULTIMODELS

An Intuitive Introduction

A multimodel is a modular model that subsumes multiple submodels that together represent the behavior of different aspects of an entity. The description may warrant a different formalism (e.g., discrete event and continuous time). Multimodels offer a very rich paradigm to represent a variety of models. At any time, at least one (i.e., one, some, or all) of the submodels is active. Several types of multimodels are possible (Yilmaz and Ören, 2004). Multimodel formalism was originally introduced by Ören (1987b, 1991) and later elaborated on by Fishwick and Zeigler (1992). In this section, multimodels are explained in an intuitive way.

Figure 9.1 is a depiction of a mixture of three aspects of water (i.e., fluid, ice, and vapor); the physical properties of each of which has to be modeled in a separate submodel. Indeed, ice and vapor are different aspects of liquid water, and depending on the pressure and heat given to or be taken from them, one, two, or all the three aspects of water can exist at a given time. Each external input (inputs to the multimodel) can be connected to each submodel. The output of each submodel (such as its mass and temperature) is the outputs of the parent multimodel. Submodels can become active depending on the input values (in this example, pressure and heat) and their current

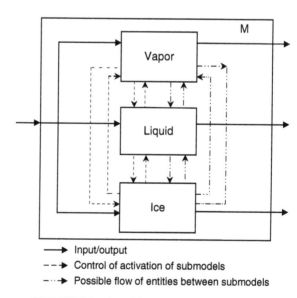

    ⟶ Input/output
    – – ▶ Control of activation of submodels
    – ·· ▶ Possible flow of entities between submodels

FIGURE 9.1   A multiaspect multimodel of water.

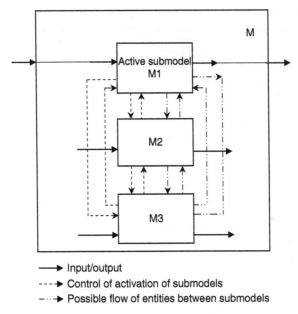

→ Input/output
---→ Control of activation of submodels
–··→ Possible flow of entities between submodels

FIGURE 9.2 A multistage multimodel.

state. Furthermore, there is a possible flow of entities between submodels. This type of multimodel is called *multiaspect multimodel*, as several aspects of an entity may exist, namely several submodels may be active, simultaneously. A multiaspect multimodel represents different aspects of an entity or a system and allows transition from a submodel to another.

In other types of multimodels, submodels may exist one at a time; for example, stages of a butterfly (such as egg, larva, pupa, and butterfly), stages of a missile, or five stages of a parachute jump such as (i) payload-free flight, (ii) parachute deployment to fully stretched rigging, (iii) inflation, (iv) deceleration and stabilization, and (v) system descent (Doherr, 2002). This type of multimodel is called *multistage multimodel*. A multistage multimodel is a multimodel where the number of submodels is fixed and submodels become activated in a sequential way. Figure 9.2 is a depiction of a multistage multimodel where there is a predefined sequence of alternate submodels. The difference between a multiaspect (Figure 9.1) and multistage (Figure 9.2) models can be seen in the couplings. The submodels in multiaspect model may all be simultaneously active, which is not the case in multistage model where they get activated sequentially.

A Systematic View

Multimodels can be categorized into three broad categories based on (i) modeling methodology, (ii) submodel activation knowledge, and (iii) submodel structure. Table 9.2 lists 30 types of multimodels based on these three major criteria.

TABLE 9.2   Types of Multimodels Based on Modeling Methodologies, Submodel Structure, and on Submodel Activation Knowledge

Based on *modeling methodologies*:
1. Atomic multimodel
2. Component multimodel
3. Continuous multimodel
4. Coupled multimodel
5. Discrete multimodel
6. Dynamically coupled multimodel
7. Memoryless multimodel
8. Mixed-formalism multimodel
9. Multiresolution multimodel
10. Nested coupled multimodel
11. Single formalism multimodel
12. Statically coupled multimodel

Based on *submodel structure*:
1. Dynamic-structure multimodel (variable structure multimodel)
2. Evolutionary multimodel
3. Extensible multimodel
4. Multiaspect multimodel
5. Multiperspective multimodel
6. Multistage multimodel
7. Mutational multimodel
8. Reducible multimodel
9. Single-aspect multimodel
10. Single-perspective multimodel
11. Static-structure multimodel

Based on *submodel activation knowledge*:
1. Acyclic multimodel
2. Constraint-driven multimodel
3. Cyclic multimodel
4. Externally activated multimodel
5. Goal-directed multimodel
6. Internally activated multimodel
7. Pattern-directed multimodel

For the sake of completeness, all types of multimodels are explained in this section. However, some of them might be sufficient to model induced emergence in multi-aspect social systems. Table 9.3 presents major categories of multimodels, criteria to distinguish them, and tables where the details are given.

Tables 9.4–9.9 are ontological dictionaries of terms related with multimodels and show their taxonomies and definitions. They are based on modeling methodologies, submodel structure, and submodel activation knowledge. Ontology-based dictionaries give the definitions as well as relationships of the terms (Ören *et al.*, 2007).

TABLE 9.3 Major Categories of Multimodels and Criteria to Distinguish Them

Multimodels Based on	Major Categories of Multimodels	Details in Table
Modeling *methodology*	*Atomic* multimodel	9.4
	Coupled multimodel	9.5
Submodel *structure*	*Number of submodels* active at a given time	9.6
	Variability of structure of submodels	9.7
Submodel *activation* *knowledge*	*Nature of knowledge* to activate submodels	9.8
	Location of knowledge to activate submodels	9.9

TABLE 9.4 Ontological Dictionary of Terms Related with Multimodels: Based on *Modeling Methodology of Submodels* (Atomic Models)

Criteria for Relationships			Type of Multimodel	Definitions
Formalism: Single	Yes		*Atomic multimodel*	A multimodel consisting of two or more submodels
		Continuous	*Continuous multimodel*	A multimodel where submodels are continuous models
		Discrete	*Discrete multimodel*	A multimodel where submodels are discrete models
		Memoryless	*Memoryless multimodel*	A multimodel where submodels are memoryless models (i.e., without state variables)
	No	Mixed formalism	*Mixed formalism multimodel*	A multimodel where at least two formalisms are used in formulating submodels (namely, two or more of continuous, discrete, or memoryless models)
Resolution	Single		*Single-resolution multimodel*	A multimodel where all submodels are of the same resolution
	Multiple		*Multiple-resolution multimodel*	A multimodel where submodels may be of different resolutions

Modeling methodology can be broadly understood through two categories: atomic and coupled. Although every model has explicit interfaces (for input and output) that allow it to connect to other models, internally, the behavior is specified as a function of structure. In an atomic model, the behavior is represented as a state space and the

TABLE 9.5   Ontological Dictionary of Terms Related with Multimodels: Based on *Modeling Methodology of Submodels* (Coupled Models)

Criteria for Relationships		Type of Multimodel	Definitions
Coupling	Resultant multimodel	*Coupled multimodel*	A multimodel consisting of two or more multimodels having input/output relationships
	Each component	*Component multimodel*	A multimodel of a coupled multimodel
Nature of coupling	Static	*Statically coupled multimodel*	A multimodel where the coupling (namely input/output relationships) of multimodels and/or component multimodels cannot change over time
	Dynamic	*Dynamically coupled multimodel*	A multimodel where the coupling (namely input/output relationships) of multimodels and/or component multimodels can change over time

TABLE 9.6   Ontological Dictionary of Terms Related with Multimodels: Based on *Structure of Submodels* (*Number* of Submodels Active at a Given Time)

Criteria for Relationships			Type of Multimodel	Definitions
Number of submodels active at a given time	Only one		*Single-aspect multimodel*	A multimodel having only one submodel active at a given time
		Single perspective	*Single-perspective multimodel*	A multimodel, based on a single perspective and having only one submodel active at a given time
		Multiple perspective	*Multiperspective multimodel*	A multimodel, based on several perspectives and having only one submodel representative of this perspective being active at a given time
	Two or more		*Multiaspect multimodel*	A multimodel having at least two submodels active at a given time

TABLE 9.7   Ontological Dictionary of Terms Related with Multimodels: Nased on *Structure of Submodels* (*Variability* of Structure of the Submodels)

Criteria for Relationships			Type of Multimodel	Definitions
Static structure	Static-structure multimodel			A multimodel where structure of submodels cannot change
	Sequential activation		Multistage multimodel	A multimodel with static structure of sub-models where number of submodels is fixed and submodels are activated sequentially
Dynamic structure	Dynamic structure multimodel (variable-structure multimodel)			A multimodel where structure of submodels can change
	Number of submodels can be	Extended	Extensible multimodel	A multimodel where number of submodels and their input/output relationships are extensible
		Restrained	Restrainable multimodel	A multimodel where some submodels and/or their relationships are restrainable
	Alteration of sub-models is possible and iterative	No	Mutational multimodel	A multimodel where the structure of submodels can be altered and alteration of submodels is not iterative
		Yes	Evolutionary multimodel	A multimodel where the structure of submodels can be altered and alteration of submodels is iterative

TABLE 9.8   Ontological Dictionary of Terms Related with Multimodels: Based on *Submodel Activation Knowledge* (Nature of the Knowledge to Activate Submodels)

Criteria for Relationships			Type of Multimodel	Definitions
Constraint driven			Constraint-driven multimodel	A multimodel where submodel activation knowledge is based on constraints
Pattern-directed	Pattern-directed multimodel			A multimodel where submodel activation knowledge is based on a pattern
	There is a cycle	No	Acyclic multimodel	A multimodel where submodel activation knowledge is based on a pattern and there is no cycle in this pattern
		Yes	Cyclic multimodel	A multimodel where submodel activation knowledge is based on a pattern and there is a cycle in this pattern
Goal-directed			Goal-directed multimodel	A multimodel where submodel activation knowledge is goal-directed

TABLE 9.9   Ontological Dictionary of Terms Related with Multimodels: Based on *Submodel Activation Knowledge* (*Location* of the Knowledge to Activate Submodels)

Criteria for Relationships	Type of Multimodel	Definitions
Location of knowledge to activate submodels		
Within the multimodel (*internal* activation of submodels)	*Internally activated multimodel*	A multimodel where the knowledge to activate the submodels is within the multimodel
Outside of the multimodel (*external* activation of submodels)	*Externally activated multimodel*	A multimodel where the knowledge to activate the submodels is outside of the multimodel

transition functions that transform the input into output, as is the case with discrete event systems (DEVSs). To have more explicit behavior, the state space and transition functions are represented as a function of *time*, as is the case with continuous systems. The DEVS formalism (Zeigler *et al.*, 2000) integrates both the discrete event and the complex continuous dynamic systems. In a coupled model, many such atomic models can be connected for a composite behavior. Mittal (2013) elaborates the use of DEVS formalism for modeling CAS (discrete, continuous, and memoryless).

MODEL COUPLING

Systems theory deals with three important concepts: *structure, behavior,* and the relationship between the two using the *closed-under-composition* principle. This principle allows explicit relationship between the system's structure and its behavior using the coupling relationships. The coupling specification allows the structure to be represented as a flat system (i.e., with no hierarchy) or a system with a hierarchical containment. The coupling relationships guarantee that the resultant behavior of a coupled system remains same even if the structure goes through metamorphosis, which is usually the case in a CAS (Mittal, 2013). The above three properties form the basis for a modular specification of the system, wherein multiple systems at different levels of hierarchy can be composed toward a larger system capable of manifesting complex behaviors, connected through explicit couplings through the component's interfaces. This is a fundamental requirement for specification of a formal multicomponent system, where each component has individual properties and exchanges information through its interfaces. To perform multimodel systems engineering, multimodel systems subscribe to systems' theoretical constructs and portray the following top-level properties:

1. Modularity
2. Strict interfaces (i.e., strong data types associated with ports)

3. State representation
4. Transition functions
5. Composition
6. Decomposition
7. Hierarchical composition

Coupling of models specify input/output relationships of models (also called *component* models) which, through coupling specification form another model called *coupled* model or *resultant* model, having inputs and outputs. Coupling is important for the use of multimodels for induced emergence for the following reasons:

1. Even submodels of a multimodel need to have (hence to represent) input/output relationships.
2. A monolithic model (i.e., a model consisting of one single model) cannot represent several aspects of a social system; hence, there is a need for modular modeling; several such models can then be combined (or coupled) to form one model to represent several aspects of a social system.
3. Hierarchical modeling can be done by coupling models, some of which being already coupled models. Hierarchical modeling can be done at several levels, where a component model of a resultant model being another coupled model.
4. Coupling is the most natural way to model systems of systems.
5. Time-varying couplings where either the input/output relationships of component models or some of the component models may change are very powerful ways to specify some of the dynamics of social systems.

It is an interesting distinction of social systems from traditional mathematical modeling that some of the inputs to some component models and/or to some submodels may be deliberately curtailed (information hiding/protecting) or synthetic inputs (such as fake news) may be used. Mathematical specifications of couplings need to be extended to include information hiding possibility. As perception is very important in social systems, disinformation activities may provide fake inputs. One of the challenges for simulation-based CSSE modeling may be development of filters to detect fake inputs.

Figure 9.3, adopted from Ören (2014), represents a coupled model Z consisting of component models Z1–Z4.

Couplings can be represented in different ways. Detailed description of coupling is given by Ören (2014). The concept of declarative specification of simulation model coupling is rather old (Ören, 1971, 1975) and based on the system theory developed by Wymore (1967) and Ören and Zeigler (2012). Recently, coupling for bio-inspired systems was elaborated on by Ören and Yilmaz (2015). Figure 9.4 represents a template for model coupling. Figures 9.5 and 9.6 are the specification of the coupling depicted in Figure 9.3.

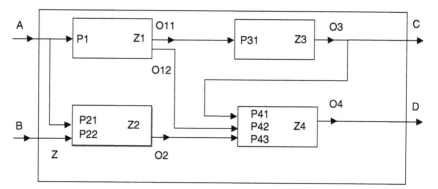

FIGURE 9.3 A coupled model Z consisting of component models Z1-Z5. (Adapted from Ören (2014)).

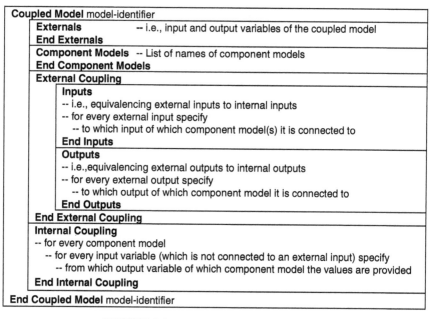

FIGURE 9.4 A template for model coupling.

VARIABLE STRUCTURE MODELS

In engineering and science, most often static models are sufficient. However, in social applications, as Greek philosopher Heraclitus said: "The only thing that is constant is change." Formalisms already exist to represent variable structure models. For example, L-system (or Lindenmayer system) models can represent growth of plants and Chaotic system models represent chaotic behavior. Another important

```
Coupled Model model-identifier
    Externals            -- i.e., input and output variables of the coupled model
        Inputs    A, B
        Outputs  C, D
    End Externals
    Component Models  -- List of names of component models
        Z1, Z2, Z3, Z4
    End Component Models
    External Coupling
        Inputs
        -- i.e.,equivalencing external inputs to internal inputs
        -- for every external input specify
            -- to which input of which component model(s) it is connected to
            Z.A → Z1.P1    -- i.e., A of Z is connected to P1 of Z1
            Z.A → Z2.P21   -- i.e., A of Z is connected to P21 of Z2
            Z.B → Z2.P22
        End Inputs
        Outputs
        -- i.e.,equivalencing external outputs to internal outputs
        -- for every external output specify
            -- to which output of which component model it is connected to
            Z.C ← Z3.O3
            Z.D ← Z4.O4
        End Outputs
    End External Coupling
    Internal Coupling
    End Internal Coupling
End Coupled Model model-identifier
```

FIGURE 9.5 Declarative specification of the coupling of the coupled model depicted in Figure 9.3 ("a → b" and "c ← d" can be read as "a is connected to b" and "c is connected from d").

variable structure system model is developed by Zeigler *et al.* (1991) and Hu *et al.* (2005).

In this section, some formalisms especially applicable for dynamic structure multimodels are presented. They may be useful to model especially in the study of emergence in social systems engineering. Mittal (2013) emphasized the use of variable structure DEVS model in formal modeling of stigmergic and CASs. Stigmergic systems have components that interact in an indirect way through the environment and CAS; on the other hand, they have multi-level structures incorporating both negative and positive feedback loops that are established using various types of couplings.

Time-Varying Couplings

In model coupling, an interesting concept is time-varying coupling (or dynamic coupling). There are two types of time-varying coupling: either component models or

```
Coupled Model model-identifier
   Internal Coupling
   -- for every component model
   -- for every input variable (which is not connected to an external input) specify
      -- from which output variable of which component model
         -- the values are provided
      Model Z1
         Z1.P1 ←Z.A        -- input is from external input
                           -- (information can be provided by the model specification environment)
      Model Z2
         Z2.P21 ← Z.A      -- input is from external input
         Z2.P22 ← Z.B      -- input is from external input
      Model Z3
         Z3.P31 ← Z1.11
      Model Z4
         Z4.P41 ← Z3.O3
         Z4.P42 ← Z1.O12
         Z4.P43 ← Z2.O2
   End Internal Coupling
End Coupled Model model-identifier
```

FIGURE 9.6 Declarative specification of the internal coupling of the coupled model depicted in Figure 9.3.

the input/output relations, or both can be time varying. (1) In a time-varying coupling, multiple models of an entity can be used. Under prespecified conditions, one of the variant models of an entity may replace the existing model. This concept was offered in early 1970s (Ören 1971, 1975). (2) The output/input relationships of models may be altered and/or external inputs to a coupled model may be connected to different component models. One of the checks to be done by a modeling system is to check dimensions of the input/output variables before accepting the coupling relationships. Another early article specified other checks that can be done by a coupling specification environment (Ören and Sheng, 1988). Challenges for dynamic model replacement and dynamic model update with multimodels were elaborated on by Yilmaz and Ören (2004). For example, dynamic model update is needed under the following conditions (Yilmaz and Ören, 2009):

Changing Scenarios: For most realistic social dilemmas, the nature of the problem changes as the simulation unfolds.

Ensembles of Models: Our knowledge about the problem (i.e., conflict) being studied may not be captured by any single model or experiment.

Uncertainty: Adaptivity in simulations and scenarios is necessary to deal with emergent conditions for evolving systems in a flexible manner.

Exploration: As simulations of complex phenomena are used to aid intuition, dynamic run-time simulation composition will help identify strategies that are flexible and adaptive.

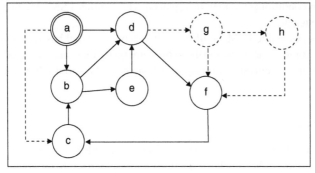

Legend:

○ ──→ States and transitions of existing finite-state machine

◌ ----→ States and transitions of extended finite-state machine

FIGURE 9.7 An extensible finite-state machine.

For these types of activities, agent-monitored simulation is very appropriate, as agents can observe, perceive, and reason about their environment to act or proact with goal-driven responses (Yilmaz and Ören, 2009).

Extensible Model

As defined in Table 9.7, an extensible model is a multimodel wherein a number of submodel and their I/O relationships are extensible. These I/O relationships are handled by coupling specification. Internally, an extensible model's behavior is modeled as an extensible finite-state machine (FSM) that incorporates new I/O functions, represented as extensible states in an existing state machine. Figure 9.7 shows a depiction of an extensible FSM. The solid circles and solid arrows represent a classical FSM with the initial state "a." It can be either a Mealy or Moore machine based on the definition of the outputs (Gill, 1962; Arbib, 1969). If at every state the output depends only on the state, then the machine is of Moore type. In both Mealy and Moore machines, the number of states is finite and does not change during the use of the model. Similarly, the transitions between states are known a priori.

If, after some conditions are satisfied or deliberately induced, a new state such as state "g" is generated, then the original FSM is extended and some new transitions to and from the new state may occur. A basic example for FSMs is a model of coin-operated soda machine (Wright, 2005). Introduction of a new coin type may be considered as an example of an extended FSM.

Similarly, after state "g," another state "h" and some other transition can occur. After some conditions satisfied or induced, some transitions, for example, from "a" to "c" can become possible. This type of FSMs can be called an *extensible finite-state machine*, which, at any given time, is a conventional FSM; however, under certain condition(s), it may be replaced with another FSM, which is a superset of the previous one.

Similar to extensible FSMs, multimodels can be extended based on emergence or induction of some conditions. Furthermore, emerged states and emerged transitions can also have fuzzy representations and each state can be a multimodel or a model to be coupled.

Restrainable Models

Restrainable models are the opposite of extensible models. An example can be the FSM having states a-h and the transitions represented both by solid and dashed arrows, as shown in Figure 9.7. After some conditions, both the states g and h and the transitions represented by dashed arrows may be restrained, therefore may not be available. Under these conditions, the initial model may be restrained to the FSM represented by circles and arrows represented by solid lines. Of course, the FSM paradigm is given just as an example. The states can represent models and multimodels, and the transitions can represent coupling of the models and multimodels. Another interpretation of restrainable models can be restraining some submodels of some multimodels. In a social system, for example, restraining some human rights by a malevolent leader (or external enemy) may limit functioning of some submodels (and may cause some others to become active).

INDUCED EMERGENCE IN COMPUTATIONAL SOCIAL SYSTEMS ENGINEERING: ROLES OF MULTIMODELS

Multimodels offer a rich paradigm to model complex dynamic systems. The definitions are deducible from the ontology presented in Computational Social Systems Engineering and Induced Emergence section where multimodels can be used to model induced emergence and methods to induce emergence are also deducible from their definitions. Tables 9.10–9.13 are a top-down decomposition of multimodels and describe how they contributed to CSS study. Table 9.14 describes how some of them (in conjunction with others in the overall multimodel) provide an opportunity to experiment with induced emergence.

BEYOND AGENT-BASED FOR MULTIMODELING AND SIMULATING EMERGENCE

Although there are various ways of doing a classification for simulation (Ören, 1987a), from one perspective, simulation can be equation-based or agent-based (Parunak et al., 1998). Both approaches employ modeling and conduct model execution on a computer. Equation-based model is mostly applied in natural sciences and engineering where the system models are formalized by mathematical equations that govern the system variables, whereas agent-based models, which are

TABLE 9.10   Multimodels Based on *Modeling Methodology of Submodels* (Atomic Models Where They can be Used for Simulation-Based CSS Studies)

Types of Multimodel	Where Can Be Used for Simulation-Based CSS Studies
Atomic multimodel	To specify basic components of social systems as multimodels that encode behavior using state variables and state machines
Continuous multimodel	To specify basic components of social systems where submodels are continuous models and require modeling of continuous phenomena such as idea spreading and disease spreading
Discrete multimodel	To specify basic components of social systems where submodels are discrete models
Memoryless multimodel	To specify basic components of social systems where submodels do not have state variables (memory variables)
Mixed formalism multimodel	To specify basic components of social systems where some submodels can be continuous, discrete, or memoryless models
Single-resolution multimodel	To specify basic components of social systems where submodels have same resolution, that is, at the same level of hierarchy
Multiple-resolution multimodel	To specify basic components of social systems where submodels have different resolutions, that is, a model is active at multiple resolutions

TABLE 9.11   Multimodels Based on *Modeling Methodology of Submodels* (Coupled Models) Where They can be Used for Simulation-Based CSS Studies

Types of Multimodel	Where Can Be Used for Simulation-Based CSS Studies
Coupled multimodel	To specify multimodels of modular systems, hierarchical systems, as well as system of systems where multimodels have input/output relationships. In the coupling, some inputs may deliberately be hidden from some component models and fake inputs may also be applied
Component multimodel	To specify each component multimodel of modular systems, hierarchical systems, as well as system of systems where coupled models become as components
Statically coupled multimodel	To specify multimodels where the input/output and the coupling of component models and component multimodels do not change over time
Dynamically coupled multimodel	To specify multimodels where some input/output relationships (couplings) of component models and/or some component models may change during a simulation study

largely employed in social systems, include agents that encapsulate the behavior of individuals that construct the system. The agents constitute their own environment, behave and interact per set of rules, and possess cognitive capabilities for information processing (Ferber, 1999).

Agent-based modeling includes identifying agents, specifying their behaviors, and capturing their interactions. Although the behavior and the interactions among the

TABLE 9.12 Multimodels Based on *Structure of Submodels* (*Number* of Submodels Active at a Given Time) Where They can be Used for Simulation-Based CSS Studies

Types of Multimodel	Where Can Be Used for Simulation-Based CSS Studies
Single-aspect multimodel	To represent current (single) aspect of a (social) system. When conditions change, another aspect may become active and can be represented by a relevant submodel
Single-perspective multimodel	To represent a (social) system with respect to a single world view (weltanschauungen) where only one submodel is active at a given time
Multiperspective multimodel	To represent a (social) system with respect to several world views, simultaneously (weltanschauungen)
Multiaspect multimodel	To represent several aspects of a (social) system as submodels that are simultaneously active. When conditions change, another set of aspects may become active and can be represented by relevant submodels

TABLE 9.13 Multimodels Based on *Structure of Submodels* (*Variability* of Structure of the Submodels) Where They can be Used for Simulation-Based CSS Studies

Types of Multimodel	Where Can Be Used for Simulation-Based CSS Studies
Static-structure multimodel	To specify multimodels where the structure of submodels do not change. The social study may be for a limited time, as the structure of social systems is rather dynamic
Multistage multimodel	To specify (social) systems with static structure of submodels where number of submodels is fixed and submodels are activated sequentially
Dynamic structure multimodel (variable structure multimodel)	To specify structure of dynamic (social) systems where the structure of the system may change, namely may extend or may be restrained. The system can change temporarily or permanently
Extensible multimodel	To specify structure of dynamic (social) systems where the structure of the system may change. For example, new dynamics can emerge or can be induced. Extensible multimodels may be appropriate especially to model-induced emergence. In this way, submodels to emerge after deliberately created conditions can be modeled
Restrainable multimodel	To specify (social) systems where some aspects of the system may become unrealizable. For example, by natural disaster(s) or by actions of a malevolent national or international leader. For example, human rights violations may restrain many other activities
Mutational multimodel	To specify structure of dynamic (social) systems where the structure of the system may change and alteration is not iterative; permanent changes may occur in the system
Evolutionary multimodel	To specify structure of dynamic (social) systems where the structure of the system may change iteratively; permanent changes may occur in the system

TABLE 9.14   Multimodels Based on *Submodel Activation Knowledge* (Nature and Location of the Knowledge to Activate Submodels) and the *Methods to Induce Emergence* for Simulation-Based CSS Studies

Types of Multimodel	Method to Induce Emergence
Constraint-driven multimodel	Constraints may cause induced emergence of some behavior. For example, allocation of resources for military spendings may contribute to reduction of budget for education and research that may have negative outcomes in the long run
Pattern-directed multimodel	A multimodel where submodel activation knowledge is based on a pattern
Acyclic multimodel	A multimodel where submodel activation knowledge is based on a pattern and there is no cycle in this pattern
Cyclic multimodel	A multimodel where submodel activation knowledge is based on a pattern and there is a cycle in this pattern. For example, there is a cycle in the life of trees
Goal-directed multimodel	By definition, goal-directed (social) events lead to desirable or undesirable induced emergence
Internally activated multimodel	Evolving systems can activate themselves as the knowledge for event-based activation resides within the model itself
Externally activated multimodel	Social systems can be activated externally to cause induced emergence. The motivation for external activation may be existence of some local wealth. The knowledge to activate is outside the multimodel and is deliberately injected

agents can be specified using equations, the common practice is to use either rules or logical operations (Helbing, 2012). It is required to have an explicit model of the agent behavior. Although agent-based modeling can be done using general purpose programming languages, there are various environments that target agent-based modeling, such as NetLogo (Tisue and Wilensky, 2004), Repast (North *et al.*, 2006), MASON (Luke *et al.*, 2004), or AnyLogic (Borshchev and Filippov, 2004).

Abar *et al.* (2017) presented a comprehensive review where they provide a concise characterization of entire spectrum of agent-based modeling and simulation tools. They note that agent-based modeling and simulation are well suited for CSS as the system under consideration is constituted by the interacting active entities that negotiate and resolve conflicts and may orchestrate gradual emerging patterns.

The categorization of Abar and colleagues based on application domains lists AgentScript, AgentSheets, AnyLogic, AOR Simulation, Ascape, GROWLab, JAS, LSD, MASS, MIMOSE, Mathematica, Modgen, NetLego, PDES-MAS, PS-I, Repast Simphony, SeSAM, SimAgent, SOARS, StarLogo, Sugarscape, Swarm, VisualBots, and VSEit as CSS tools. Most of these tools are developed in Java and they support various ways of developing agents from visual drag and drop conversational programming practices (AgentScript, Anylogic) to general purpose programming languages such as Java or C++ (OSED-MAS, PS-I, and VSEit), from domain-specific languages (MASS, MIMOSE, NetLogo, and Repast Simphony) to scripting languages such as Python (AgentScript, Repast Simphony, SOARS, and SugerScript). Classification regarding the model development effort groups Agent Script, AgentSheets, NetLogo PS-I, SeSAM SOARS, StarLogo, SugarScape, and VisualBots as simple; AnyLogic, AOR Simulation, Ascape, GROWLab, JAS, LSD, MASS, Mathematica, MIMOSA, and Modgen as moderate; and PDES-MAS, Repast Symphony, and SimAgent is hard. In terms of modeling strengths and simulation models' scalability, AgentScript, AgentSheets, GROWLab, PS-I, SOARS, StarLogo, SugarScape, and VisualBots are categorized as light weight and small scale; JAS, Mathematica, MIMOSA, Modgen NetLogo, Repast Symphony, and VSEit as medium scale; AnyLogic, AOR Simulation, Ascape, LSD, NASS, SeSAM, and SimAgent as high and large scale.

Regarding its capability in modeling strong and complex network interactions among the elements of system, which may lead to counter-intuitive behavior, the agent-based approach has been regarded as suitable in simulating emergence (Helbing and Balietti, 2011). It is crucial that the selection of the appropriate modeling technique is utmost importance. Currently, agent-based approach plays an important role in CSS with its promise to represent individual interactions from which social patterns emerge (Bouanan et al., 2016). However, it is not only the interaction network of individuals that is changing its structure, but modeling the individual agents also requires multimodeling. Then the computed emergence is not only due to variability of interaction network, but also induced as a result of structural variability of agent models.

Pawletta et al. (2002) classifies modeling and simulation of variable structure systems into three categories. Besides agent-based approaches, they introduce local structure variability that can be handled by switching equations and complex structural changes that change subsystem (subagent) structure and couplings, so requires an overall model update. Although the earlier kind of variability is supported in various equation-based modeling environments, the later one is still an active research area of multimodeling. Recent efforts from the simulation of technical systems are Pawletta et al. (2016) and Durak et al. (in preparation).

An important advancement in agent-based simulation is the consideration of the full synergy of agents and simulation. This full synergy has been studied, as agent-directed simulation, by Yilmaz et al. (2016), Yilmaz and Ören (2010), and Ören (2001). In the full synergy of agents and simulation, there are three important possibilities (Yilmaz and Ören, 2009):

1. Contribution of simulation to agents. This is simulation of systems modeled by using agents or agent-based simulation as most studies are done.

2. Contribution of agents to simulation which itself offers two other important possibilities: agent-supported simulation and agent-monitored simulation (Ören and Yilmaz, 2012).

3. Agent-monitored simulation is especially important in monitoring submodels of multimodels as well as dynamic couplings during the execution of simulation.

Agent-based modeling in the context of multimodeling needs to be explored in further detail and continues to be a subject of ongoing research.

CONCLUSIONS AND FUTURE STUDIES

Social systems, as very important complex systems, have been studied intensively by social scientists. Society with all its complexities requires a multidisciplinary approach for social systems engineering. Incorporating modeling and simulation in social systems engineering is a preferred methodology due to the benefits that modeling and simulation provide at various levels. However, modeling and simulation of a social system is a hard problem and an active area of research. Although modeling deals with multidisciplinary and multi-paradigm approaches, the simulation aspects bring computation into the mix. CSSE will bring a new era for the future management of social systems and will aid social scientists with advanced computational techniques. Instead of performing experiments on the social systems and gaining on-the-job experience, simulation can be used for decision-making in complex social issues. For example, pilots are trained on simulators, before they are allowed to fly airplanes; future decision makers for complex social systems may benefit from similar trainings. A methodology to conceive and model complex reality, namely multimodeling methodology, is discussed in this chapter for simulation-based CSSE. Thirty types of models are presented and their usage in CSS engineering is described. Some of them can be used for inducing emergence in the overall CSS model. We also discussed the state-of-the-art in agent-based modeling toolset and found that most of them do not yet exploit multimodeling as a means to model induced emergence. Therefore, the field is open for advancement of agent technology by multimodels and dynamic couplings. Implementations at several aspects may provide more adequate tools to understand induced emergence.

DISCLAIMER

The author's affiliation with the MITRE Corporation is provided for identification purposes only and is not intended to convey or imply MITRE's concurrence with, or support for, the positions, opinions, or viewpoints expressed by the author. Approved for Public Release, Distribution Unlimited (Case: PR_17-3254-4).

REFERENCES

Abar, S., Theodoropoulos, G.K., Lemarinier, P., and O'Hare, G.M. (2017) Agent based modelling and simulation tools: a review of the state-of-art software. *Computer Science Review*, **24**, 13–33.

Arbib, M.A. (1969) *Theories of Abstract Automata*, Prentice-Hall, Englewood Cliffs, NJ.

Ashby, W.R. (1956) *An Introduction to Cybernetics*, Wiley.

Borshchev, A. and Filippov, A. (2004) Anylogic—multi-paradigm simulation for business, engineering and research. The 6th IIE Annual Simulation Solutions Conference, Orlando, FL.

Bouanan, Y., Zacharewicz, G., and Vallespir, B. (2016) DEVS modelling and simulation of human social interaction and influence. *Engineering Applications of Artificial Intelligence*, **50**, 83–92.

Cioffi-Revilla, C. (2014) *Introduction to Computational Social Science*, Springer-Verlag, London.

Conte, R. (2012) Manifesto of computational social science. *The European Physical Journal Special Topics.*, **214** (1), 325–346.

De Wolf, T. and Holvoet, T. (2005) Emergence versus self-organization: different concepts but promising when combined. *Lecture Notes in Artificial Intelligence*, **3464**, 1–15.

Doherr, K.-F. (2002) Parachute Flight Dynamics and Trajectory Simulation. DocSlide, http://docslide.net/documents/10-parachute-flight-dynamics-and-trajectory-simulation-doher.html (accessed: 04 September 2017)

Durak, U., Pawletta, T., and Ören, T. (in preparation) Simulating variable system structures for engineering emergence, in *Engineering Emergence: A Modeling and Simulation Approach* (eds L. Rainey and M. Jamshidi), CRC Press.

Durant, W. (1935) The Story of Civilization, vol. 1 – Our Oriental Heritage. Simon and Shuster, https://archive.org/stream/storyofcivilizat035369mbp/storyofcivilizat035369mbp_djvu.txt (accessed 12 November 2017).

ED-Emergence: Etymological Dictionary – Emergence (2000), http://www.etymonline.com/index.php?allowed_in_frame=0&search=emergence (accessed: 04 September 2017).

Ferber, J. (1999) *Multi-Agent Systems: An Introduction to Distributed Artificial Intelligence*, 1st edn, Addison-Wesley Longman Publishing Co., Inc., Boston, MA.

Fishwick, A.P. and Zeigler, B.P. (1992) A multimodel methodology for qualitative model engineering. *ACM Transactions on Modeling and Simulation*, **2** (1), 52–81.

Foo, N.Y. and Zeigler, B.P. (1985) Emergence and computation. *International Journal of General Systems*, **10** (2–3), 163–168. doi: 10.1080/03081078508934879

Furnham, A., Richards, S.C., and Paulhus, D.L. (2013) The dark triad of personality: A 10 year review. *Social and Personality Compass*. doi: 10.1111/spc3.12018

Gill, A. (1962) *Introduction to the Theory of Finite-State Machines*, McGraw-Hill.

Helbing, D. (2012) Agent-based modeling, in *Social Self-Organization*, Springer, Berlin, Heidelberg, pp. 25–70.

Helbing, D. and Balietti, S. (2011) How to do agent based simulations in the future: From modeling social mechanisms to emergent phenomena and interactive systems design. SFI Working Paper: 2011-06-024.

Hu, X., Zeigler, B.P., and Mittal, S. (2005) Variable structure in DEVS component-based modeling and simulation. *Simulation*, **81** (2), 91–102.

Lazer, D. *et al.* (2009) Computational social science. *Science*, **323** (5915), 721–723.

Luke, S., Cioffi-Revilla, C., Panait, L. and Sullivan, K. (2004) Mason: a new multi-agent simulation toolkit. 2004 Swarmfest Workshop, Ann Arbor, MI.

Mittal, S. (2013) Emergence in stigmergic and complex adaptive systems: a discrete event systems perspective. *Journal of Cognitive Systems Research*.

Mittal, S. (2014a) Model Engineering for Cyber Complex Adaptive Systems, European M&S Symposium, Bordeaux, France.

Mittal, S. (2014b) Attention-focusing in activity-based Intelligent Systems, Workshop on Activity-based Modeling and Simulation, Zurich, Switzerland.

Mittal, S. and Cane, S.A. (2016) Contextualizing emergent behavior in system of systems engineering using gap analysis. Symposium on M&S of Complexity in Intelligent, Adaptive and Autonomous Systems, Spring Simulation Multi-conference, Pasadena, CA.

Mittal, S. and Martin, J.L.R. (2017) Simulation-based complex adaptive systems, in *Guide to Simulation-Based Disciplines: Advancing our Computational Future* (eds S. Mittal, U. Durak, and T. Ören), Springer, UK.

Mittal, S. and Rainey, L. (2015) Harnessing emergence: the control and design of emergent behavior in system of systems engineering. SummerSim: Summer Simulation Multi-conference 2015, Chicago, USA, 26–29 July, 2015.

Mittal, S. and Zeigler, B.P. (2017) Theory and practice of M&S in cyber environments, in A. Tolk and T. Ören, *The Profession of Modeling and Simulation*, Wiley & Sons.

Mittal, S., Durak, U., and Ören, T. (eds) (2017) *Guide to Simulation-Based Disciplines: Advancing our Computational Future*, Springer.

Mogul JC. Emergent (mis)behavior vs. complex software systems. HP Labs Tech Reports, http://www.hpl.hp.com/techreports/2006/HPL-2006-2.pdf

North, M.J., Collier, N.T., and Vos, J.R. (2006) Experiences creating three implementations of the repast agent modeling toolkit. *ACM Transactions on Modeling and Computer Simulation (TOMACS)*, **16** (1), 1–25.

Ören, T.I. (1971) GEST: a combined digital simulation language for large-scale systems. Proceedings of the Tokyo 1971 AICA (Association Internationale pour le Calcul Analogique) Symposium on Simulation of Complex Systems, Tokyo, Japan, September 3–7, pp. B-1/1–B-1/4.

Ören, T.I. (1975) Simulation of time-varying systems, in *Advances in Cybernetics and Systems* (ed. J. Rose), Gordon & Breach Science Publishers, England, pp. 1229–1238.

Ören, T.I. (1987a) Simulation: taxonomy, in *Systems and Control Encyclopedia* (ed. M.G. Singh), Pergamon Press, Oxford, England, pp. 4411–4414. (Chinese translation appeared in: Acta Simulata Systematica Sinica, 1989, 1, 60-63).

Ören T.I. (1987b) Model update: a model specification formalism with a generalized view of discontinuity. Proceedings of the Summer Computer Simulation Conference, Montreal, Quebec, Canada, 1987 July 27–30, pp. 689–694.

Ören, T.I. (1991) Dynamic templates and semantic rules for simulation advisors and certifiers, in *Knowledge-Based Simulation: Methodology and Application* (eds P.A. Fishwick and R.B. Modjeski), Springer-Verlag, Berlin, Heidelberg, New York, Tokyo, pp. 53–76.

Ören, T.I. (2001, Invited Paper) Impact of data on simulation: from early practices to federated and agent-directed simulations (eds A. Heemink, *et al.*) Proceedings of EUROSIM 2001, June 26–29, 2001, Delft, the Netherlands.

Ören, T.I. (2014) Coupling concepts for simulation: a systematic and comprehensive view and advantages with declarative models. *International Journal of Modeling, Simulation, and Scientific Computing (IJMSSC)*, **5** (2), 1430001–14300017 (article ID: 1430001). doi: 10.1142/S179396231 43000015 (online version 2014-01-21) (Invited review paper).

Ören, T.I. and Sheng, G. (1988) Semantic rules and facts for an expert modelling and simulation system. Proceedings of the 12th IMACS World Congress, Paris, France, July 18–22, 1988, vol. 2, pp. 596–598.

Ören, T.I. and Yilmaz, L. (2012) Agent-monitored anticipatory multisimulation: A systems engineering approach for threat-management training. Proceedings of EMSS'12 – 24th European Modeling and Simulation Symposium, (eds F. Breitenecker, A. Bruzzone, E. Jimenez, F. Longo, Y. Merkuryev and B. Sokolov), September 19–21, 2012, Vienna, Austria, pp. 277–282. ISBN 978-88-97999-01-0 (Paperback), ISBN 978-88-97999-09-6 (PDF).

Ören, T.I. and Yilmaz, L. (2015, Invited article) Awareness-based couplings of intelligent agents and other advanced coupling concepts for M&S. Proceedings of the 5th International Conference on Simulation and Modeling Methodologies, Technologies and Applications (SIMULTECH'15), Colmar, France, July 21–23, 2015, pp. 3–12.

Ören, T.I. and Zeigler, B.P. (2012) System theoretic foundations of modeling and simulation: a historic perspective and the legacy of A. Wayne Wymore. *Special Issue of Simulation – The Transactions of SCS*, **88** (9), 1033–1046. doi: 10.1177/0037549712450360

Ören, T.I., Ghasem-Aghaee, N. and Yilmaz, L. (2007) An ontology-based dictionary of understanding as a basis for software agents with understanding abilities. Proceedings of the Spring Simulation Multiconference (SpringSim'07). Norfolk, VA, March 25–29, 2007, pp. 19–27. (ISBN: 1-56555-313-6).

Ören, T., Mittal, S., and Durak, U. (2017) The evolution of simulation and its contributions to many disciplines. Chapter 1 of:, in *Guide to Simulation-Based Disciplines: Advancing our Computational Future* (eds S. Mittal, U. Durak, and T. Ören), Springer, pp. 3–24.

Parunak, H.V.D., Savit, R. and Riolo, R.L. (1998) Agent-based modeling vs. equation-based modeling: a case study and users' guide. International Workshop on Multi-Agent Systems and Agent-Based Simulation. Springer, Berlin Heidelberg, pp. 10–25.

Pawletta, T., Lampe, B., Pawletta, S., and Drewelow, W. (2002) A DEVS-based approach for modeling and simulation of hybrid variable structure systems, in *Modeling, Analysis, and Design of Hybrid Systems* (eds S. Engel, G. Frehse, and E. Schnieder), Lecture Notes in Control and Information Sciences 279, Springer.

Pawletta, T., Schmidt, A., Zeigler, B.P. and Durak, U. (2016) Extended Variability Modeling Using System Entity Structure Ontology within MATLAB/Simulink. 49th Annual Simulation Symposium, Pasadena, CA.

Tisue, S. and Wilensky, U. (2004) Netlogo: a simple environment for modeling complexity. International Conference on Complex Systems, Boston, MA.

Vinerbi, L., Bondavalli, A. and Lollini, P. (2010) Emergence: A new Source of Failures in Complex Systems. Third International Conference on Dependability.

Wikipedia (2015) http://en.wikipedia.org/wiki/Computational_social_science.

Wright, D.R. (2005) Finite-State Machines, CSC2015 Class Notes, North Carolina State University, http://www4.ncsu.edu/~drwrigh3/docs/courses/csc216/fsm-notes.pdf (accessed 07 July 2017).

Wymore, A.W. (1967) *A Mathematical Theory of Systems Engineering: The Elements*, Krieger, Huntington, NY.

Yilmaz, L. and Ören, T. (2004) Dynamic model updating in simulation with multimodels: a taxonomy and a generic agent-based architecture. Proceedings of SCSC 2004 – Summer Computer Simulation Conference, July 25–29, 2004, San Jose, CA, pp. 3–8.

Yilmaz, L. and Ören, T.I. (2009 – All Chapters by Invited Contributors) Agent-directed Simulation and Systems Engineering. Wiley Series in Systems Engineering and Management, Wiley-Berlin, Germany. 520p.

Yilmaz, L. and Ören, T. (2010) Intelligent agent technologies for advancing simulation-based systems engineering via agent-directed simulation. *SCS M&S Magazine. SCS M&S Magazine*, July (Invited paper).

Yilmaz, L., Ören, T., Madey, G., Sierhuis, M. and Zhang, Y.. (2016) Proceedings of the Symposium on Agent-Directed Simulation, SpringSim'16 – 2016 Spring Simulation Multiconference, Pasadena, CA, USA, April 3–6, 2016, SCS, San Diego, CA.

Zeigler, B.P. (2016) A note on promoting positive emergence and managing negative emergence in systems of systems. *Journal of Defense Modeling and Simulation: Applications, Methodology, Technology*, **13** (1), 133–136. doi: 10.1177/1548512915620580

Zeigler, B.P., Kim, T.G., and Lee, C. (1991) Variable structure modelling methodology: an adaptive computer architecture example. *Transactions of the Society for Computer Simulation*, **7** (4), 291–319.

Zeigler, B.P., Kim, T.G., and Praehofer, H. (2000) *Theory of Modeling and Simulation*, 2nd edn, Academic Press, New York.

10

APPLIED COMPLEXITY SCIENCE: ENABLING EMERGENCE THROUGH HEURISTICS AND SIMULATIONS

Michael D. Norman[1], Matthew T.K. Koehler[1], and Robert Pitsko[2]

[1] *The MITRE Corporation, Bedford, MA, USA*
[2] *The MITRE Corporation, McLean, VA, USA*

INTRODUCTION – A COARSE-GRAINED LOOK

Overview

This chapter is an attempt to create a practical synthesis of complexity science and traditional engineering: introduces a set of heuristics to engineer for emergence, explores a number of them with a simulation of unmanned vehicle (UxV) swarms, and provides taxonomy to understand the levels of autonomy in engineered UxVs.

The chapter is organized into five main sections:

- *Introduction – A Coarse-Grained Look section*[1] introduces the fundamental concepts of complex systems, many of which occur in the natural world, and how they may be united with traditional engineering concepts by leveraging observations and research from the field of complexity science,

---

[1] In the Section 1 introduction, technically specific terms will be used without a clear definition given. The precise definitions for these terms can be found in Section 2. If starting with a detailed description of terms is important to the reader, the second section should be read first, then the introduction, which provides a coarse-grained perspective.

Emergent Behavior in Complex Systems Engineering: A Modeling and Simulation Approach,
First Edition. Edited by Saurabh Mittal, Saikou Diallo, and Andreas Tolk.
© 2018 John Wiley & Sons, Inc. Published 2018 by John Wiley & Sons, Inc.

- *Definitions and a Taxonomy for Applied Complexity Science section* provides definitions for concepts such as complex systems and emergence and presents a framework for using simulation to inform systems engineering (SE),
- *Heuristics for Applying Complexity Science to Engineer for Emergence section* introduces heuristics for applying complexity science to engineer systems for emergence,
- *Unmanned Autonomous Vehicle (UxV) Swarms section* presents an unmanned vehicle swarm case as an exemplar for the integration of simulation and engineering, and
- *Operational UxV Swarms section* discusses the state of the art of engineering for emergence in the domain of UxV swarms, motivates the application of applied complexity science, and suggests a path forward.

Complex Systems and Emergence: A Path to Resilience

Why are we interested in complex systems? We are interested in them because of their remarkable abilities to adapt, evolve, and produce aggregate dynamics and/or effects that are more than simply the sum of their components' contributions or effects. These aggregate dynamics or effects are *emergent*, and they exert influences on the system and its environment that may lead to surprising system performance (positive or negative).

These aggregate dynamics or effects cannot be fully understood via examination of components in isolation. Often the emergent properties of a system will persist in a way that component interchange, adaptation, removal, or addition is of little consequence; here, we find the "ghost in the machine" (Ryle, 1949). A standing wave in a river is an example. Even though the water molecules constantly change and debris flows along with the water, the wave persists. Emergence is a fascinating topic and a unimaginably difficult problem for engineering because it is unapproachable using the traditional scientific perspective of reductionism (Kauffman, 2010). General work in applying complexity science to traditional engineering practices has been done in the fields of complex SE (Norman and Kuras, 2006; Norman, 2015) and enterprise SE (Rebovich and White, 2010). In this chapter, we try to make progress on the practical integration of complexity science and traditional engineering.

Often the concept of emergence, much like complexity, is perceived in either a positive or a negative light. The emergent and often dynamic patterns that are brought into existence can be useful, for example, the symphony of variably phased oscillations of neuronal activity that are the signature of a healthy brain. They can also be destructive, as is the case when complex networks of neurons all become entrained to oscillate in phase, creating an epileptic seizure (Jirsa *et al.*, 2014).

Not only do complex systems adapt and evolve, they do so in a self-organized manner, without any central control mechanism (Holland, 1992; Strogatz, 2004; Taleb, 2012); this is found in even one of the most complex systems, the human brain. Although cognitive neuroscientists have discovered cortical subnetworks that are relatively specialized (Tononi *et al.*, 1994), there is no homunculus instructing the brain on how to organize itself (Bar-Yam, 2002). This self-organization can come from two basic mechanisms, a self-evolutionary manner (Bak *et al.*, 1987) and from external design or engineering (Carlson and Doyle, 1999).

Adaption and evolution are not typical capabilities of complicated systems. Informally defined, a complicated system is one that is composed of many components, in which each component is only connected to a relatively small number of the other components; each component's behavior is almost entirely prescriptive, and the whole system can be understood through reductionist methods (e.g., linear analysis). Complicated systems can be easily recognized because they are engineered, deployed, and employed in discrete steps using linear analytical methods (e.g., superposition in circuit analysis).

Aristotle's eloquently put, "the whole is greater than the sum of its parts" (Von Bertalanffy, 1972), has become a hallmark of emergence and a signal of underlying complexity. Dissecting a complex system to study it loses information about the relationships that existed between the components. Many readers will undoubtedly have examples of such systems readily at hand: economies, natural ecosystems, and social networks all share a set of remarkably similar properties; most of which we are still just beginning to identify and understand. A thorough discussion of the shared properties of organically formed complex systems (Holland, 1995; Bar-Yam, 1997; Wilson, 2000; Miller, 2016) is beyond the scope of this chapter.

One shared property of complex systems that is of interest to those who would like to study and design them is their potential resilience (Gao *et al.*, 2016) to particular classes of perturbation and, in some cases, their antifragility (Taleb, 2012). In the sections that follow, we turn our attention to the process of designing and engineering systems that have emergent properties.

Engineering Resilience

Traditional Systems Engineering Traditional SE has little to offer in solving problems of complexity (Norman, 2015). The reader is encouraged to review the foundational SE literature (Norman and Kuras, 2006; Rebovich and White, 2010; Pitsko, 2014) relevant to the topic. Suffice to say, the SE process of gathering all requirements before any system development or field testing (Blanchard *et al.*, 1990) is neither realistic nor especially helpful while designing systems that need to perform in dynamic, perturbed environments (Pitsko, 2014; Norman and Koehler, 2017).

Perturbing Complex Systems Whenever we make a change to an organic or designed complex system, we are perturbing it. Examples of perturbing complex systems include introducing, changing, or removing economic; environmental; or social policies. These perturbations can come from agents within the system, emergent effects, environmental input or outside forces that are changing the composition or "physics" of the system, or applying some sort of stress to it.

Designing Complex Systems Designing complex systems, sometimes referred to as "complex systems engineering" (Norman and Kuras, 2006; Sheard and Mostashari, 2009), is not a common human skill, but the few successes have resulted in disruptions. Technological examples of disruptive, designed complex systems include the TCP/IP stack (Norman, 2015), blockchain-based technology (e.g., ethereum) (Brito and Castillo, 2013), peer-to-peer networking (Schollmeier, 2001), and Wikipedia (http://www.wikipedia.org).

A non-technological example of a designed complex system would be the United States Government as defined by the constitution. The decision by the U.S. founders to create a republic was based on the intuitive understanding of a pure democracy's emergent, self-destructive properties if executed at a large-enough scale (Bovard, 1995). Their solution for avoiding the negative emergent effects of a large-scale direct democracy was a republic of many smaller states. Thus, the founders sought to create a system that avoided the tragedy of the commons and enabled perturbations to be dealt with at a proportional scale. The founders also knew of the fragility that comes from a strict hierarchy due to its inability to deal with multi-scale complexity (Bar-Yam, 2004b). Although it may or may not be the case that the given system's designers intended to create a complex system, that is what resulted in the above examples.

Our simple goal for this chapter is to document the process of designing a set of rules, which will produce desirable emergent effects when computationally modeled. It is often the case that minimalist changes trump more interventionist ones. The next section describes the lexicon of Applied Complexity Science in explicit detail.

DEFINITIONS AND A TAXONOMY FOR APPLIED COMPLEXITY SCIENCE

Rigorous Definitions

This section provides rigorous definitions for the complexity science concepts being explored. These definitions are made explicit to properly ground the computational system design.

Complex Adaptive System

Although there is no universally agreed definition of a complex adaptive system, there are a number of commonalities that suffice for our purpose (Miller and Page, 2009; Holland, 1995):

- some number of discrete entities
- heterogeneity
- interactions
- a meaningful space
- learning/adaptation (sometimes).

So, given these components and our purpose, we define a complex (adaptive) system as a system that contains many, heterogeneous, interacting entities that exist in a meaningful space and may change their behavior over time based on their experiences. One important element of these systems is that based on the interactions among entities (positive and negative feedback) and the adaptation or change that may occur, the dynamics of the system will likely be nonlinear especially when reacting to a perturbation (Lorenz, 2000). Moreover, emergence makes these systems especially interesting.

Emergence

Emergence is often defined as the manifestation of a surprising feature of a complex system. In fact, the field of complex systems has even been characterized as the "science of surprise" (Casti, 1994). Unfortunately, the term "surprise" remains difficult to describe rigorously and remains a function of the observer. Therefore, while we will not attempt to propose a definitive definition of emergence in this chapter, we will use a working definition: an emergent phenomenon is a system-level dynamic not seen in any one, isolated component and, therefore, is not "easily" discovered, a priori, through a reductionist approach (e.g., linear superposition).

As an example, even though ants are very simple creatures, an ant colony can produce optimal foraging paths, build structures, farm, go to war, and so on (Hölldobler and Wilson, 1983). It would be difficult, if not impossible, to discover these features of the colony via a careful examination of an individual ant in isolation. It takes a number of ants, interacting within an environment, to create these colony-level features. This raises the issue of scale; how many ants are needed to create these emergent properties? Although this issue has been discussed (Anderson, 1972), it has received little systematic attention. We will discuss this in the following sections.

Often a complex system must be computationally modeled to understand the rules of interaction (and thus the agent decision space) that drive the emergent phenomenon. The advent of modern computing capabilities has enabled generative and non-reductionist approaches to modeling emergence.

It should be noted that one of the most striking features of emergent phenomena is its ability to exert influence on the system's constituent components. Consider, as an example, the myriad agents making market decisions to buy/sell a security leading to that security's dynamic market price. According to our working definition of emergence, the value of a security is an emergent entity with its own top-down influence on the agents that make up the market: dynamic price action directly influences a person's decision to buy, sell, or hold a security.

Complicated System

In contrast to complex systems, a complicated system is a sum of the constituent components, which behave in a linear manner. Here, reductionist approaches are perfectly fruitful. Moreover, the behavior of the system is predictable through a careful examination of the components and their stable combinations. A passenger car is an example of a complicated system. By examining the brake pedal, an observer can understand that it is designed to increase the pressure in the brake lines via a hydraulic system, which, correspondingly, increases the pressure of the brake pads against the rotors. Finally, this pressure causes friction dissipating kinetic energy as heat and eventually brings the vehicle to rest. Importantly, unless there is a failure of some sort, the brake pedal will only do that. The order of operations of other systems in the passenger car does not matter to the brake.

Engineering

The Academy of Engineering (www.nae.edu) stresses that engineering is the constraint-based application of science. Importantly, the academy also stresses that "to 'engineer' a product means to construct it in such a way that it will do exactly what you want it to, without any unexpected consequences" (https://www.nae.edu/About/FAQ.aspx). Clearly, this implies the engineering of linear systems, whose dynamics are predictable.

If engineering is all about the design and creation of systems that behave predictably, how then can engineering possibly make use of complexity science where a defining characteristic is that system-level dynamics are unpredictable based on an examination of components in isolation? We contend, with modern computing hardware and simulation software, one can complement existing engineering tools and methods to potentially explore the emergent properties of a system *as it is being designed*. Existing engineering tools and techniques can be used to design individual components of a system and the representation and interaction of these components can be explored with tools and techniques of complexity science to understand the corresponding emergent system-level properties.

As depicted in Figure 10.1, we see the union of the tools and techniques of engineering and complexity science as necessary for the practice of complex SE. These tools and techniques complement each other. Traditional engineering can be used to create components and then complexity science can be used to evaluate the performance of the overall system (the components in aggregate). This, in turn, may suggest changes for the individual components to improve system-level performance and the cycle begins again. Note that this process can be initiated from either the engineering or complexity science domains; system design can begin with an exploration of dynamics as a way of informing prototypical engineering efforts that seek to explicitly enable emergence.

Utility Function

It takes sufficient understanding of the overall system utility function (what does success look like) to enable and make use of emergence in complex systems. The overall system utility function is closely related to the utility functions of the individual components. As each component tries to maximize its own utility, the overall system utility function increases (for early work to create a foundation for a science of complex systems design, see (Tumer and Wolpert, 2004)). In a truly zero sum environment, this may not be possible, meaning that leveraging emergence cannot be done in all cases. As discussed by Tumer and Wolpert (2004), leveraging the emergence of

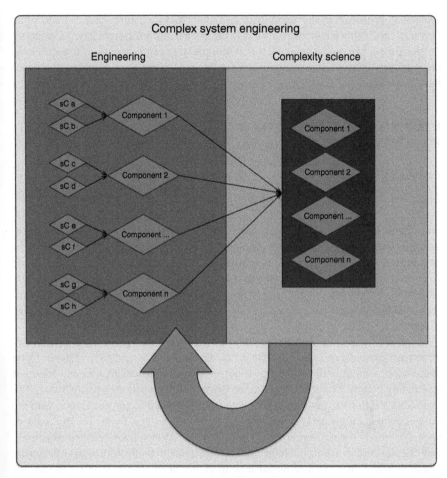

FIGURE 10.1 Complex systems engineering. The feedback between traditional engineering practice and complexity science. The emergent phenomenon exhibited at the largest scale of the system's existence (the right side of the figure) has a top-down influence on the system's components (created via traditional engineering of subcomponents (sC a–h) that are combined into larger system components (Component 1 ... n)). Traditional systems engineering practices do not pay mind to this cross-scale feedback loop, the study of which is a primary focus of complexity science.

collective dynamics can be difficult and is driven in part by the scale of the system, the control within or over the system, and system-level information available to the components.

Tumer and Wolpert (2004) continue with two more important points: the collective can be designed via model-based design (meaning the components use a model of the system for decision-making) or via a model-free design (meaning the components react to their environment without an understanding of the overall system). Finally, Tumer and Wolpert (2004) characterize the human challenge as that of a "forward" problem (trying to analyze and understand existing systems) and of an "inverse" problem (where the desire is to design a complex system from scratch). Here, we focus on the latter. Before discussing those, it is necessary to explore the decentralized system design heuristics that various pursuits in complexity science have uncovered.

HEURISTICS FOR APPLYING COMPLEXITY SCIENCE TO ENGINEER FOR EMERGENCE

Enabling Emergence

There are a number of taxonomies of complex system and emergence "types" (Maier, 1996; Bar-Yam, 2004b; Miller and Page, 2009). No matter what taxonomy one chooses to use, to date, designed complex systems fall to the less emergent and adaptive end of the spectrum compared to natural ones and represent just the beginning of a new era of design. This chapter describes a novel perspective and terminology to allow us to look further into the future of complexity and automation than we have previously been able to do. It is a synthesis of the seemingly antithetical worlds of engineering and complexity science, which we refer to as *Applied Complexity Science*.

Applied Complexity Science seeks to *engineer for emergence* in the pursuit of effectively perturbing existing systems or designing new complex systems using evidence-based methods. It should be noted that we consider making any changes an existing complex system, for better or worse, to be classified as perturbation. The methods used to these ends include computational modeling and simulation with the goal of growing antifragile complex systems in the spirit of Taleb. The hallmark of such systems is their inherent ability to not only "bounce back" from exceedingly rare or catastrophic events but to become stronger due to them. Although engineering disciplines have not yet discovered how to create developed complex systems, we propose that the most promising path is through the application of complexity science. The following heuristics are offered as design-time and implementation-time considerations to enable emergence.

Scale Matters   Things that work at one scale do not necessarily work at another. Recognition of scale leads to recognition of emergence (Anderson, 1972). Often feedback loops exist between scales. Always maintain *scale discipline* when designing and modeling complex systems. Inquiry (perturbing or designing) into complex systems often requires the scale of the question being asked to be explicitly stated, as the modeling efforts must be cognizant of and resultant solution space must accommodate that scale, as well. Scale in this context is inclusive of both size and granularity of perspective. A strict interpretation of the nuances between differing definitions of scale, scope (Ryan, 2007), and complexity (Bar-Yam, 2004c) is outside the bounds of this chapter.

A system's ability to cope with changing environmental conditions while maintaining or improving its functionality must be understood at both the scale of the component and the scale of the aggregate, and the mapping between the two must be made explicit. Strict hierarchical control systems create fragile enterprises (Norman, 2015) if the complexity at lower scales is greater than the comprehension of the leaders at the highest scale (Bar-Yam, 2004a). This has the effect of severely limiting emergent behavior and the agility, adaptability, and evolution, which enable complexity. Moreover, this "control disconnect" can lead to poor decision-making based on a misunderstanding of the actual causal relationships at lower levels of the hierarchy.

Focus on Relationships and Interactions   One central theme that ties man-made complex systems together is that they are created from the perspective of a macroscale system existing as a functionally emergent property of the relationships of the microscale components. A flock emerges from a group of birds because of their interactions. Similarly, an ant colony can farm, build structures, find optimal foraging paths, and go to war not because any given ant is a brilliant tactician, architect, or mathematician but rather because they interact in meaningful ways.

Decentralize by Default   Centralized systems are prone to developing single points of failure. By decentralizing and creating some level of redundancy, the likelihood that a system fails as the result of an attack, a natural hazard, or simply by bad luck affecting the central single point of failure is significantly reduced, so the system becomes more stable. Natural complex systems are fundamentally decentralized. It is important to note the distinction between a distributed system and a decentralized system. Simply put, a decentralized system is one in which each of the major components is under individual control; in a distributed system, this in not necessarily the case. When designing complex systems, we want to decentralize as much as is reasonably possible. Any of the complex systems previously discussed can lose virtually any single component and continue to function because the components can interact with themselves and their environment, making decisions at their scale of interaction using locally available information. If a poor decision is made, the consequences of that decision do not propagate (typically) through the system unless other components (which may henceforth be referred to as *agents*) (Holland, 2006) decide to continue its propagation. If, however, decision-making across components is centralized, one bad decision can cause a systemic failure (Taleb, 2012). In this way, complex systems are resilient to perturbation precisely because they are decentralized.

Cascading failures in complex systems can be avoided through decentralization; consider the prosumer's (a producer and consumer of a given property) act of producing electricity via solar panels to protect against the extreme complications built into the architecture of the power grid (i.e., centralized production and many hierarchical levels of distribution). The prosumer is now able to function without the rest of the system if those complications tip into complexity and lead to cascading failures and emergent blackouts. An electric grid where every consumer is also a producer creates a basic complex system that has an emergent property *of* electricity.

Allow Ambiguity   The only way to gain traction in the midst of complexity, especially when pursuing modeling efforts, is to take a coarse-grained view of the system, allowing fine-grained details to go unrealized unless, or until, they are required for capturing the emergent properties of interest (Barry *et al.*, 2009). The search for the proper level of abstraction that captures enough details of the system's relationships to reproduce the emergent properties of interest is, in itself, an art form.

As shown in "Reference modelling in support of M&S—foundations and applications," (Tolk *et al.*, 2013), for nontrivial systems, the incompleteness theorem is applicable, which means that consistent formal representations are necessarily incomplete, and complete representations are inconsistent. The recommended solution is a complete, but inconsistent, reference model that is used to derive a set of incomplete but consistent (simulation) systems. They form a set of orchestrated alternatives that allow focus on diverse facets and, like the hurricane map in the weather forecasts, present a whole picture based on the set of ambiguous models. Ambiguity allows one to consistently cover all known aspects of a system completely.

Like the "NATO Code of Best Practice for C2 Assessment/Selecting an Orchestrated Set of Tools" observes already (p. 199):

> The natural tendency of an analyst is to simplify a problem. Part of that simplification is to select a tool, preferably only one, which will meet the analysis requirements. In the analysis of [Command and Control (C2)] combat operations, this may be possible if the analysis is properly scoped. In the analysis of [Operations Other Than War] C2, the issues are typically too numerous, the variables are too confounding and the scope is too broad for one tool to satisfy all analysis requirements. An orchestrated set of complementary tools will normally be required.

Decide Locally Using a Model-Free Design To enable emergence, components need to make decisions without a model of the entire system; we refer to this as model-free design. This means that data that is available in the immediate environmental context is all that is required for an agent to make decisions.

More data means more aggregation time. If data must be transmitted to and then processed in a centralized location, where results of processing then have to be retransmitted to be made actionable, value is lost due to the delays across transmission, aggregation, and action. When thinking about data usage in a designed complex system, decentralized data fusion (Durrant-Whyte *et al.*, 1990), in which data is stored and accessed in a completely federated manner, may be required. Furthermore, increased data may simply result in more noise, more non-constructive redundancy and either increased cognitive load on an operator or agent or further layering of machine-based assumptions and may result in making all parts worse off (Tumer and Wolpert, 2000).

Incorporate Dynamic Modeling (As Part of the Complex System Design Process) The system itself should be designed in an executable modeling environment, so that the effects of decentralized decision-making, changing environmental conditions, nested feedback loops, and other dynamics that display hysteresis and path dependence will become apparent and may be analyzed. The modeling effort is not done to make predictions about the future, but to build insights into behavior and performance, as well as to actively probe the fragility of the system under design.

Certain situations may even require the use of active modeling at run-time within the complex system itself; consider as biological analog to this the emergent complexity of a human brain anticipating a social event.

Understand Emergent Properties (i.e., Resilience) Resilience is often described as the ability of a system to bounce back from a shock (Zolli and Healy, 2013). Although this could be an apt description of the aggregate or emergent behavior we see as a complex system interacts with its environment, it is not a useful description if the goal is to design a resilient system. As we begin to prepare tomorrow's systems engineers to directly grapple with the inevitable emergent properties of the systems they create, they may be able to leverage emergent properties to acquire functionality that would otherwise be unavailable. These properties may include inter alia, resilience/robustness to certain classes of stressor/perturbation, and antifragility.

Complexity Is Not Always the Answer

To be clear, if a problem space is of a limited scale, that is, occasional system-wide failures are acceptable, then a solution does not have to take the form of a complex system. A complication-based solution may be far more cost effective given global resource-scarcity conditions and is acceptable when a failure in the system would not propagate any further into the complex network in which the system is already situated so that the emergent properties of the aggregate network are not compromised. Hierarchical centralization of control in a complex system is perfectly suitable so long as it is nested at the scale of an expendable component or set of components (Norman, 2015).

A Starting Point

No matter where a system's boundaries are drawn, the starting point for designing a system that functions without centralized control is not clear. What is clear is the need to capture the relationships of consequence to the questions being explored by modeling the structure and dynamics. We are now nearing the edge of the possible, and the rest of this chapter will explore one method of using simulation to leverage emergence, the essence of designing a complex system by applying complexity science. Table 10.1 represents an attempt to create a complexity-based taxonomy of autonomous systems to aid in further discussion.

The first column of Table 10.1 indicates the level of autonomy. The second column explains the properties a system must possess to achieve that given level. The third column lists the Applied Complexity Science design heuristics that would need to be employed in the creation of a decentralized system capable of achieving that level of autonomy.

In the examples that follow, we focus on the design of unmanned vehicle swarms (UxVs). We assume that the engineering of a specific component of Figure 10.1 is successful. The focus of this chapter is on the use of the tools and techniques of complexity science and the corresponding feedback into the traditional engineering

TABLE 10.1   A Taxonomy of Applied Complexity Science

Level of Autonomy	Properties of System	Heuristics Employed in System Design
0	• Micromanagement of systems • Distributed design 　■ Preprogrammed travel paths	None
1	• Emergence levered via local decision-making for navigation of agents in system space 　■ Tactical goals established externally	1–3
2	• L1 + decentralized design at large scale (including decentralized decision-making using local information)	1–7
3	• L2 + agents have limited ability to vary parameters of their own rules of interaction • May be based on competition of internal agent models to guide *adaptation* (self-induced change of the agent)	1–7
4	• Antifragile complex systems capable of designing themselves • L3 + agents have ability to change themselves and propagate/fuse those changes to/with others • Rate of system *evolution* (self-induced change of a system-wide/emergent property) may be artificially accelerated (by either internal or external influence); "policy vote" 　■ Altering aggregate goals in response to environmental perturbations • Generate a collective solution when confronting a problem at a collective scale; that is, *emergent computation* • Future swarm strategies are formed and tested via emergent computation across the agent base	1–5, 6 (applied recursively), 7

process that may occur. We also assume that the system must be able to scale and that we have little control over the environment. This means that each member of the UxVs will not be able to have a model of the larger system available for local decision-making. Therefore, the UxVs will need to be designed in a model-free way, with the swarm components reacting to each other and their environment without global knowledge and without the ability to communicate with all other members of their swarm. This presents a challenge, as we must create a set of rules and incentives for the members of the swarm so that as selfish maximizers, they will create a high-performance collective capable of emergence.

UNMANNED AUTONOMOUS VEHICLE (UxV) SWARMS

The example complex system designed in this chapter is a swarm of autonomous unmanned vehicles (UxVs). The agent-based modeling platform NetLogo 6.0.1 is used for this purpose. Although we use the UxV example to explore many of the ideas discussed above, we cannot cover them all. The examples in this chapter explore the following heuristics of Applied Complexity Science: scale matters, focus on the relationships, decentralize by default, allow ambiguity, use dynamic modeling, understand emergent properties, and decide locally using model-free design.

The future operational environment will make use of increasingly large numbers of increasingly autonomous entities. As scale increases, it is not possible for a human to command each entity directly. With the highly prototypical Perdix effort, the Defense Advanced Research Projects Agency (DARPA) has signaled the research community's inability to scale manpower across a swarm and the desire to enable swarm autonomy (Perdix Fact Sheet). Available artifacts around the Perdix program dated as recently as January 2017 indirectly claim that the system exhibits two emergent properties: the ability to adapt to some amount of swarm population dynamics and the ability for swarm agents to navigate without prespecified paths but with individually prespecified goal locations.

Based on qualitative analysis of publicly available videos of recent tests of the Perdix swarm and associated command and control, we believe that the Perdix program achieves autonomy somewhere between levels 0 and 1 on the Taxonomy of Autonomy in Applied Complexity Science, as presented in Table 10.1. Any further specifics on higher levels of autonomy planned for the Perdix program are currently unknown.

As a one-to-many mapping of operator-to-swarm agents seems to be the direction of current research interest, one of our goals is to discover what that mapping looks like and begin to ask systematic and informed questions around operational teaming.

Emergence in Swarms - Why Is a Swarm Resilient?

There exists a tendency to treat a swarm, which is decentralized, as a distributed system. A swarm is decentralized because its components possess agency. We define agency to mean the capacity for an actor to act in a given environment, and implicitly an agent is endowed with certain degrees of freedom and decision-making responsibilities within them. At scale, this agency drives emergent properties that transcend properties of individual membership. This means the swarm is resilient to almost any localized perturbation. This is brought about by the fact that each agent is operating against only a local data picture (i.e., it is a model-free system), accepting (or oblivious to) the continuous uncertainty of nonlocal events; this creates space for emergence to be leveraged and avoids the problem of a system that can fall out of synchronization.

One important emergent property of swarms is resilience: disruptions at the local level do not lead to disruptions at the aggregate level. If the tipping point for swarm behavior has been reached (e.g., density of agents), the entities will begin to interact and make increasingly autonomous decisions about mission execution and these emergent properties of the swarm will persist and demonstrate path dependence.

To design a swarm that can demonstrate useful emergent properties, we start by determining how many swarm agents are required and what their relative density must be for emergent properties to be observed.

Simulating Emergence and Exploring Applied Complexity Science Heuristics

The design of individual UxVs is generally understood and consistent with current system engineering tools and methods. The design of swarms of UxVs, on the other hand, is less generally understood. This section focuses on developing the understanding required to begin to engineer how UxVs interact to enable emergence and set conditions for the swarm to be resilient to perturbations. The Applied Complexity Science heuristics are noted in bold font as this example is developed.

Modeling a swarm in NetLogo (*incorporate dynamic modeling*) is an exercise in *allowing ambiguity* because we are abstracting the engineering specifics of the component and resultant system to get our hands around the relationships and consequent dynamics (e.g., a radio connection between two devices may be modeled as a logical link so long as the questions being asked of the system do not pertain to spectrum management).

In the initial simulation, we test a set of canonical relationships (*focus on the relationships*) used to create a "boid flock" (Reynolds, 1987). This was used as a starting point because of the emergent pattern formation that occurs as a result of these simple rules. This minimalist approach to modeling collective behavior has produced dynamics with remarkably similar characteristics to the patterns displayed by natural flocks of birds, so it is assumed that the essence of the emergent flocking behavior has been captured, despite the intentional absence of other bird-specific details that are not believed to have an impact on the birds' relationships to one another.

We construct our model around these relationship rules: cohesion (the tendency of the UxVs to cluster together), separation (the tendency of the UxVs to not get too close together), and alignment (the tendency for the UxVs to move in the same direction). Each agent makes decisions for itself with information only about its immediate environment (*decide locally*).

The flocking behavior that the model produces is a result of completely decentralized and model-free sensing and decision-making (*decentralize by default*); the flock is an emergent entity and as such should be resilient to particular classes of perturbation (*understand emergent properties*) such as the removal of any given UxV.

Swarm coherence is defined here to be represented by the variance of headings across the group; it is assumed that this is a dynamic metric whose value will vary over time. A high swarm coherence would be represented by a low variance and a low swarm coherence by a high variance. It should be noted that a swarm's variance should never be too close to zero, as the dynamics of a swarm would be destroyed in such a situation, given the fluid nature of swarm behavior within a finite environment.

Using these rules, we then explore how the density of UxVs impacts the creation of a coherent swarm (defined as most swarm elements moving in the same direction). One UxV does not make a swarm (*scale matters*). On the other hand, if there are too many UxVs, then it will be too hard for them to find a common direction and not get in each other's way. For this experiment, we held the number of UxVs constant at 250 and varied the size of the landscape from 31×31 patches to 101×101 patches (here a "patch" is a square unit of terrain). We find that swarms form very quickly at a density of 250 UxVs in a 51×51 patch size environment. As the size of the environment is increased, it takes the swarm longer and longer to cohere. As would be expected, this general trend is seen across all runs for a given environment size, and overlap may exist when comparing performance of two individual runs from slightly different-sized environments.

Figure 10.2 depicts the resulting swarm dynamics in a collective of 250 agents as the region size is increased. This baseline helps us understand the potential impacts of design decisions. Figure 10.2 shows how changes in the density of UxV impact the ability of the UxVs to form a coherent swarm. Intuitively, we suspect that denser groups will produce emergent swarms more quickly than sparser groups. Here, we define a coherent swarm when the variance of UxV headings is very low or very high (high variance is an artifact created when the swarm is heading north, causing headings to fluctuate between 0 and 360 degrees). In all the diagrams below, the UxVs exist in a toroidal landscape that varies from 31×31 patches to 101×101 patches in 10 patch increments. Heading variances below the gray horizontal line indicate a coherent swarm.

What happens when we perturb the system? Building on the UxVs example, we use the simulation to understand the impact of perturbations on swarm performance. For example, what happens if there are obstacles in the swarm's environment? We now know that a swarm forms very quickly at a density of 250 UxVs in a 51×51 patch. Using that density, what happens as we increase the density of obstacles? At what point does the swarm break down?

After understanding the engineering constraints revolving around building a swarm in isolation, an exploration of the swarm in a representative environment is performed by adding obstacles. As we add obstacles that take up 1.1% of the space, it very quickly begins to destroy the swarming dynamics. By about 5%, the swarm headings are largely random. Very low variance indicates a coherent swarm.

If we envision a military application of this swarm, it is not difficult to imagine that 5% of the airspace in which the swarm would be operating could be restricted (airports, civilian centers, other governmental facilities, etc.). This being the case, it is worth noting that the current set of rules may not create a coherent swarm under those circumstances. However, more careful analysis of the potential obstacles should be undertaken.

In the first few simulations, obstacles are placed randomly within the landscape. We find that variance tends to increase with the number of obstacles. Highly coherent swarms are not found if four or more obstacles are introduced, as is indicated by the horizontal gray line in Figure 10.3.

In Figure 10.4, we see the impact of obstacles that are highly organized. Having organized obstacles (i.e., obstacles placed in a line) significantly disrupts the coherence of the swarm more quickly than do the unorganized obstacles (i.e., obstacles placed randomly within the environment). However, unexpectedly, as the organized object grows in size, new structures develop in the swarm dynamics. Looking at obstacle sizes 2-7, there is very little structure in the heading plots (Figure 10.4, c3–h3). However, as the size of the obstacle becomes large enough to effectively create a barrier within the swarm's area of operations, new structures emerge within the swarm dynamics as indicated by the exaggerated vertical components of both the variance and the heading plots. This is an emergent effect of the swarm avoiding the large obstacles and turning to move into free space.

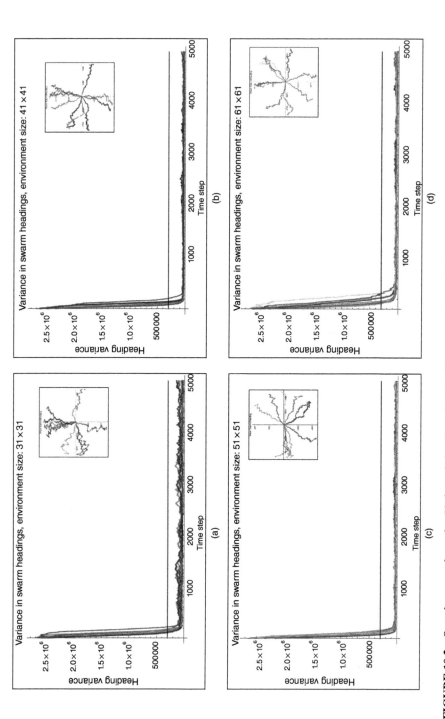

FIGURE 10.2 Sample swarm dynamics. Variance in heading of the U × V swarm as density is decreased by holding swarm size constant and varying environment from a 31 × 31 patch size to a 101 × 101 patch size in increments of 10. Swarms form very quickly at a density of 250 U × Vs in a 51 × 51

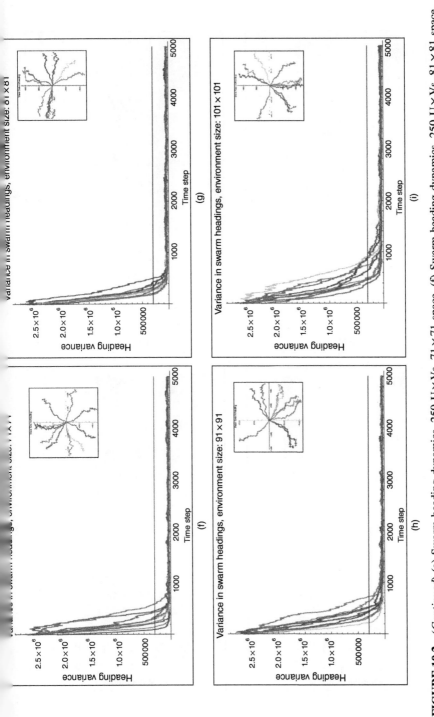

FIGURE 10.2 (*Continued*) (e) Swarm heading dynamics, 250 U × Vs, 71 × 71 space. (f) Swarm heading dynamics, 250 U × Vs, 81 × 81 space. (g) Swarm heading dynamics, 250 U × Vs, 91 × 91 space. (h) Swarm heading dynamics, 250 U × Vs, 101 × 101 space.

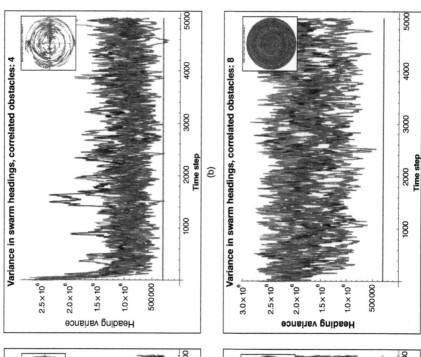

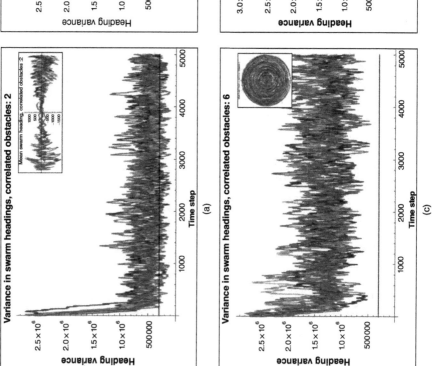

What is interesting from this exploration of swarming and obstacles is that different "types" of obstacles impact the swarm differently at different scales. Unorganized obstacles act like noise monotonically causing the coherence of the swarm to break down. Organized obstacles, on the other hand, more quickly cause degradation in swarm coherence, but as their scale continues to grow the obstacles, create new coherence within the swarm.

By way of a final experiment, suppose the goal is to follow a moving object and have the UxVs spread out around it. This might be the case if there was a need to create a mesh network around a set of ground forces to relay communications in a very cluttered environment. An engineer might first try biasing the swarm by first having them all set their heading toward the ground unit and then run the swarm algorithm. Unfortunately, this rule is too strong and collapses the swarm, at best creating a line of UxVs trailing the ground units. To fix this, a rule could be added to the UxVs to spread out so that they are no closer to each other than some fixed amount. Unfortunately, such a rule does not preclude UxVs from leaving the area, thus limiting the potential of the desired UxV-based communications network. An obvious fix, therefore, is to add another rule stating that each agent should endeavor to keep at least one UxV in communications range. However, once simulated, it becomes clear that this "fix" does not preclude pairs of UxVs from leaving the group together.

The above solutions are examples of how subtle micromanagement of an emergent phenomenon can add fragility to the system. Rather than adding to the swarm's rules, fine-tune adjustments to the existing rules may be able to be made that engender desired emergent behavior while not adding unforeseen fragility to the system.

Assuming that the UxVs move faster (in these simulations, they are required to move approximately four times, or more, faster) than the forces they are following, then the current swarm rules are usable.

First, the direction of the ground forces is included as part of the swarm heading. Second, the minimum separation distance is set to be quite high (this, in turn, may require changes to the sensor platform of the UxVs as they must be able to "see" farther than their separation distance). Finally, the maximum turn the swarm can make is set to align to roughly double that of cohere and separate turns. This generates a robust swarm that follows the ground forces and does a "good" job of spreading out around them to create a mesh network. It does not, however, guarantee coverage always as the swarm may overshoot the ground units and need to circle back.

By creating this swarm ruleset, emergence is leveraged to engineer a complex system.

FIGURE 10.3 Highly coherent swarms. The impact of unorganized obstacles on swarm dynamics, as can be seen, variance tends to increase with the number of obstacles. Highly coherent swarms are not found if 4 or more obstacles are introduced, as is indicated by the horizontal gray line. (a) Variance of heading, 250 U $\times$ Vs in a 51 $\times$ 51 world, correlated obstacle of size 2. (b) Variance of heading, 250 U $\times$ Vs in a 51 $\times$ 51 world, correlated obstacle of size 4. (c) Variance of heading, 250 U $\times$ Vs in a 51 $\times$ 51 world, correlated obstacle of size 6. (d) Variance of heading, 250 U $\times$ Vs in a 51 $\times$ 51 world, correlated obstacle of size 8.

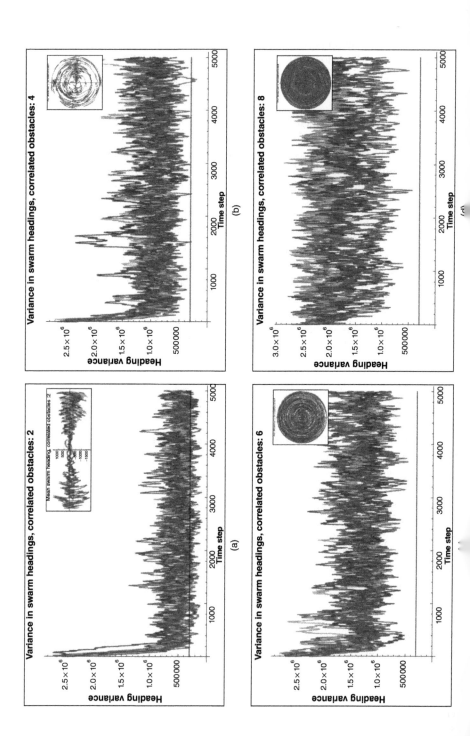

OPERATIONAL UxV SWARMS

Near-operational military swarming capabilities in the true sense of the term are nonexistent. Experimental military functions are limited to centrally controlled or distributed configurations that are micromanaged by one or more human operators or limited in scale to tens of vehicles (Wolf et al., 2017). These distributed systems assume lossless transmission, complete information, shared decision spaces (maintaining a shared decision space is an incredible burden), and limitless power supplies; none of which are available in the real world. As such, emergence has not been *designed* into any operational UxV teams, although one prototypical research effort, Perdix, may be working toward that goal.

Other governmental efforts to create "swarms" have been distributed systems of very small scale, not decentralized swarms as presented in this chapter. If we were to design a system from the perspective of traditional engineering a distributed system, we would want to put identical decision-making software on each agent and give each agent access to the identical sensor/data set for continuity of decision-making, with communication channels between all agents, carrying data about everything known by all of the agents. Each agent would require this global data picture for the system to function as intended, and real-world complications such as time delay and environmental perturbations expose the fragility of a distributed system design that does not enable emergence. In a system such as this, almost any perturbation could cause the unified situational picture to become corrupted and individual agent perspectives to drift, creating divergence in the group's planning due to wrongly perceived differences in the environment. Relying on a unified global data picture is a fragile design decision. Government command and control architectures were not designed with emergent swarm behavior in mind.

DARPA's Perdix effort's contribution to leveraging emergence, if descriptions are accurate, is summarized as follows:

> Controlling 100 drones individually would be overwhelming, so much like a sport coach, operators call "plays" (e.g., surveilling a field) and Perdix decides how best to run them. Because Perdix cannot change their plays, operators can predict the swarm's behavior without having to micromanage it.

Bees are believed to participate in emergent computation during colony migration planning events (Makinson and Beekman, 2014) to guide their aggregate behavior.

FIGURE 10.4   Highly organized obstacles. The impact of organized obstacles on swarm dynamics. We can see that as the organized obstacles become large enough, new emergent swarm dynamics form within the system's new boundaries. (a) Variance of heading, 250 U × Vs in a 51×51 world, correlated obstacle of size 2. (b) Variance of heading, 250 U × Vs in a 51×51 world, correlated obstacle of size 4. (c) Variance of heading, 250 U × Vs in a 51×51 world, correlated obstacle of size 6. (d) Variance of heading, 250 U × Vs in a 51×51 world, correlated obstacle of size 8.

Expert opinion based on the available information is that the Perdix do not decide where to navigate because of decentralized design and/or emergent computation, but are given specific coordinates and specific operational objectives as part of their play-calling. This means the next clear goal for Applied Complexity Science in this space will be to increase swarm autonomy to a solid level 1, and then to level 2, as shown in Figure 10.1.

DISCUSSION AND CONCLUSIONS

Modern computing hardware and simulation software can complement existing engineering tools and methods and, potentially, be used to explore the emergent properties of a system *as it is being designed.*

Autonomous swarms must be imbued with the agency to make decisions in a decentralized manner to truly be autonomous and leverage the problem-solving power of emergent behavior, as well as perhaps glean the benefits of resilience and antifragility.

As scale increases, a complex system inevitably goes through transitions of both structure and function. It is difficult to believe that a swarm of two hundred and fifty will behave the same as a swarm of one million. Consider as an analogy the government of Switzerland. It is speculated that Switzerland's decentralized, direct democracy is what has allowed it to remain so stable while still existing at the scale of a nation-state, which has historically not been the most stable (Taleb, 2012). The Swiss system is empowered to make highly autonomous decisions at multiple scales due to Swiss Canton sovereignty. As a result, perturbations are handled at their own scale. The system adapts to the perturbations locally and prevents perturbations from cascading through the system. Therefore, the systemic fragility created by a strong centralized control mechanism never arises. An extremely large swarm should be something like Switzerland. Dynamic modeling is required to begin to implement such a system. Currently, there are no true swarm capabilities being researched or implemented by any governments or militaries we are aware of.

To fully enable and leverage emergence for swarms, one of the requirements is the need to transition from a distributed design to a decentralized design as a default (level 2 of Table 10.1). Distributed design does not necessarily involve more than one control entity, whereas decentralized design necessitates that decisions be made based only on locally available data, which means the perspective of one agent should not be identical to the perspective of another. The emergent coherence of the swarm is a result of myriad decisions made at the scale of each agent, the complexity of which becomes difficult to understand completely.

For the sake of discussion, swarm coherence is defined here to be represented by the variance of headings across the group; it is assumed that this is a dynamic metric whose value will vary over time. A high swarm coherence would be represented by a low variance and a low swarm coherence by a high variance. It should be noted that a swarm's variance should never be too close to zero, as the dynamics of a swarm would be destroyed in such a situation, given the fluid nature of swarm behavior within a finite environment.

TABLE 10.2 Swarm Coherence as a Function of Density for Environment Sizes 30×30 to 100×100

Environment Size	Minimum Heading Variance (Average Over 10 Runs)	Earliest Time Step (t) for Minimum Heading Variance (Average Over 10 Runs)
31×31	21 289.05	3183.6
41×41	18 719.23436	2338.3
51×51	17 792.29474	2697.6
61×61	18 808.29094	2565.6
71×71	19 366.58371	2914.7
81×81	20 670.87091	3850.5
91×91	21 400.06905	3082.2
101×101	22 387.4731	3853.3

The first experiment involves setting the basic foundations for exploring the swarm coherence dynamics of the canonical boids implementation. Here, the focus is on discovering the impact of swarm density on swarm coherence, without introducing internal obstacles/perturbations. From Table 10.2, it is clear that the size of the environment has a huge impact on swarm coherence. With an environment size of 100 patches, there is no convergence to a globally coherent pattern. As the swarm's density is decreased by holding the number of agents constant at 256, but the size of the environment is increased to 400 patches square, there is an unusual pattern that seems to be indicative of phase transition in swarm dynamics. Realistically, this is interpreted to be just enough space for some swarming but with enough other agents in the way that the "separate" function adds a lot of noise. As the size of the environment is increased to 30×30, it is clear that the change in swarm density is causing another phase transition in the system's dynamics to occur. For the first time, we see coherence across all experimental runs with these parameters. Coherence is specifically defined here as a swarm heading variance less than 250 000, and it is shown in Figure 10.2 as a horizontal gray line. This trend of convergence tightens until it peaks in a 50×50 environment, where the heading coherence of the entire 250 agent swarm dips below 18 000 in only 2697.6 time steps. It could also be argued that the 40×40 patch set of runs produced a highly performant swarm due to the incredibly tight variance of 18.7 K achieved in only 2338.3 time steps. As environment size grows from 60×60 onward, swarm convergence time and variance increase almost monotonically.

The major takeaway is the evidence that a "sweet spot" for swarm density exists for achieving rapid swarm coherence. High-density swarms do not function well, and low-density swarms take much longer to form.

As noted earlier, this "sweet spot" changes based on the types of environmental perturbations present. Future work will look to enable adaptation (level 3 of Table 10.1) of the agents so that goal swarm coherence metrics can be achieved across varied environmental conditions.

The system demonstrates remarkable resilience to perturbation because it is *decentralized*. This experimentation demonstrates that swarm coherence is destroyed when all swam members are ordered to follow a single target, even temporarily. It was found that nth order effects on swarming coherence are least adversely impacted by engineering efforts if the parameters of the canonical boid relationships are altered, rather than introducing new relationships to the system. In effect, preserving simplicity at the individual level, the system exhibits more stable complex behavior at the aggregate level. Future experimentation will examine the effects of increasing levels of autonomy and the implementation of decentralization of higher order functions to ascend through the levels shown in Table 10.1.

It is worth noting that during modeling, it was decided to allow ambiguity and not concern the initial effort with the details of physical propulsion, for example. Fine-grained attributes will be introduced once further paths for uncovering value in application are discovered.

Local data were used to make local decisions (*decide locally*) in a model-free context; the swarm agents never receive control signals from other agents; they merely make decisions based on the activity in their immediate vicinity.

All of this was made possible with *dynamic modeling*, which demonstrates that the emergent swarm itself is resilient to perturbation, even though individual swarm agents are not (*understand emergent properties*).

Unanticipated emergent effects are apparent in the initial attempts to influence and guide aggregate functionality. A better understanding of emergent properties is needed to approach this. Currently, the best methods are trial and error, but future work will address the need to create a repeatable process around this activity. This experimentation created a follower mesh network, requiring rule changes, rather than additions or removal to achieve stability of function.

Future work will involve the development of a more rigorous definition of swarms – one that incorporates more than variance/heading correlations. This will provide more rigorous metrics to evolve toward higher levels of autonomy. We anticipate being able to use this technology for various effects and plan to develop novel rulesets for communications jamming (e.g., cluttering up a radar tower) as well as protection (e.g., protect fighter from surface-to-air missile array).

DISCLAIMER

The authors' affiliation with The MITRE Corporation is provided for identification purposes only and is not intended to convey or imply MITRE's concurrence with, or support for, the positions, opinions, or viewpoints expressed by the authors. Approved for public release: Case: 17-2641.

REFERENCES

Anderson, P.W. (1972) More is different. *Science*, **177** (4047), 393–396.

Bak, P., Tang, C., and Wiesenfeld, K. (1987) Self-organized criticality: an explanation of the 1/f noise. *Physical Review Letters*, **59** (4), 381.

Barry, P.S., Koehler, M.T. and Tivnan, B.F. (2009) Agent-directed simulation for systems engineering. Proceedings of the 2009 Spring Simulation Multiconference. Society for Computer Simulation International, p. 15.

Bar-Yam, Y. (1997) *Dynamics of Complex Systems*, vol. 213, Addison-Wesley, Reading, MA.

Bar-Yam, Y. (2002) General features of complex systems, in *Encyclopedia of Life Support Systems (EOLSS)*, UNESCO, EOLSS Publishers, Oxford, UK.

Bar-Yam, Y. (2004a) *Making Things Work*, Knowledge Industry.

Bar-Yam, Y. (2004b) A mathematical theory of strong emergence using multiscale variety. *Complexity*, **9** (6), 15–24.

Bar-Yam, Y. (2004c) Multiscale variety in complex systems. *Complexity*, **9** (4), 37–45.

Blanchard, B.S., Fabrycky, W.J., and Fabrycky, W.J. (1990) *Systems Engineering and Analysis*, vol. 4, Prentice Hall, Englewood Cliffs, NJ.

Bovard, J. (1995) *Lost Rights: The Destruction of American liberty*, Palgrave Macmillan.

Brito, J. and Castillo, A. (2013) *Bitcoin: A Primer for Policymakers*, Mercatus Center at George Mason University.

Carlson, J.M. and Doyle, J. (1999) Highly optimized tolerance: a mechanism for power laws in designed systems. *Physical Review E*, **60** (2), 1412.

Casti, J.L. and Ford, J. (1994) Complexification. *Nature*, **371** (6495), 296.

Durrant-Whyte, H.F., Rao, B.Y.S. and Hu, H. (1990, May) Toward a fully decentralized architecture for multi-sensor data fusion. Robotics and Automation, 1990. Proceedings, 1990 IEEE International Conference on, IEEE, pp. 1331–1336.

Gao, J., Barzel, B., and Barabási, A.L. (2016) Universal resilience patterns in complex networks. *Nature*, **530** (7590), 307–312.

Holland, J.H. (1992) Complex adaptive systems. *Daedalus*, **121** (1), 17–30.

Holland, J.H. (1995) *Hidden Order: How Adaptation Builds Complexity*, Basic Books.

Holland, J.H. (2006) Studying complex adaptive systems. *Journal of Systems Science and Complexity*, **19** (1), 1–8.

Hölldobler, B. and Wilson, E.O. (1983) Queen control in colonies of weaver ants (Hymenoptera: Formicidae). *Annals of the Entomological Society of America*, **76** (2), 235–238.

Jirsa, V.K., Stacey, W.C., Quilichini, P.P. *et al.* (2014) On the nature of seizure dynamics. *Brain*, **137** (8), 2210–2230.

Kauffman, S.A. (2010) *Reinventing the Sacred: A New View of Science, Reason and Religion*, Basic Books, New York.

Lorenz, E. (2000) The butterfly effect. *World Scientific Series on Nonlinear Science Series A*, **39**, 91–94.

Maier, M.W. (1996, July) Architecting principles for systems-of-systems. INCOSE International Symposium, vol. 6, no. 1, pp. 565–573.

Makinson, J.C. and Beekman, M. (2014) Moving without a purpose: an experimental study of swarm guidance in the Western honey bee (Apis mellifera Linnaeus). *Journal of Experimental Biology*, **217**, 2020–2027 10.1242/jeb.103283.

Miller, J. (2016) *A Crude Look at the Whole: The Science of Complex Systems in Business, Life, and Society*, Basic Books.

Miller, J.H. and Page, S.E. (2009) *Complex Adaptive Systems: An Introduction to Computational Models of Social Life*, Princeton University Press.

Norman, M.D. (2015) Complex systems engineering in a federal IT environment: lessons learned from traditional enterprise-scale system design and change. Systems Conference (SysCon), 2015 9th Annual IEEE International, IEEE, pp. 33–36.

Norman, M.D. and Koehler, M.T.K. (2017) Cyber defense as a complex adaptive system: a model-based approach to strategic policy design. In the Conference of the Computational Social Science Society of the Americas (CSS 2017).

Norman, D.O. and Kuras, M.L. (2006) Engineering complex systems, in *Complex Engineered Systems* (eds D. Braha, A. Minai, and Y. Bar-Yam), Springer, Berlin Heidelberg, pp. 206–245.

Pitsko, R. (2014) Principles for Architecting Inherently Adaptable Complex Systems. Ph.D. Dissertation at the Stevens Institute of technology, Hoboken, NJ.

Rebovich, G. Jr. and White, B.E. (eds) (2010) *Enterprise Systems Engineering: Advances in the Theory and Practice*, CRC Press.

Reynolds, C.W. (1987) Flocks, herds and schools: a distributed behavioral model. *ACM SIGGRAPH Computer Graphics*, **21** (4), 25–34.

Ryan, A.J. (2007) Emergence is coupled to scope, not level. *Complexity*, **13** (2), 67–77.

Ryle, G. (1949) *The Concept of Mind*, Barnes & Noble, New York.

Schollmeier, R. (2001, August) A definition of peer-to-peer networking for the classification of peer-to-peer architectures and applications. Peer-to-Peer Computing, 2001. Proceedings. First International Conference on, IEEE, pp. 101–102.

Sheard, S.A. and Mostashari, A. (2009) Principles of complex systems for systems engineering. *Systems Engineering*, **12** (4), 295–311.

Strogatz, S. (2004) *Sync: The Emerging Science of Spontaneous Order*, Penguin, UK.

Taleb, N.N. (2012) *Antifragile: Things That Gain from Disorder*, vol. **3**, Random House.

Tolk, A., Diallo, S.Y., Padilla, J.J., and Herencia-Zapana, H. (2013) Reference modelling in support of M&S—foundations and applications. *Journal of Simulation*, **7** (2), 69–82.

Tononi, G., Sporns, O., and Edelman, G.M. (1994) A measure for brain complexity: relating functional segregation and integration in the nervous system. *Proceedings of the National Academy of Sciences*, **91** (11), 5033–5037.

Tumer, K. and Wolpert, D. (2000, July). Collective intelligence and Braess' paradox. AAAI/IAAI, pp. 104–109).

Tumer, K. and Wolpert, D.H. (2004) *Collectives and the Design of Complex Systems*, Springer Science & Business Media.

Von Bertalanffy, L. (1972) The history and status of general systems theory. *Academy of Management Journal*, **15** (4), 407–426.

Wilson, E.O. (2000) *Sociobiology*, Harvard University Press.

Wolf, M.T., Rahmani, A., de la Croix, J.P., Woodward, G., Vander Hook, J., Brown, D., Schaffer, S., Lim, C., Bailey, P., Tepsuporn, S., Pomerantz, M., Nguyen, V., Sorice, C., Sandoval, M. and Pomerantz, M. (2017) CARACaS multi-agent maritime autonomy for unmanned surface vehicles in the Swarm II harbor patrol demonstration. Proceedings of SPIE 10195, Unmanned Systems Technology XIX, , 5 May 2017, pp. 101950O–101950O, doi: 10.1117/12.2262067.

Zolli, A. and Healy, A.M. (2013) *Resilience: Why Things Bounce Back*, Simon and Schuster.

SECTION III

ENGINEERING EMERGENT BEHAVIOR IN COMPUTATIONAL ENVIRONMENTS

11

TOWARD THE AUTOMATED DETECTION OF EMERGENT BEHAVIOR

Claudia Szabo and Lachlan Birdsey

School of Computer Science, The University of Adelaide, Adelaide, SA 5005, Australia

INTRODUCTION

Complex systems often exhibit behavior that cannot be reduced to the behavior of their individual components alone and require thorough analysis once unexpected properties are observed (Davis, 2005; Johnson, 2006; Mogul, 2006). These *emergent properties* are becoming crucial as systems grow both in size (with respect to the number of components and their behavior and states) and in coupling and geographic distribution (Johnson, 2006; Mogul, 2006; Bedau, 1997; Holland, 1999). Due to significant research interest in the past decades, a plethora of emergent properties examples, from flocks of birds, ant colonies, to the appearance of life and of traffic jams, has been observed and identified. In software systems, connection patterns have been observed in data extracted from social networks (Chi, 2009) and trends often emerge in big data analytics (Fayyad and Uthurusamy, 2002). More malign examples of emergent behavior include power supply variation in smart grids due to provider competition (Chan *et al.*, 2010), the Ethernet capture effect in computer networks (Ramakrishnan and Yang, 1994), and load-balancer failures in a multi-tiered distributed system (Mogul, 2006). As emergent properties may have undesired and unpredictable consequences (Mogul, 2006; Ramakrishnan and Yang, 1994; Floyd and Jacobson, 1993), systems that exhibit such behaviors become less credible and difficult to manage.

Emergent Behavior in Complex Systems Engineering: A Modeling and Simulation Approach,
First Edition. Edited by Saurabh Mittal, Saikou Diallo, and Andreas Tolk.
© 2018 John Wiley & Sons, Inc. Published 2018 by John Wiley & Sons, Inc.

Although emergent properties have been the focus of research since the 1970s (Bedau, 1997; Holland, 1999; Cilliers, 1998; Gardner, 1970; Seth, 2008), very few methods for their identification, classification, and analysis exist (Seth, 2008; Chen *et al.*, 2007; Kubik, 2003; Szabo and Teo, 2012; Brown and Goodrich, 2014). Moreover, existing methods are usually employed only on simplified examples that are rarely found in real life. For example, the flock of birds model suggests that flocks result from the birds obeying three rules, as opposed to the myriad rules that affect flocking in real life. Approaches can be classified broadly from two orthogonal perspectives. In the first perspective, approaches propose to identify emergence as it happens (Kubik, 2003; Szabo and Teo, 2012) and aim to use formal or meta-models of calculated composed model states. Toward this, a key issue remains in the identification of variables or attributes that describe the system components, or the *micro-level*, and the system as a whole, or the *macro-level*, and the relationships and dependencies between these two levels. These definitions allow the specification of emergence as the set difference between macro-level and the micro-level but are difficult to capture and computationally expensive to calculate.

In contrast, the second perspective uses a definition of a known or observed emergent property and aims to identify its cause, in terms of the states of system components and their interaction (Seth, 2008; Chen *et al.*, 2007). A key issue when using this *post-mortem* perspective is that a prior observation of an emergent property is required and that emergent properties need to be defined in such a way that the macro-level can be reduced or traced back to the micro-level. Moreover, current approaches (Seth, 2008; Chen *et al.*, 2007; Kubik, 2003; Brown and Goodrich, 2014) are demonstrated using simple models such as Flock of Birds or Predator-Prey, which have limiting assumptions and constraints when applied to more complex systems. For example, most approaches do not consider mobile agents (Kubik, 2003), assume unfeasible a priori specifications and definitions of emergent properties (Szabo and Teo, 2012), or do not scale beyond models with a small number of agents (Teo *et al.*, 2013). In the multi-agent systems community, approaches focus more on the engineering of systems to exhibit beneficial emergent behavior and less on its identification (Bernon *et al.*, 2003; Jacyno *et al.*, 2009; Salazar *et al.*, 2011). Moreover, approaches that engineer emergent behavior do not ensure that no other side effects occur as a consequence.

In this chapter, we propose architecture to identify and analyze potential emergent behaviors in multi-agent systems as they happen. We analyze a multi-agent system as it is executing and record system snapshots. These snapshots are then compared with previously recorded instances of emergent behavior according to our pre-defined distance functions. If current snapshots are similar, then the system state is highlighted to the system expert for further analysis. Our proposed architecture is extensible and permits the collection and aggregation of several metrics that have been shown in the literature to indicate emergence (Kubik, 2003; Chan, 2011a; Shalizi, 2006; Szabo and Teo, 2013; Brown and Goodrich, 2014). Examples of such metrics include measures of interaction (Chan, 2011a) and Statistical Complexity (Shalizi, 2006) among others. In this chapter, we focus on three metrics, namely, Hausdorff distance (Birdsey and Szabo, 2014), Active Hausdorff distance, and Statistical Complexity.

In this chapter, we define and implement three metrics, namely, Hausdorff distance, Active Hausdorff distance, and Statistical Complexity, all of which can detect emergence. Hausdorff distance (Huttenlocher *et al.*, 1993) and Active Hausdorff distance determine the similarity of interaction graphs (IGs), with Active Hausdorff taking advantage of the interactions between agents. Statistical Complexity is used to determine the amount of information entropy in an interaction graph. To test Hausdorff distance, Active Hausdorff distance, and Statistical Complexity, we implemented the Flock of Birds, Game of Life, and Predator-Prey models and performed experiments over several configurations of these systems.

OVERVIEW OF EXISTING WORK

Complex systems are omnipresent in today's world. They exist in nearly every facet of life, from biological systems (Odell, 1998) to the technological (Odell, 1998; Johnson, 2006). Over the last few decades, a significant body of research has been undertaken to classify many of the unexpected behaviors that a complex system produces (Johnson, 2006; Mogul, 2006; Odell, 1998; Fromm, 2006). Early research into emergence carried out by Holland (1999) and Bedau (1997) has assisted in defining key terms such as *weak* and *strong* emergence as well as defining such terms in different degrees (Holland, 2007). In this chapter, we consider *weak* emergence as being the *macro-level* behavior that is a result of *micro-level* component interactions and *strong* emergence as the *macro-level* feedback or causation on the *micro-level*.

Bedau (1997) states that an emergent property can be defined as "a property of assemblage that could not be predicted by examining the components individually." Emergence can be seen in many real-world systems such as technological and nature-driven systems. For example, the neurons in the brain individually fire impulses but together form an emergent state of consciousness (Odell, 1998). The flocking of birds is a well-known example of emergent behavior in nature. Independent birds aggregate around an invisible center and fly at the same speed for flock creation. The birds come together to create something that would be entirely indiscernible by studying only one or two birds. Two key examples of systems where emergent behavior is caused by interactions are the Flock of Birds model (Reynolds, 1987) and the cellular automata Game of Life model (Gardner, 1970). The former achieves its emergent properties through each bird flocking around a perceived flock center, whereas in the latter model, the emergent properties are achieved by the patterns that are formed by the cells' transitions between states. Studies that propose various processes of detecting and identifying emergent behavior mainly use either one or both of these systems to prove the validity of their proposed approach (Chan *et al.*, 2010; Seth, 2008; Szabo and Teo, 2013; Chan, 2011a).

Multi-agent systems are a useful formalism to model complex systems. The components present in a complex system can be modeled as agents that perform their respective actions and interactions. The modeling of these components as agents allows for unnecessary information to be abstracted away leaving only the actions and interactions needed for a particular outcome. These agent-based models are then used in simulations to assist with research and analysis (Johnson, 2006). Multi-agent systems can be engineered to exhibit emergent properties (Fromm, 2006; Savarimuthu *et al.*, 2007). Several formalisms have been proposed to obtain or engineer emergent behavior, such as the DEVS extension proposed by Mittal (2013) and Birdsey *et al.* (2016), but they have yet to be employed in practice. By creating models where emergence is an easily attainable product derived from agents' interactions, users are relieved from having to model every aspect of the complex system under study. Multi-agent systems that have been designed to exhibit emergence are usually engineered to focus on self-organization and co-operation between agents. These systems generally rely on a system expert to identify the emergent behavior (Jacyno *et al.*, 2009; Salazar *et al.*, 2011; Savarimuthu *et al.*, 2007). For example, human societies and the myriad ways that emergent properties can arise are generally modeled using this approach in order to study aspects such as norm emergence (Jacyno *et al.*, 2009; Savarimuthu *et al.*, 2007).

Chan *et al.* (2010) highlights that agent-based simulation is the most suitable method for modeling systems containing unexpected or emergent behaviors because it emphasizes that the actions and interactions between agents are the main causes for emergent behaviors. Several works support the use of agent-based modeling for studying emergent behaviors (Salazar *et al.*, 2011; Fromm, 2006; Pereira *et al.*, 2012; Serugendo *et al.*, 2006). In addition to the Flock of Birds and Game of Life models, Chan *et al.* (2010) show that other complex systems such as social networks and electricity markets, implemented within an agent-based simulation, can exhibit emergent properties, which can then be identified. The methods in Chan *et al.* (2010) for detecting emergence rely upon the presence of a system expert, who can identify the emergent behavior.

Considerable research has been done in developing methods for the detection of emergence, and as discussed above, existing methods assess emergence in either a *post-mortem* setting or a *live* setting (Szabo and Teo, 2012). *Post-mortem* analysis methods are applied after the system under study has finished executing and use data that were recorded during the execution (Szabo and Teo, 2012, 2013). In contrast, *live* analysis methods are used while the system under study is executing (Szabo and Teo, 2012; Chan, 2011a). Most existing works focus on *post-mortem* analysis methods (Szabo and Teo, 2013; Chen *et al.*, 2009; Tang and Mao, 2014). In addition to *post-mortem* and *live* analysis, methods can be classified into three main types (Teo *et al.*, 2013): *grammar-based* (Szabo and Teo, 2013; Kubik, 2003), *event-based* (Chen *et al.*, 2007), or *variable-based* (Seth, 2008; Szabo and Teo, 2013; Tang and Mao, 2014).

Some forms of *live* analysis involve *grammar-based methods*. These attempt to identify emergence in multi-agent systems by using two grammars, L_{WHOLE} and L_{PARTS}. Kubik (2003) defines that L_{WHOLE} describes the properties of the system as a whole and L_{PARTS} describes the properties obtained from the reunion of the parts and, in turn, produces emergence as the difference between the two solutions. L_{WHOLE} and L_{PARTS} can be easily calculated as the sets of words that are constructed from the output of agent behavior descriptions. This method does not require a prior observation of the system in order to identify possible emergent properties or behaviors, which therefore makes it suitable for large-scale models where such observations are notoriously difficult (Teo *et al.*, 2013). However, as grammars require a formation of *words*, the process through which these words are formalized can suffer badly as the model grows in scope, leading to computational issues, especially for large-scale systems (Teo *et al.*, 2013; Kubik, 2003). To address this, some works attempt to identify *micro*-level properties and model interaction and performing reconstructability analysis on this data (Szabo and Teo, 2013); however, this analysis is required to take place in a *post-mortem* context.

Some forms of *post-mortem* analysis involve *event-based methods*, in which behavior is defined as a series of both simple and complex events that changed the system state, as defined by Chen *et al.* (2007). Complex events are defined as compositions of simple atomic events where a simple event is a change in state of specific variables over some non-negative duration of time. These state changes, or state transitions, are also defined by a set of rules. Each emergent property is defined manually by a system expert as a complex event. It is the particular sequence of both complex and simple events in a system that leads to emergence occurring in the system. However, this method relies heavily on the system experts and their specific definitions. Furthermore, it can suffer from both agent and state-space explosion, making it unsuitable for large systems.

In *variable-based methods*, a specific variable or metric is chosen to describe emergence. Changes in the values of this variable signify the presence of emergent properties (Seth, 2008). The center of mass of a bird flock could be used as an example of emergence in bird flocking behavior, as shown in Seth (2008). Seth's approach uses Granger causality to establish the relationships between a macro-variable and micro-variables and proposes the metric of G-Emergence, a near-*live* analysis method. This has the advantage of providing a process for emergence identification that is relatively easy to implement. However, the approach requires system expert knowledge as observations must be defined for each system. Szabo and Teo (2013) proposed the use of reconstructability analysis to determine which components interacted to cause a particular emergent property (defined through a set of variables). They identified the interactions that cause birds to flock (Reynolds, 1987), the cells that cause the glider pattern in Conway's Game of Life (Gardner, 1970), and the causes of traffic jams. However, their method is heavily dependent on the choice of the variable set that represents the micro- and macro-levels and requires the intervention of a system expert.

Variable-based methods from other fields, such as information theory and machine learning, have been adapted with the goal of emergence detection. Information theory approaches for detecting emergence have also been proposed by using such techniques as Shannon Entropy (Gershenson and Fernandez, 2012; Prokopenko *et al.*, 2009; Tang and Mao, 2014) and variety (Holland, 2007; Yaneer, 2004). These have advantages over other *variable-based methods* in that they can process large amounts of data efficiently. Tang and Mao (2014) proposed measures of relative entropy that depend on the main emergent property of a system under study. However, these methods require the input of a system expert because they rely on the emergent property of a system being classified along with a specific function to be defined for that particular property. Machine learning classification techniques have also been proposed as a way of detecting emergence. A variant of Bayesian classification (Brown and Goodrich, 2014) has been used to successfully detect swarming and flocking behavior in biological systems such as the flock of birds model (Reynolds, 1987). This approach involves identifying *key features* of an agent, such as how many neighbors an agent has, and uses this information to determine the likelihood that a random set of agents is exhibiting emergence. Other methods from machine learning have been utilized, such as Conditional Random Fields and Hidden Markov Models in Vail *et al.* (2007), but with the goal of activity recognition in domain-specific contexts. Vail *et al.* used Conditional Random Fields and Hidden Markov Models somewhat successfully to determine if agents were performing a particular distinct action based on their relational position to other agents.

In summary, a substantial number of methods for detecting emergence both in a *live* or *post-mortem* capacity exists. However, prior research has three main shortcomings. Firstly, many of the devised methods rely on a system expert for a definition of emergent behavior. This implies that a true autonomous method for detection has not yet been achieved. Secondly, the testing and validation of these methods is always performed on a restricted number of models, usually the Flock of Birds model and the Game of Life model. Although these are two important models in the study of emergence, there are many types of multi-agent systems that possess extremely different properties and therefore the methods may prove to be limited. Thirdly, although there has been research into developing an integrated approach for detection of emergence (Teo and Szabo, 2008), it has focused on *post-mortem* methods. It is clear that *live* methods possess benefits over *post-mortem* methods. Aside from being able to detect emergence as it is occurring, *live* methods do not require a prior observation of an emergent property in the system under study and require very little system expert input.

IDENTIFYING EMERGENCE USING INTERACTION GRAPHS

Our proposed architecture for identifying emergent behavior as it happens focuses on employing a wide variety of metrics in the analysis of unexpected behavior. The proposed architecture relies on the use of multi-agent system simulation as the fundamental methodology for system analysis and is divided into three main components, namely, (i) *Modeling and Simulation*, (ii) *Metric Collection*, and (iii) *Analysis and Visualization*, as shown in Figure 11.1.

Modeling and simulation: The system under study is implemented as an agent-based model for use in agent-based simulation software tools, for example, Repast (2017). This requires the explicit definition of agents, agent states, and agent interactions. Connected to the simulation software is a tool, the *SnapshotTaker*, that takes snapshots at certain specified time periods.

Metric collection: The *Metric Collection* modules are responsible for the acquisition and interpretation of snapshot data collected by the *SnapshotTaker*. A meta-model describing the data and how it is collected and calculated

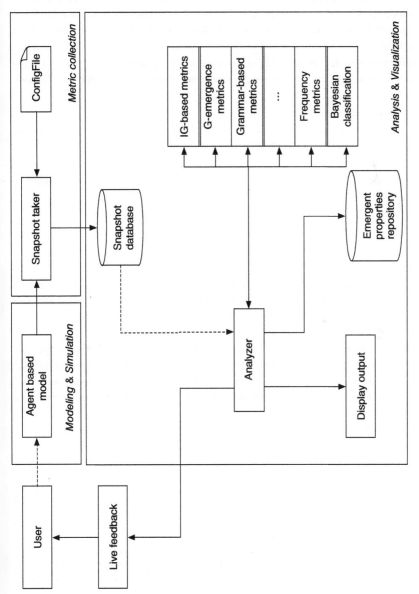

FIGURE 11.1 Architecture overview.

drive this process. As our architecture is designed to be used with different simulation software, the *Metric Collection* phase requires a configuration file. This file dictates how the data received by the *SnapshotTaker* is to be transformed into an appropriate format to be used by the *Analysis and Visualization* modules. For example, if the simulation software outputs read-only log data, the configuration file informs the *SnapshotTaker* how to transform the data to conform to the system meta-model.

Interaction graphs: In any multi-agent system, it is possible to discern and place interactions into three categories independent of the system under study (Chan, 2011b):

1. *Information poll*:

   Agent a_i needs to obtain information about a_j to make a calculation. a_i acquires the information without a_j being informed.

2. *Communication*:

   Agent a_i sends a message to a_j. Upon receiving this message, agent a_j may either:

   (1) Perform some computation leading to a state change, or

   (2) Perform some computation without changing state.

   Both a_i and a_j are aware of the interaction.

3. *Indirect*:

   An agent a_i may perform an action on the Environment En, which in turn will affect agent a_j. En may be represented as an agent. En can also perform interactions on any agent. The two agents indirectly interacted may or may not be aware of each other.

For a multi-agent system M comprised of n agents a_i, we define an interaction graph to capture the interactions between agents over a given time interval T^s, where s is the size of the interval in time units and remains the same for a simulation run. The term *interactions* will capture all types of interactions as defined above. An interaction graph is a directed graph where each vertex represents an agent, $a \in M$, and each arc represents a interaction between two agents, $a_i \to a_j$, and carries a weight w_{ij}. This is formally defined as:

$$IG_{T^s}(M) = < V_{T^s}, E_{T^s} > \qquad (11.1)$$

$$V_{T^s} = \{a_i | a_i \in M, i = 1, \ldots, n\} \qquad (11.2)$$

$$E_{T^s} = \{(a_i, a_j, w_{ij}) | a_i, a_j \in V_{T^s}, w_{ij} \in Z^+\} \qquad (11.3)$$

The weight w_{ij} of the arc $a_i \to a_j$ is incremented every time an interaction between a_i and a_j takes place. Currently, this calculation increments the weight, regardless of what type of interaction occurs between agents, but future formulae can be refined to assign different weights to various interaction types. This could be the case, for example, if more importance was assigned to indirect interaction as a cause of emergence (Szabo and Teo, 2013).

Snapshot Calculation: Snapshots consist of information about agents, the environment, and instances of the specified metric formalism over the time interval

T_k, where k is the time when the snapshot was taken. As such, the snapshot for time T_k contains all interaction information from $[T_{k-s}, T_k]$. The snapshots are metric agnostic, thus allowing for the calculation of various emergent behavior metrics without the need to re-execute the simulation run. Each snapshot will include interaction graphs and all agent and environment states. This allows the system expert to perform various additional calculations in the analysis and visualization stage over a particular snapshot of interest. For example, if the distance calculation over two interaction graphs shows an emergent property, the system expert may want to analyze the Statistical Complexity of that snapshot as it has evolved over the previous snapshots. Formally,

$$S_{T_k}(M) = \{a_i, IG_{T_k}, \ldots |a_i \in M\} \tag{11.4}$$

$$S(M) = \{S_{T_k} | T_k \in T\} \tag{11.5}$$

where $S_{T_k}(M)$ defines the snapshot for the time interval T_k and $S(M)$ is the set of all collected snapshots.

Analysis and Visualization

In the *Analysis and Visualization* step, snapshots from the executing system are analyzed using the selected metrics to determine if an emergent property is present. Each metric requires access to previously acquired snapshots that are known to contain emergent properties, which are stored in the *Emergent Properties Repository*. The *Analysis and Visualization* step can be further divided into two steps, namely, *Comparison* and *Analysis*. The *Comparison* step applies distance functions to the collected metrics. For example, when using an interaction metric and its associated interaction graph, the distance function calculates the similarity between the interaction graph as captured in the current snapshot and interaction graphs of systems that have been previously shown to exhibit emergent behavior. The *Analysis* step highlights the snapshot to the user if any significant similarities are detected.

We define three metrics for detection of emergence in multi-agent systems: Hausdorff Distance, Active Hausdorff Distance, and Statistical Complexity.

Hausdorff Distance

The Hausdorff distance *(HD)* is a metric that is used to determine the similarity between two graphs with respect to shape (Huttenlocher *et al.*, 1993). For two interaction graphs $IG(A)$ and $IG(B)$, the Hausdorff distance is defined as

$$HD(A, B) = \max\{h(A, B), h(B, A)\} \tag{11.6}$$

where

$$h(A, B) = \max_{a \in A}\{\min_{b \in B}\{d(a, b)\}\} \tag{11.7}$$

and d is the distance between vertices a and b, with $a \in A$ and $b \in B$, respectively. For points in a two-dimensional Euclidian space, the distance d could be calculated as the Euclidian distance between points $a(x_a, y_a)$ and $b(x_b, y_b)$ as $d(a, b) = \sqrt{(x_a - x_b)^2 + (y_a - y_b)^2}$, where (x_a, y_a) and (x_b, y_b) are the Cartesian coordinates of points a and b, respectively. However, any distance can be used. It is important to highlight that the use of this definition implies that the vertices of the interaction graphs need to be assigned coordinates in a two-dimensional Euclidian space. This can be performed by assigning coordinates to the interaction graph nodes without any semantic meaning. If the system under study already employs some coordinate abstraction, then the associated distance calculation can be employed. For example, in a bird flocking model defined in "Flock of Birds", the coordinates could be those of the birds. Intuitively, the Hausdorff distance between interaction graphs A and B measures how close A and B are to each other, where "close" means that every vertex in A is close to some vertex in B, using some distance function such as the two-dimensional Euclidian distance. When other distances are used, domain ontologies can be employed to further refine the meaning of similarity.

The Hausdorff distance focuses on the position of the agents and only considers agents that are interacting, that is, are included in the interaction graph. This implies that this metric will not give useful results in cases where agents are stationary or agent interactions are sparse. In cases where the agents are stationary, the agents' positions across all interaction graphs are unchanging, which would lead to results with constant values. If agent interactions are sparse, the interaction graphs would contain wildly different information at each interval as few agents would be captured. This could be remedied by careful selection of the snapshot interval size. Moreover, as the coordinate information is recorded at the end of the snapshot interval and the metric does not consider interaction beyond the presence of the nodes in the IG, the distance function ignores cases in which the emergent behavior happens in the middle of the interval. This further makes the metric dependent on the size of the snapshot interval. For example, if a flock formed at the start of a snapshot interval but was broken midway through the snapshot interval, the agents' positions at the end of the snapshot would likely not represent a flock.

Active Hausdorff Distance

To address the drawbacks of HD identified above, we define a variation of HD, namely the Active Hausdorff Distance (HDA), which is calculated in a similar manner as the HD, but following a pre-processing step: $HDA(A, B) = HD(A', B')$, where A' is obtained from A using a pre-processing algorithm.

The pre-processing algorithm aims to move agents closer to the agents with which they had been interacting the most, as shown in Figure 11.2. Edges in the interaction graph are sorted in descending order of their weights (line 1). Nodes are then moved toward other agents by considering the inverse strength of their interaction (line 6). This is done by looking at the node's positions and determining the correct direction in which the move should take place (line 8 – onward). The purpose of this pre-processing step is to position agents closer together geometrically based on the frequency and strength of the interactions of each agent over the snapshot interval.

```
pre_process(A)
1. sort(A)
2. for Edges e in A
3.      a = e.startNode; b=e.endNode; w=e.weight
4.      xDiff = Math.abs(a.x - b.x)
5.      yDiff = Math.abs(a.y - b.y)
6.      xOffset = xDiff / (2 * w);
7.      yOffset = yDiff / (2 * w)
8.      if (a.x < b.x)
9.          a.x += (xDiff/2) - xOffset
10.         b.x -= (xDiff/2) + xOffset
11.     else if (a.x > b.x)
12.         a.x -= (xDiff/2) + xOffset
13.         b.x += (xDiff/2) - xOffset
14. 15.if (a.y < b.y)
16.         a.y += (yDiff/2) - yOffset
17.         b.y -= (yDiff/2) + yOffset
18.     else if (a.y > b.y)
19.         a.y -= (yDiff/2) + yOffset
20.         b.y += (yDiff/2) - yOffset
```

FIGURE 11.2   Pre-processing agents based on their interactions.

HDA can be modified further to penalize interactions that actively discourage emergent behavior. For example, in the flock of birds model, if a large proportion of the birds were to use the separation interaction over a given interval, it would discourage emergent behavior from happening.

Statistical Complexity   Statistical Complexity (SC) is a metric that determines the amount of information entropy, or complexity, in a particular graph. As the entropy of a system decreases, it can be shown that collective behavior increases (Tang and Mao, 2014). As emergence can be construed as being a collective behavior, it makes sense to consider measures of entropy to capture emergence. The simplest way to determine Statistical Complexity is by using Shannon Entropy (Gershenson and Fernandez, 2012; Prokopenko *et al.*, 2009), which is defined as

$$SC_{SE} = -\sum_{i=0}^{n} p(a_i)\log_{10}p(a_i) \tag{11.8}$$

where $p(a_i)$ is the probability mass function with respect to agent a_i, and n is the total number of agents. While generic probabilistic mass functions can be defined to cover all systems, more accurate results may be obtained by specifying it for each particular system. For example, for the Flock of Birds model the probability mass function, $p(a_i)$, could be defined as:

$$p(a_i) = \frac{\sum_{j=0}^{N}\{w_{jk}|j \vee k = i\}}{(n^2 - n) \times s \times \text{types of interactions}} \tag{11.9}$$

where $w_{jk} \in E$, s is the snapshot size, n is the total number of agents in the snapshot, N is the total number of interactions that occurred in the snapshot, and *types of interactions* is the number of interaction rules for the system. This particular $p(a_i)$ calculation represents the proportion of total interactions across the snapshot interval in which agent a_i is involved.

As statistical complexity operates on a single interaction graph, a threshold or series of thresholds must be defined so as to allow for comparison between a candidate interaction graph and a reference interaction graph. A simple way of achieving this is by taking the difference of the result for the candidate IG and the reference interaction graph (IG_e).

$$D(\text{IG}, IG_e) = \text{SC}_{SE}(IG_e) - \text{SC}_{SE}(IG) \tag{11.10}$$

By taking the difference, the change in information entropy between the candidate and reference interaction graphs becomes evident. If we were to use the individual $\text{SC}_{SE}(IG)$ result, it may not highlight an emergent behavior, as some systems do not show emergent behavior when this result tends to zero (Tang and Mao, 2014).

This method, unlike the Hausdorff methods, does not focus on the position of the agents. Instead it relies entirely on interactions between agents and discrete representations of those interactions. Previous applications of Shannon Entropy in multi-agent systems have focused on the states of the agents and environment, rather than the interactions between the agents (Tang and Mao, 2014; Cavalin *et al.*, 2014).

EXPERIMENTAL ANALYSIS

To analyze the suitability of Hausdorff distance, Active Hausdorff distance, and Statistical Complexity, we apply them to well-known models that have been shown to exhibit emergence, namely, Flock of Birds, Game of Life, and Predator-Prey models. The Flock of Birds and Game of Life models are suitable for assessing the qualities of the proposed metrics as they are both simple models and also exhibit direct and indirect interactions. The Predator-Prey model is more complicated as it features two distinct types of mobile agents, both of which have rules for being removed or added to the system. For each experiment, we compare between the snapshots acquired during the current execution and the snapshots of the same model exhibiting emergence. As discussed above, these snapshots are stored in the *Emergent Properties Repository*.

Table 11.1 presents an overview of each model, focusing on agent type, initial population size, environment, and interaction types. We focus on both mobile and stationary agents that have interactions of type information poll and communication.

In the following, we present a brief overview of our prototype implementation followed by an overview of the three models we used for our experiments: Flock of Birds, Game of Life, and Predator-Prey.

TABLE 11.1   Differences in Model Features

Model	Agent Type	Initial Population Sizes	Environment	Interaction Types
Flock of Birds	Mobile	20, 50	Unbounded, closed	Information poll
Game of Life	Stationary	225, 400	Bounded, closed	Information poll
Predator-Prey	Mobile	12, 16	Unbounded, open	Information poll, communication

Prototype Implementation

The current prototype implementation of the architecture consists of three separate components: *Modeling and Simulation*, the *MetricAggregator*, and the *MetricSuite*. All the three components are written in Java. The visualization of graphs is currently achieved using the graphing tool (Medusa, 2017). The *Modeling and Simulation* component includes the Repast simulation suite (Repast, 2017), which is the execution environment for our models. Repast's back-end handles the agent simulation, including time management, while also having a visual representation of the currently running simulation. Repast's scheduling system allows for methods to be run at arbitrary times, which is an approach that is highly suitable for a time-based snapshot system. The *MetricAggregator* captures and stores the snapshots. The *MetricSuite* consists of implementations for the three metrics and our *Emergent Properties Repository*. This component allows for the *Emergent Properties Repository* to execute selected metrics depending on snapshots that have been received and to compare them with the previously annotated snapshots stored in the repository. If a metric returns a result indicating that a new snapshot possesses emergence, the *Emergent Properties Repository* annotates that snapshot and stores it for future comparisons.

Model Overview

Flock of Birds   The Flock of Birds model (Reynolds, 1987) captures the motion of bird flocking and is a seminal example in the study of emergence. At the macro-level, a group of birds tends to form a flock, as shown in Figure 11.4b. Flocks have aerodynamic advantages, obstacle avoidance capabilities, and predator protection, regardless of the initial positions of the birds. At the micro-level, each bird obeys three simple rules (Reynolds, 1987):

1. separation – steer to avoid crowding neighbors
2. alignment – steer toward average heading of neighbors
3. cohesion – steer toward average position of neighbors

We model this as a multi-agent system in which each bird is an agent that has the three-movement rules defined above. Other bird attributes include initial position and initial velocities. In our experiments, the initial bird positions can be either fixed or assigned randomly at start up. Bird velocities are assigned randomly. The model parameters can also influence emergent behavior analysis. As such, we collect and analyze interaction graphs of Flock of Birds models with sizes of 20 and 50 birds, with fixed and randomly assigned position values and randomly assigned velocity values.

Game of Life   Conway's Game of Life model (Gardner, 1970) represents cells that interact with their neighbors to determine in what state they should be at the next time step, either alive or dead. At the macro-level, patterns emerge between groups of cells, such as the Pulsar pattern as shown in Figure 11.3. At the micro-level, the rules for each cell are as follows, where X, Y, and Z are the parameters for the Game of Life model (Gardner, 1970; Chan *et al.*, 2010):

1. A live cell with at least X and at most Y live neighbors will remain alive in the next time step.
2. A dead cell with exactly Z live neighbors will become alive in the next time step.
3. Otherwise, the cell will die in the next time step where $0 \leq X, Y, Z \leq \epsilon$ and $X \leq Y$, where ϵ is the maximum number of neighbors, which in a two-dimensional configuration is 8.

Certain combinations of X, Y, and Z settings can reveal emergent behavior such as patterns, like the glider (Szabo and Teo, 2013), and shapes appearing in the cellular structure (Chan *et al.*, 2010).

We model the Game of Life as a multi-agent system where each cell is an agent. A two-dimensional grid of cells, of size $n \times n$, is established and the initial state of each cell is either fixed or chosen at random on start up. The attributes recorded for each cell are the cell state and the states of the cell's eight neighbors at the start of the time step. Snapshots are taken of Game of Life models with sizes of both 20×20 cells and 15×15 cells are collected and analyzed, each cell having a total of two possible states. Their initial states are either randomized or set to allow the creation of a particular pattern (Chan *et al.*, 2010). For our experiments, we followed Conway's initial X, Y, and Z rules, which are 2, 3, and 3, respectively.

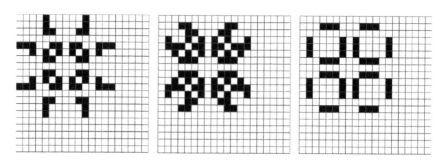

FIGURE 11.3   Game of life with pulsar pattern.

Predator-Prey The Predator-Prey model has been used in studying emergent behavior, with varying rule sets (Chen *et al.*, 2007). The Predator-Prey model has two types of agents, namely, Predators and Prey. As in nature, both types of agents wish to survive to create a new generation. Survival for the Predators is to eat, whereas survival for the Prey is to avoid being eaten. The main difference of this model from the previous two is that agents can be added to or removed from the system during execution and that the different agent types obey different rules. Specifically, Predators obey four rules, whereas the Prey obeys two rules. The rules of the Predators are as follows:

1. If a Prey is detected within distance d, kill the Prey with probability $p(predKill)$. If successful, the Prey is removed from the system immediately.
2. If some Prey is killed, a new Predator is born at the Prey's location, after one time step has passed.
3. If a Predator is not within distance d, it dies with probability $p(predDeath)$.
4. Move one step in any random direction if no Prey was killed and the Predator has not died.

The rules of the Prey are:

1. Move one step in any random direction.
2. Give birth to a new Prey at a random location with probability $p(preBirth)$.

In our multi-agent system implementation, Predator and Prey are assigned initial random positions. Their velocities are also randomized at initialization as well as for each time step. The three-rule probabilities, *predKill*, *predDeath*, and *preBirth* are decided by the system expert. Both agent types have a common set of states, life, and procreation, but the Predator has extra states to indicate if it has seen and/or killed some prey. If an agent is removed from the simulation, that is, upon death, their life state is set to "dead." This is to ensure that no information generated by the system is lost as it may prove valuable to a particular metric.

We collect snapshots of Predator-Prey systems with varying grid sizes and number of Predators. The variation of the grid sizes is shown by Chen *et al.* (2007) to have an impact on the occurrence of emergent behaviors. Due to the stochastic nature of the Predator-Prey model with respect to agent destruction and procreation, creating a baseline for comparison is somewhat problematic. Chen *et al.* (2007) define emergence occurring in the Predator-Prey model as when starvation affects a Predator that is caused by either one or more Predators "over-hunting."

Experimental Results

We present key results from our experiments alongside figures of the interaction graph that achieved the best result for three of our metrics: Hausdorff, Active Hausdorff, and Statistical Complexity. Each result is achieved by comparing each candidate graph (IG) with a reference interaction graph (IG_e) that shows an emergent state.

Each experimental result shows the minimum, median, mean, standard deviation, and average runtime.

The experiments use three different interval sizes, namely, $s = 2$, $s = 5$, and $s = 10$ ticks, over a simulation run of 1000 ticks. The snapshots that are considered for comparison occur at ticks 100, 500, and 1000. These are chosen because they represent the system shortly after beginning, mid-way through its execution, and at the end of its execution, respectively. The time interval T_k refers to the time interval that finished at tick k. Our comparisons involve snapshots that contain differing interval and population sizes and properties exclusive to each system.

Hausdorff Distance We use the Euclidian distance function as the secondary distance function across all the experiments. We study the Flock of Birds and Predator-Prey models. As the Hausdorff Distance only considers an agent's position in space, the results derived for the Game of Life model, in which all the agents are stationary, are meaningless for detecting emergence.

Flock of Birds: Table 11.2 presents the values of HD for a model of 20 birds, $IG(B_{20})$, and Figure 11.4a is the companion interaction graph with the best result. The

TABLE 11.2 HD($IG(B_{20})$, $IG_e(B_{20})$): 20 Birds, $s = 5$

Time Interval	Min	Median	Mean	σ	Runtime (ms)
T_{100}	109.77	161.62	165.62	31.34	0.43
T_{500}	110.49	173.93	203.15	84.86	0.28
T_{1000}	154.44	273.62	381.78	211.88	0.31

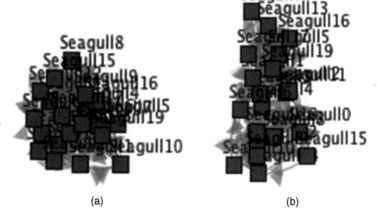

(a)   (b)

FIGURE 11.4 Comparison between $IG(B_{20})$ and $IG_e(B_{20})$. (a) Flock of birds: $IG(B_{20})$ with minimum HD: 109.77. (b) Flock of birds: reference graph: $IG_e(B_{20})$

experiment compares interaction graphs at different time steps with $IG_e(B_{20})$, shown in Figure 11.4b, which, for this experiment, is taken to be the interaction graph of a model with 20 birds that have the same starting positions. For both models, the initial velocities are randomly assigned. It is important to note here that a validation run in which $IG_e(B_{20})$ represents exactly the same model, that is, the velocities have the same values, leads to distance values of zero. The large variance shown in this particular experiment, and across many of the experiments, is caused by the initialization of bird velocities always being random. As the Flock of Birds experiments take place in an unbounded environment, it is highly likely that a small number of birds may be situated a large distance away from the flock, therefore influencing the distance calculations.

Table 11.3 presents the result for a comparison where both the reference and candidate graphs contain 50 birds. Figure 11.5 presents both the candidate interaction graph with best result and the reference graph, with birds in dark gray and interactions in light gray. Visual comparison between candidate graph with the best result and the reference clearly shows that emergence is present in multiple subsets of agents.

TABLE 11.3 $HD(IG(B_{50}), IG_e(B_{50}))$: 50 Birds, $s = 5$

Time Interval	Min	Median	Mean	σ	Runtime (ms)
T_{100}	143.13	171.79	15.32	188.51	0.08
T_{500}	101.33	139.84	58.60	249.96	0.08
T_{1000}	86.84	215.83	88.08	346.39	0.08

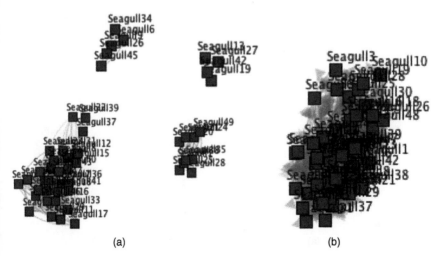

(a)   (b)

FIGURE 11.5 Comparison between $IG(B_{50})$ and $IG_e(B_{50})$. (a) Flock of birds: $IG(B_{50})$ with minimum HD: 86.84. (b) Flock of birds: reference graph: $IG_e(B_{50})$.

Predator-Prey: Tables 11.4 and 11.5 present the values of HD for a Predator-Prey model with 4 predators and 8 prey, and a grid size of 10×10 and 20×20, respectively. The interaction graphs, shown in Figure 11.6 (dark gray – lion, light gray – antelope), contain the best results presented in these tables (5.657 and 14.142, respectively) and do not resemble the reference graph at all.

TABLE 11.4 HD(IG($P_{4,8}$), IG$_e$($P_{4,8}$)): 4 Predators, 8 Prey, 10×10 Grid, $s = 5$

Time Interval	Min	Median	Mean	σ	Runtime (ms)
T_{100}	5.657	10.607	11.172	3.798	0.093
T_{500}	5.657	12.021	17.536	15.495	0.1
T_{1000}	8.485	16.971	33.375	34.424	0.129

TABLE 11.5 HD(IG($P_{8,8}$), IG$_e$($P_{4,8}$)): 8 Predators, 8 Prey, 20×20 grid, $s = 5$

Time Interval	Min	Median	Mean	σ	Runtime (ms)
T_{100}	14.142	23.335	24.89	10.246	0.145
T_{500}	33.941	60.811	59.821	17.877	0.196
T_{1000}	55.154	76.368	78.206	23.533	0.464

(a)   (b)

FIGURE 11.6 Comparison between IG($P_{4,8}$) and IG$_e$($P_{4,8}$). (a) Predator-Prey: IG($P_{4,8}$) with minimum HD: 5.657. (b) Predator-Prey: reference graph: IG$_e$($P_{4,8}$).

TABLE 11.6 HDA(IG(B_{20}), IG$_e$(B_{20})): 20 Birds, $s = 5$

Time Interval	Min	Median	Mean	σ	Runtime (ms)
T_{100}	64.21	103.03	99.85	26.60	1.16
T_{500}	81.31	213.69	221.42	134.02	0.74
T_{1000}	3.58	91.06	161.85	212.47	1.41

Active Hausdorff Distance Although normal Hausdorff is capable of detecting when emergence has occurred for some systems, it has two major flaws. First, it can only detect emergence that is a result of the agents' positions in space. Second, it only considers the positions recorded at the end of the snapshot interval. Active Hausdorff distance (HDA) utilizes a pre-processing phase that re-positions the agents based on the quantity and type of interactions they were involved in over the snapshot interval. To highlight the benefits of HDA over HD, we also perform comparisons where the snapshot interval sizes are different between the comparison and reference graph.

Flock of Birds: Table 11.6 presents the results for HDA for a model of 20 birds. Figure 11.7a is the interaction graph with the minimum HDA value. This result clearly shows that we are also able to capture the system state in which only a subset of the birds are flocking.

Similar results are obtained for models with 50 birds, in that HDA is able to identify emergence when comparing between $IG(B_{50})$ and $IG_e(B_{50})$. Moreover, HDA achieves great results when comparing models of different sizes and this can be seen in Figure 11.7c, which shows the interaction graph for the snapshot with the smallest HDA for $IG(B_{50})$.

This is an important advance as this emergent behavior might have been missed had only metrics captured at the end of the interval been used. Moreover, our experiments show that HDA is less dependent on the size of the snapshot interval than HD. Specifically, in a comparison where the snapshot interval size is significantly different between the comparison and reference graphs, HDA is capable of identifying emergence. We are also able to show that emergence is occurring in a comparison graph representing a system that contains two types of birds, whereas the reference graph still contains one type. Figure 11.8 presents the graph with the best result for a comparison graph with two types of birds with a total population of 50.

Predator-Prey: Table 11.7 presents the values of HDA for a Predator-Prey model with 4 predators, 8 prey, and a grid size of 10×10.

The results generated from the Active Hausdorff distance experiments are identical to the ones generated in the Hausdorff distance experiments. This suggests that despite containing mobile agents, the interactions between agents are far too infrequent to provide the Active Hausdorff pre-processing phase with any information. This is consistent across all the normal Hausdorff and Active Hausdorff Predator-Prey experiments. The 10×10 grid size is the most condensed space configuration used in our experiments and as the grid size increases, we see that the Hausdorff and Active Hausdorff distances grow in proportion to the grid size. This

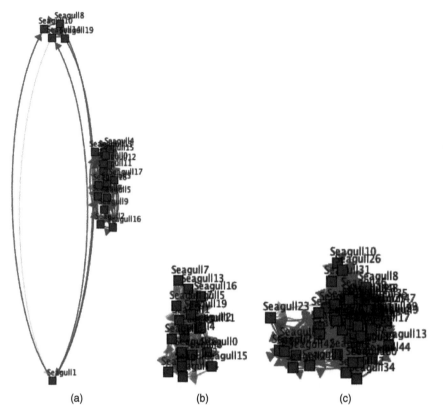

(a)   (b)   (c)

FIGURE 11.7   Comparison between Flock of Birds $IG(B_{20})$, $IG(B_{50})$ and $IG_e(B_{20})$. (a) Flock of Birds: $IG(B_{20})$ with minimum HDA: 3.58. (b) Flock of Birds: reference graph: $IG_e(B_{20})$. (c) Flock of Birds: $IG(B_{50})$ with minimum HDA: 72.35.

is caused by the agents being able to traverse across more space, thereby making the interactions more infrequent.

Statistical Complexity   Statistical Complexity (SC) does not consider the agent's position in space but instead focuses the interactions between the agents. SC can be used in two main ways. First, the SC of an interaction graph gives details about the quantity of information of interactions across the entire snapshot interval. This can signify the presence of self-organization in the system. However, a large or small quantity of information does not give us a clear indication of emergence present. To alleviate this issue, the SC of two interaction graphs is compared. This will show us how similar the information present in a candidate snapshot is compared to the information present in a reference snapshot. We use the same probability mass function across these experiments:

$$p(a_i) = \frac{\sum_{j=0}^{N} \{w_{jk} | j \vee k = i\}}{(n^2 - n) \times s \times \textit{types of interactions}} \tag{11.11}$$

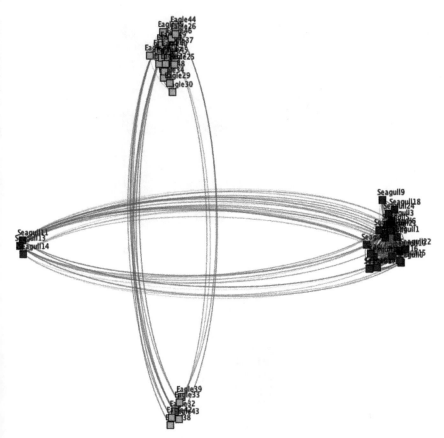

FIGURE 11.8 IG(B_{50}) with minimum HDA: 43.06.

TABLE 11.7 HDA(IG($P_{4,8}$), IG$_e$($P_{4,8}$)): 4 Predators, 8 Prey, 10 × 10 Grid, $s = 2$

Time Interval	Min	Median	Mean	σ	Runtime (ms)
T_{100}	4.123	9.899	10.029	3.451	0.55
T_{500}	4.123	17.678	26.434	26.14	0.551
T_{1000}	4.123	35.355	45.667	41.277	0.706

Flock of Birds: Figure 11.9a presents the 20 bird model that has the smallest difference between itself and the reference graph, $IG_e(B_{20})$. It is clear that several subsets of birds have moved away from each other. Despite this, the numerous edges (interactions) between the subsets and the main flock suggest that at some point over the snapshot interval, all the subsets were part of one flock. Table 11.8 presents the results for the comparison between $IG(B_{50})$ and $IG_e(B_{20})$. Figure 11.9b presents the 50 bird model

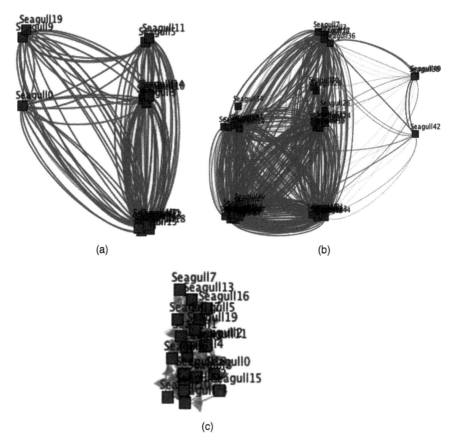

(a)   (b)

(c)

FIGURE 11.9   Comparison between $IG(B_{20})$, $IG(B_{50})$ and $IG_e(B_{20})$. (a) $IG(B_{20})$ with SC_{SE} closest to 0: −0.53. (b) $IG(B_{50})$ with SC_{SE} closest to 0: 0.003. (c) Flock of Birds: reference graph: $IG_e(B_{20})$.

TABLE 11.8 $SC(IG(B_{50}), IG_e(B_{20}))$: 50 Birds, $s = 10$

Time Interval	Min	Median	Mean	σ	Runtime (ms)
T_{100}	0.805	1.154	1.136	0.202	0.42
T_{500}	−0.177	0.531	0.542	0.566	0.64
T_{1000}	−0.948	−0.056	0.078	0.773	0.95

that has the smallest difference between itself and the reference graph, $IG_e(B_{20})$. Similar to the comparison involving $IG(B_{20})$, the large number of interactions indicates that the birds were flocked at some point during the snapshot interval.

Table 11.9 presents the results of a comparison where the comparison graph is derived from a larger snapshot interval and contains more birds and more bird types

TABLE 11.9 $SC(IG(B_{50}), IG_e(B_{20}))$: 50 Birds, $s = 10$

Time Interval	Min	Median	Mean	σ	Runtime (ms)
T_{100}	0.933	1.621	1.572	0.281	0.21
T_{500}	0.352	1.052	0.964	0.282	0.42
T_{1000}	0.07	0.934	0.892	0.42	0.45

FIGURE 11.10 $IG(B_{50})$ for two bird types with SC_{SE} closest to 0: 0.07.

than the reference graph. Figure 11.10 displays the graph that obtained a result closest to zero; that is, the Statistical Complexity of the comparison graph is as close to the Statistical Complexity of the reference as possible. It is visually discernible that several flocks have occurred.

Game of Life: Table 11.10 contains the SC_{SE} values obtained from the same models as used in the HD and HDA experiments. These values are more significant than the values acquired from executing HD and HDA on the same comparison. The probability mass function, $p(a_i)$, used in SC_{SE} is the same as the probability mass function used in the Flock of Birds experiments. Similar to the SC_{SE} calculations in the Flock of Birds model, we present the difference in SC_{SE} between the reference and comparison graph. Table 11.11 contains the SC_{SE} values obtained from a comparison where the candidate graph is taken from a model where another emergent pattern, the glider

TABLE 11.10   SC(IG(G_{20}), IG$_e$(G_{20})): 20 × 20 Cells, $s = 5$; Random Initialization

Time Interval	Min	Median	Mean	σ	Runtime (ms)
T_{100}	−431.515	−382.405	−371.529	52.203	4.064
T_{500}	−431.515	−404.496	−400.075	36.462	1.829
T_{1000}	−431.515	−404.496	−401.085	34.177	1.866

TABLE 11.11   SC(IG(G_{20}), IG$_e$(G_{20})): 20 × 20 Cells, $s = 5$; Glider Pattern

Time Interval	Min	Median	Mean	σ	Runtime (ms)
T_{100}	−431.515	−431.515	−431.515	0.000	1.539
T_{500}	−431.515	−431.515	−431.515	0.000	1.553
T_{1000}	−431.515	−431.515	−431.515	0.000	1.661

pattern, occurs. The minimum results attained in Tables 11.10 and 11.11 are possible when the reference snapshot contains no interactions. This suggests that all the cells are dead at these time steps.

Tables 11.12 and 11.13 contain the SC$_{SE}$ values obtained from a comparison where the comparison graph was extracted from a model that was designed to force the pulsar pattern to occur. Figure 11.11 displays the graph with the SC result closest to zero. These results confirm our idea that the information stored in a snapshot is enough to determine that emergence is occurring in these particular models, despite a significantly different number of agents in the case of the $IG(G_{15})$ comparison graph. The zero-valued results of these experiments also highlight that the models in the comparison graphs were showing emergence in the form of the pulsar pattern and also that the proportion of interactions to the snapshot interval size was identical between the comparison and reference graphs. We found similar results in comparisons where both the comparison and the reference graphs contained the glider model.

TABLE 11.12   SC(IG(G_{20}), IG$_e$(G_{20})): 20 × 20 Cells, $s = 5$; Pulsar Pattern

Time Interval	Min	Median	Mean	σ	Runtime (ms)
T_{100}	0.000	0.000	0.000	0.000	2.913
T_{500}	−106.286	−106.286	−106.286	0.000	2.923
T_{1000}	0.000	0.000	0.000	0.000	2.948

TABLE 11.13 SC(IG(G_{15}), IG$_e$(G_{20})): 15 × 15 Cells, $s = 5$; Pulsar Pattern

Time Interval	Min	Median	Mean	σ	Runtime (ms)
T_{100}	0.000	0.000	0.000	0.000	2.156
T_{500}	−106.286	−106.286	−106.286	0.000	2.159
T_{1000}	0.000	0.000	0.000	0.000	2.101

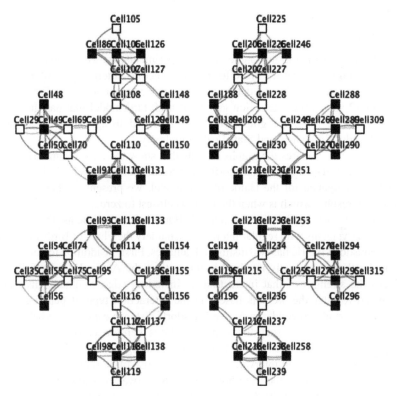

FIGURE 11.11 Game of life: $IG(G_{20})$ with SC$_{SE}$ closest to 0: 0.000.

Predator-Prey: As both HD and HDA have not succeeded to show the presence of emergence in our Predator-Prey model, likely due to the large separation between all agents, considering only the interactions may highlight when emergence is occurring. Table 11.14 presents the values of SC$_{SE}$ for a Predator-Prey model with 4 predators and 8 prey, and a grid size of 20 × 20. We find that the Statistical Complexity metric has solely indicated that there is more activity present within the $IG(P_{4,8})$ than the $IG_e(P_{4,8})$. This suggests that the Statistical Complexity metric cannot detect emergence for our Predator-Prey model.

TABLE 11.14 $SC(IG(P_{4,8}), IG_e(P_{4,8}))$: 4 Predators, 8 Prey, 20×20 Grid, $s = 5$

Time Interval	Min	Median	Mean	σ	Runtime (ms)
T_{100}	−5.111	−4.182	−4.461	0.449	0.008
T_{500}	−6.04	−4.182	−4.368	0.587	0.008
T_{1000}	−10.684	−5.111	−6.318	2.319	0.018

Discussion

Across the Flock of Birds and Game of Life experiments, we were able to show that our architecture and metrics are effective at detecting emergence. Unfortunately, our metrics were not able to determine emergence in the Predator-Prey model. Figure 11.12 shows the interaction graphs of a 20 bird model that has the closest similarity to the reference graph, $IG_e(B_{20})$, obtained using Hausdorff distance, Active Hausdorff distance, and Statistical Complexity.

Figure 11.13 compares the graphs of cells that show the closest similarity to the reference graph, $IG_e(G_{20})$. As Hausdorff distance and Active Hausdorff distance cannot detect emergence for the Game of Life model, we present the best Statistical Complexity results, which is when the result is closest to zero.

For the Active Hausdorff and normal Hausdorff experiments that consider non-stationary agents with regular interactions, namely the Flock of Birds model, the Active Hausdorff metric has two distinct advantages. First, in addition to considering only nodes that are actually involved in interaction, the Active Hausdorff distance analyzes the strength of that interaction by including edge weights. This can be customized to alter the weights based on the interaction type, depending on the system under study. Second, the Active Hausdorff distance captures the behavior of the system over the *entire* snapshot interval, and it is therefore capable of identifying whether the system has been in an emergent state during the interval and has ceased at the end of the snapshot interval. This makes the use of an interaction metric less dependent on the size of the snapshot interval. Moreover, as shown in the runtime results, the calculation of the Active Hausdorff distance is only marginally slower, due to the amount of pre-processing required.

For the Predator-Prey model that contains agents that interact infrequently, the Hausdorff and Active Hausdorff distance metrics were ineffective. This was still the case when the agent space was reduced. There are two main reasons why this may have occurred. First, in the designing of the model, we only considered five types of interactions that could occur, as well as five possible states for the Predator and two for the Prey. There could have been extra states and interactions to be considered. Second, in our aim to run the simulation for a long period of time, we had to reduce the possibility of a population burst from the Prey. This led to the reduction of the initial population that further led to the reduction of possible interactions from occurring.

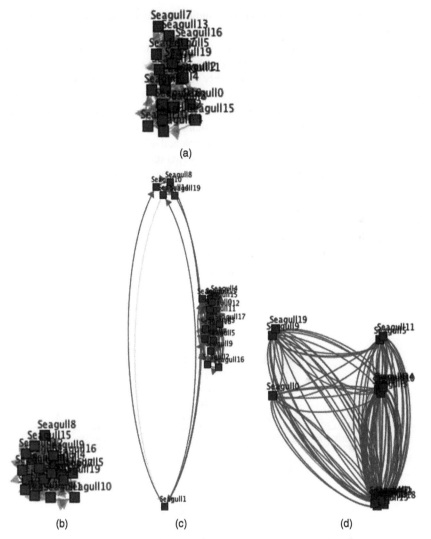

FIGURE 11.12 Best results from HD, HDA, and SC$_{SE}$ when compared to the reference graph $IG_e(B_{20})$. (a) Reference graph: $IG_e(B_{20})$. (b) HD: IG(B_{20}). (c) HDA: IG(B_{20}). (d) SC$_{SE}$: IG(B_{20}).

The Statistical Complexity experiments show that an agent's position in space may not necessarily be the main contributing factor to a system exhibiting emergent behavior. It highlights the knowledge that on the *micro*-level, emergent behavior is derived from interactions. In the case of the Game of Life model, Statistical Complexity was the only metric that gave any suggestion of emergence occurring. For other systems where this is also the case, such as with social networks and power-grids, we believe that Statistical Complexity would provide indications as to whether emergence was occurring.

For the Flock of Birds model, we successfully detected emergence across several model configurations using a reference graph that contained only 20 birds. Despite the different model configurations ostensibly very similar semantically, they are very distinct from an automated emergence identification perspective with respect to numbers of birds, number of types, and their position and velocity values. For the Game of Life model, we were able to detect a particular type of emergence occurring, the pulsar pattern, using a reference graph that contained 400 cells. As the Game of Life model features stationary agents, each with only two states, each particular configuration is quite similar to each other. Despite this, we were able to single out a particular type of emergent behavior occurring.

We also found that graphs that were obtained at different time intervals could be compared successfully. In extreme cases, we theorize that the interval gap would be a problem, but use of certain metrics may lessen this issue as well as how much information is present in the comparison and reference graphs. We also found that comparisons where the reference graph and candidate graph represented different

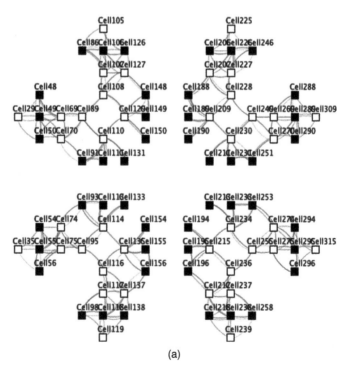

(a)

FIGURE 11.13   Best results from SC_{SE} for Game of Life models with different agent amounts when compared to the reference graph $IG_e(G_{20})$. (a) Reference graph: $IG_e(G_{20})$. (b) SC_{SE}: $IG(G_{15})$. (c) SC_{SE}: $IG(G_{20})$.

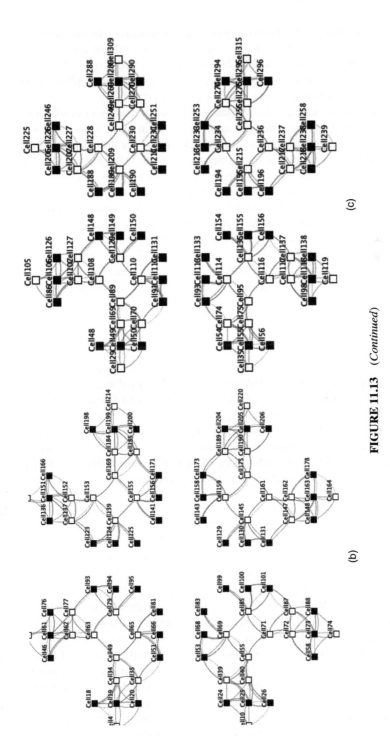

FIGURE 11.13 *(Continued)*

257

numbers of agents did not influence the results significantly. In the Flock of Birds experiments, we were able to detect emergence with a reference graph of just 20 birds in a system that contained 2 types of birds and 50 birds overall. In the Game of Life model, we were able to detect emergence with a reference graph that contained nearly double the amount of cells than in the comparison graph.

As Hausdorff distance and Active Hausdorff distance employ a Euclidian distance, they rely on correctly representing an agent's position in space. Thus, it is crucial that the positional data are validated through checking whether the triangle inequality is maintained between all points in the interaction graph. The agent-based implementations of our three models are of course subject to validation concerns, as are any future agent-based models used with our architecture.

Our proposed metrics rely on comparisons between candidate and reference graphs and thus our *Emergent Properties Repository* is required to contain at least one reference graph before any calculations can occur. The initial reference graphs in the repository must have their emergent properties validated before they can be considered for use in analysis. This could be achieved using methods such as Complex Events (Chen *et al.*, 2007) or reconstructability analysis (Szabo and Teo, 2013).

CONCLUSION

This chapter studies the assumption that emergence is determined by the interactions that take place between the *micro-level* agents and thus *weak* emergence occurs. This assumption is based on previous findings and discussions (Bedau, 1997; Holland, 2007, 1999; Gore and Reynolds, 2008; Pereira *et al.*, 2012; Chan, 2011a). We did not identify or utilize indirect interactions and interactions that involved the environment. Despite these types of interactions being able to lead to emergence, considering them lies outside the scope of our architecture. Although we could have modeled indirect interactions as chains of direct interactions, this has been seldom covered by previous research and also lies outside the scope of our research.

Identifying emergent behavior as it happens in multi-agent systems, without prior knowledge of emergence, is a challenging yet intriguing problem that could bring significant benefits. In this chapter, we propose an architecture that is able to identify emergent behavior as it happens by comparing snapshots of the current system with those taken in previous executions of other systems following existing insight (Birdsey and Szabo, 2014; Mittal and Rainey, 2015). The snapshots of the execution of the multi-agent system are collected at various snapshot intervals and those significantly similar to reference snapshots are highlighted to the user. To facilitate an exhaustive comparison, and as no metric has been shown to perfectly capture emergent behavior, our architecture is designed to include a variety of metrics and secondary distance functions. By analyzing the interactions between agents by way of our metrics, our proposed interaction graphs and associated differences have been used to showcase our proposed approach. The metrics with which we performed our experiments have shown promising results. Hausdorff and Active Hausdorff were able to successfully show various stages of flocking in systems with differing population sizes as well as bird types. Statistical Complexity was able to precisely show when the cells in the Game of Life model achieved a particular pattern. Future work includes further analysis of our approach by using models from different domains and different metrics. It may prove useful to store a snapshot as a binary string representing the system and agents states that could therefore allow measures such as information entropy and the Levenshtein distance (Navarro, 2001). The Levenshtein distance determines how different two strings are by counting the minimum number of operations it would take to turn one into the other. This could be used as a secondary distance function for a particular metric in a system that places great reliance on emergence being a product of agent states. Several validation concerns remain as formal proofs of the correctness of the Active Hausdorff distance, Statistical Complexity, and other proposed metrics in identifying emergence are needed. Lastly, enhancing the *Emergent Properties Repository* to enable complete self-population of reference snapshots, thereby alleviating the initial input of a system expert, is a major challenge that would greatly improve our architecture and may be solved using more complex emergence detection methods.

REFERENCES

Bedau, M. (1997) Weak emergence. *Nous*, **31** (11), 375–399.

Bernon, C., Gleizes, M.-P., Peyruqueou, S., and Picard, G. (2003) ADELFE: a methodology for adaptive multi-agent systems engineering, in *Engineering Societies in the Agents World III. ESAW 2002*, Lecture Notes in Computer Science, vol. **2577** (eds P. Petta, R. Tolksdorf, and F. Zambonelli), Springer-Verlag, Berlin, Heidelberg, pp. 156–169.

Birdsey, L. and Szabo, C. (2014) An architecture for identifying emergent behavior in multi-agent systems. Proceedings of the 13th International Conference on Autonomous Agents and Multiagent Systems, pp. 1455–1456.

Birdsey, L., Szabo, C., and Falkner, K. (2016) CASL: a declarative domain specific language for modeling complex adaptive systems. Winter Simulation Conference (WSC), 2016, IEEE, pp. 1241–1252.

Brown, D.S. and Goodrich, M.A. (2014) Limited bandwidth recognition of collective behaviors in bio-inspired swarms. Proceedings of the 13th International Conference on Autonomous Agents and Multiagent Systems, pp. 405–412.

Cavalin, P., Neto, S.B., Pinhanez, C., Gribel, D., and Appel, A.P. (2014) Large-scale multi-agent-based modeling and simulation of microblogging-based online social network, in *International Workshop on Multi-Agent Systems and Agent-Based Simulation*, Lecture Notes in Computer Science, vol. **8235**, Springer-Verlag, Berlin, Heidelberg, pp. 17–33.

Chan, W.K.V. (2011a) Interaction metric of emergent behaviors in agent-based simulation. Proceedings of the Winter Simulation Conference, Phoenix, AZ, USA, pp. 357–368.

Chan, W.K.V. (2011b) Interaction metric of emergent behaviors in agent-based simulation, in *Winter Simulation Conference* (eds S. Jain, R.R. Creasey Jr., J. Himmelspach, K.P. White, and M.C. Fu), IEEE.

Chan, W., Son, Y.S., and Macal, C.M. (2010) Simulation of emergent behavior and differences between agent-based simulation and discrete-event simulation. Proceedings of the Winter Simulation Conference, pp. 135–150.

Chen, C., Nagl, S.B., and Clack, C.D. (2007) Specifying, detecting and analysing emergent behaviours in multi-level agent-based simulations. Proceedings of the Summer Computer Simulation Conference.

Chen, C.C., Nagl, S.B., and Clack, C.D. (2009) A formalism for multi-level emergent behaviours in designed component-based systems and agent-based simulations, in *From System Complexity to Emergent Properties*, edited by Aziz-Alaoui, M. A., and C. Bertelle, Springer, pp. 101–114.

Chi, L. (2009) Transplating social capital to the online world: insights from two experimental studies. *Journal of Organizational Computing and Electronic Commerce*, **19**, 214–236.

Cilliers, P. (1998) *Complexity & Postmodernism*, Routledge, London.

Davis, P. (2005) New paradigms and challenges. Proceedings of the Winter Simulation Conference, Orlando, FL, USA.

Fayyad, U. and Uthurusamy, R. (2002) Evolving data into mining solutions for insights. *Communications of the ACM*, **45** (8). doi: 10.1145/545151.545174.

Floyd, S. and Jacobson, V. (1993) The synchronization of Periodic Routing Messages. Proceedings of Special Interest Group on Data Communication, pp. 33–44.

Fromm, J. (2006) On engineering and emergence. arXiv preprint nlin/0601002.

Gardner, M. (1970) Mathematical games: the fantastic combinations of John Conway's new solitaire game "Life". *Scientific American*, **223**, 120–123.

Gershenson, C. and Fernandez, N. (2012) Complexity and information: measuring emergence, self-organization, and homeostatis at multiple scales. *Complexity*, **18** (2), 29–44.

Gore, R. and Reynolds, P. (2008) Applying causal inference to understand emergent behavior. Proceedings of the Winter Simulation Conference, Miami, USA, pp. 712–721.

Holland, J. (1999) *Emergence, From Chaos to Order*, Basic Books.

Holland, O.T. (2007) Taxonomy for the modeling and simulation of emergent behavior systems. Proceedings of the 2007 Spring Simulation Multiconference, pp. 28–35.

Huttenlocher, D.P., Klanderman, G.A., and Rucklidge, W.J. (1993) Comparing images using the Hausdorff distance. *IEEE Transactions on Pattern Analysis and Machine Intelligence*, **15**, 850–863.

Jacyno, M., Bullock, S., Luck, M., and Payne, T.R. (2009) Emergent service provisioning and demand estimation through self-organizing agent communities. Proceedings of the International Conference on Autonomous Agents and Multiagent Systems, vol. 1, pp. 481–488.

Johnson, C.W. (2006) What are emergent properties and how do they affect the engineering of complex systems? *Reliability Engineering & System Safety*, **12**, 1475–1481.

Kubik, A. (2003) Towards a formalization of emergence. *Journal of Artificial Life*, **9**, 41–65.

Medusa (2017) Medusa-Visualization, https://sites.google.com/site/medusa3visualization/home (accessed 13 November 2017).

Mittal, S. (2013) Emergence in stigmergic and complex adaptive systems: a formal discrete event systems perspective. *Cognitive Systems Research*, **21**, 22–39.

Mittal, S. and Rainey, L. (2015) Harnessing emergence: the control and design of emergent behavior in system of systems engineering. Proceedings of the Conference on Summer Computer Simulation, SummerSim '15, San Diego, CA, USA, Society for Computer Simulation International, pp. 1–10.

Mogul, J.C. (2006) Emergent (mis)behavior vs. complex software systems. Proceedings of the 1st ACM SIGOPS/EuroSys European Conference on Computer Systems, New York, USA, pp. 293–304.

Navarro, G. (2001) A guided tour to approximate string matching. *ACM Computing Surveys*, **33** (1), 31–88.

Odell, J. (1998) Agents and emergence. *Distributed Computing*, **12**, 51–53.

Pereira, L.M., Santos, F.C. et al. (2012) The emergence of commitments and cooperation. Proceedings of the 11th International Conference on Autonomous Agents and Multiagent Systems, vol. 1, International Foundation for Autonomous Agents and Multiagent Systems, pp. 559–566.

Prokopenko, M., Boschetti, F., and Ryan, A.J. (2009) An information-theoretic primer of complexity, self-organization and emergence. *Complexity*, **15**, 11–28.

Ramakrishnan, K.K. and Yang, H. (1994) The ethernet capture effect: analysis and solution. Proceedings of the IEEE Local Computer Networks Conference, Minneapolis, MN, USA.

Repast (2017) The Repast Suite, http://repast.sourceforge.net/ (accessed 13 November 2017).

Reynolds, C. (1987) Flocks, herds, and schools: a distributed behavioral model. Proceedings of ACM SIGGRAPH, pp. 25–34.

Salazar, N., Rodriguez-Aguilar, J.A., Arcos, J.L., Peleteiro, A., and Burguillo-Rial, J.C. (2011) Emerging cooperation on complex networks. Proceedings of the International Conference on Autonomous Agents and Multiagent Systems, pp. 669–676.

Savarimuthu, B., Purvis, M., Cranefield, S., and Purvis, M. (2007) Mechanisms for norm emergence in multiagent societies. Proceedings of the 6th International Joint Conference on Autonomous Agents and Multiagent Systems, AAMAS '07, pp. 173:1–173:3.

Serugendo, G.D.M., Gleizes, M.P., and Karageorgos, A. (2006) Self-organisation and emergence in MAS: an overview. *Informatica (Slovenia)*, **30** (1), 45–54.

Seth, A.K. (2008) Measuring emergence via nonlinear granger causality. Proceedings of the Eleventh International Conference on the Simulation and Synthesis of Living Systems, pp. 545–553.

Shalizi, C.R. (2006) Methods and techniques of complex systems science: an overview, in *Complex Systems Science in Biomedicine*, Springer, New York, pp. 33–114.

Szabo, C. and Teo, Y. (2012) An integrated approach for the validation of emergence in component-based simulation models. Proceedings of the Winter Simulation Conference, pp. 2412–2423.

Szabo, C. and Teo, Y.M. (2013) Post-mortem analysis of emergent behavior in complex simulation models. Proceedings of the 2013 ACM SIGSIM Conference on Principles of Advanced Discrete Simulation, ACM, pp. 241–252.

Tang, M. and Mao, X. (2014) Information entropy-based metrics for measuring emergences in artificial societies. *Entropy*, **16**, 4583–4602.

Teo, Y. and Szabo, C. (2008) CODES: an integrated approach to composable modeling and simulation. Proceedings of the 41st Annual Simulation Symposium, Ottawa, Canada, pp. 103–110.

Teo, Y.M., Luong, B.L., and Szabo, C. (2013) Formalization of emergence in multi-agent systems. Proceedings of the 2013 ACM SIGSIM Conference on Principles of Advanced Discrete Simulation, ACM, pp. 231–240.

Vail, D.L., Veloso, M.M., and Lafferty, J.D. (2007) Conditional random fields for activity recognition. Proceedings of the 6th International Joint Conference on Autonomous Agents and Multiagent Systems, AAMAS '07, pp. 235:1–235:8.

Yaneer, B.-Y. (2004) A mathematical theory of strong emergence using multiscale variety. *Complexity*, **9**, 15–24.

12

ISOLATING THE CAUSES OF EMERGENT FAILURES IN COMPUTER SOFTWARE

Ross Gore

Virginia Modeling Analysis and Simulation Center, Old Dominion University, Norfolk, VA 23529, USA

SUMMARY

Faults in computer software can be either deterministic or non-deterministic. For deterministic faults, a variety of approaches exist for automated debugging including statistical approaches, state-altering approaches, and general approaches. Although these approaches are effective, they assume the fault in the software is deterministic; the activation of the fault is reproduced with a certain set of inputs to the system. However, for many faults, the assumption that a fault is deterministic is not true. The activation and propagation of the fault are not always reproducible. These types of faults are emergent. Emergent faults are caused by (i) employing stochastics in software and (ii) the effect of the internal environment of the system on fault activation conditions. In the first case, the output of software that employs stochastics can include some natural random variance. In these cases, it is difficult to determine if the software passes or fails a given test case because there is variance in the output. In the second case, the states of a system (e.g., hardware, operating system behavior, and application) executing the program can impact defect activation. For each of these cases, we present specific approaches to account for the non-determinism needs to be introduced to enable existing automated debugging approaches to be applied. These techniques, used in combination with traditional automated debugging approaches, represent different methodologies available to handle emergent faults in software.

Emergent Behavior in Complex Systems Engineering: A Modeling and Simulation Approach,
First Edition. Edited by Saurabh Mittal, Saikou Diallo, and Andreas Tolk.
© 2018 John Wiley & Sons, Inc. Published 2018 by John Wiley & Sons, Inc.

INTRODUCTION

When a failure is first observed in software, the prospect of localizing the fault caus-ing it can be daunting. A software fault is an invalid token or bag of tokens in a program statement that will cause a failure when the compiled code for the program that implements the source code is executed (Munson *et al.*, 2006). A software failure occurs when the output of an executed test case, supplied for the program, does not produce the output specified by the requirements (Laprie *et al.*, 1992).

Common practice is to apply classic debugging techniques to identify the program statements and interactions that lead to the fault. This practice is largely manual and it can consume weeks, months, and even years of effort resulting in the consumption of an enormous amount of resources. As a result, any improvement in the process can greatly decrease the cost. This has motivated the development of a variety of different automated debuggers that assist in fault localization by automating part of the process of searching for faults (Harrold *et al.*, 1998; Renieris and Reiss, 2003; Vessey, 1986; Jones and Harrold, 2005; Jones *et al.*, 2002).

Although these debuggers are effective, they assume the fault in the software is deterministic. Deterministic faults are activated and propagated with a certain set of inputs every time the software runs. In other words, they are always reproducible. However, for many faults, this assumption is not true. Over multiple runs, the activa-tion and propagation of the fault cannot always be reproduced. These types of faults are emergent. Emergent faults primarily occur in two classes of software.

The first class of software employs floating-point computations and continuous stochastic distributions to represent, or support the evaluation of, an underlying model. This class of software includes exploratory simulations where the test cases for the software are defined by an analytical solution. The stochastics create at least two difficult questions for users tasked with localizing emergent faults: (i) does an output have to match an analytical solution exactly to be correct and (ii) if not, how close must an output be to the analytical solution to not be considered a fault?

The second class of software is implemented in an environment where the schedul-ing of internal tasks (e.g., hardware, operating system behavior, and application) that execute the software can influence how and when within execution a fault is acti-vated. Reproducing a specific execution sequence for this class of software is difficult because events are scheduled by the operating system, not the software being tested. To debug a fault, the programmer must predict the order of execution that caused the fault. Furthermore, many faults activated by effects of the internal system environ-ment crash the system itself, ending the debugging process prematurely.

The remainder of the chapter proceeds as follows. First, we provide an overview of the different automated debugging approaches that can be applied for non-emergent faults. Then, we define emergent faults and discuss how they arise in the two afore-mentioned classes of software. For each of these classes of software, we present spe-cific techniques to account for the absent determinism. Finally, we describe how these techniques provide infrastructure to enable existing automated debugging approaches to be applied to enable effective localization of emergent faults.

ISOLATING THE CAUSE OF DETERMINISTIC FAILURE

In this section, the advantages and drawbacks of existing fault localization approaches for deterministic faults are reviewed. Deterministic faults are reproduced with a certain set of inputs to the system every time the inputs are present. First, statistical approaches to deterministic fault localization are discussed, followed by state-altering approaches. Then, fault-specific approaches and static approaches to deterministic fault localization are reviewed.

Statistical Debugging Approaches

There has been considerable research on using statistical approaches for deterministic fault localization. These approaches, referred to as statistical debuggers, require test inputs, corresponding execution profiles, and a labeling of the test executions as either succeeding or failing. The execution profiles reflect the coverage of program elements. The approaches employ an estimate of suspiciousness to rank the program elements.

Developers examine program elements in decreasing order of the suspiciousness estimate until the fault is discovered. Program elements refer to individual program statements or the truth values of conditional propositions represented by branches or inserted predicates. Statistical approaches at the statement level are more efficient but less effective than approaches at the predicate level because they only consider statement coverage and not the values of variables within statements. Statement-level statistical approaches are reviewed in Statement-Level Approaches section, then predicate-level statistical approaches are summarized in Predicate-Level Approaches section.

Statement-Level Approaches Statement-level statistical fault localization research focuses on the discovery that failing program executions are likely to have different statement coverage patterns compared to passing program executions (Harrold *et al.*, 1998). Initial research in this area centered around identifying statistics of statement coverage to effectively estimate the suspiciousness of statements.

The goal was to create a statistic that maximized the suspiciousness estimate for the program statement containing the fault. In the nearest neighbor approach, a failing execution is paired with a larger number of passing executions. The passing execution that is most similar to the failing execution is identified and differences between the two are labeled as suspicious (Renieris and Reiss, 2003).

Set-union and set-intersection approaches have been explored as well. The set-intersection approach identifies suspicious statements by computing the set difference between the statements that are present in every passed execution and the statements in a single failing execution (Vessey, 1986). The set-union approach identifies suspicious statements by removing the union of all statements (or entities) in passing executions from the statements (or entities) within a failing execution (Vessey, 1986). Formally, given a set of passing test cases P containing individual

$$E_{\text{initial}} = E_f - \bigcup_{p \in P} E_p \qquad E_{\text{initial}} = \bigcap_{p \in P} E_p - E_f$$

Set union approach   Set intersection approach

FIGURE 12.1   The set union and set intersection approaches to statistical debugging.

passed test cases p_i, and a single failing test case f, the set of coverage entities executed by each p is E_p, and the coverage entities (E_{initial}) executed by f is E_f, the two approaches are shown in Figure 12.1.

Tarantula is the first statistical approach to recognize that statements exercised more often by failing runs than by passing runs are likely to be faulty (Jones and Harrold, 2005; Jones et al., 2002). Tarantula calculates the fraction of failed executions that execute a statement, over all passed executions. It uses the former fraction over the sum of the two fractions as the statement's suspiciousness estimate. The Tarantula suspiciousness estimate measures specificity (or precision).

Figure 12.2 shows the program, mid(), and its test suite. The program mid() takes three integers as input and is required to output the median value. The function fails to properly identify the median number for some inputs because there is a fault in Statement 7. Statement 7 should read $m = x$; however, it reads $m = y$.

Figure 12.2 illustrates the process of employing statistical debugging to localize this fault. The debugger begins by executing mid() for each of the test inputs shown at

Program source code		3, 3, 5	1, 2, 3	3, 2, 1	5, 5, 5	5, 3, 4	2, 1, 3	Suspiciousness	Rank
mid() {				Execution trace					
1	read("Enter 3 numbers:", x, y, z);	●	●	●	●	●	●	.16	7
2	m = z;	●	●	●	●	●	●	.16	7
3	if (y < z)	●	●	●	●	●	●	.16	7
4	if (x < y)	●	●			●	●	.25	3
5	m = y;			●				.00	13
6	else if (x < z)	●				●	●	.33	2
7	m = y; // THIS IS A BUG!	●					●	.50	1
8	else				●	●		.00	13
9	if (x > y)				●	●		.00	13
10	m = y;				●			.00	13
11	else if (x > z)					●		.00	13
12	m = x;							.00	13
13	print("Middle number is: ", m);	●	●	●	●	●	●	.16	7
}				Actual vs. specified					
		3 vs. 3	2 vs. 2	2 vs. 2	5 vs. 5	4 vs. 4	1 vs. 2		
				Pass/fail label					
		P	P	P	P	P	F		

FIGURE 12.2   Statistical debugging example using Tarantula.

the top of the figure. The execution of mid() for each test input is traced to record the statements that are executed. The columns below the test inputs reflect each execution trace: a black dot signifies that the statement was executed, and the lack of a black dot signifies that the statement was not executed.

Once mid() is executed for a given test input, the actual output of the program is compared to the specified output. The actual and specified outputs for each test input are shown immediately below the execution trace. The actual output is written in italics and the specified output is underlined. These outputs determine if the corresponding execution trace is labeled as passing or failing. If the actual output matches the specified output, then the execution trace passes; otherwise, it fails. The result of applying this labeling process to each execution trace of mid() is shown in the bottom row of the figure.

Labeling each execution trace as passing or failing enables the suspiciousness and rank of each statement in the source code of mid() to be computed. Suspiciousness measures how likely it is that a statement contains a fault. It is calculated by computing the ratio of the number of failing execution traces that include the statement to the number of total execution traces that include the statement. The suspiciousness of each statement is shown in second right-most column of Figure 12.2. The eclipse plugin for Tarantula is shown in Figure 12.3, note the program being debugged is different from the one shown in Figure 12.2.

Other statement suspiciousness estimates balance Tarantula's specificity measure with a sensitivity (or recall) measure. For example, in statement-level bug isolation (SBI), Yu et al. combine the proportion of executions including the statement that are faulty and the proportion of faulty executions in which the statement appears together via their harmonic mean to yield a suspiciousness estimate balancing both sensitivity and specificity (Yu et al., 2008). Similarly, the Ochiai measure balances sensitivity and specificity via their geometric mean (Abreu et al., 2007).

Over time, Tarantula, SBI, and other similar approaches were shown to be more effective than set-union and set-intersection (Jones and Harrold, 2005; Yu et al., 2008; Abreu et al., 2007). As a result, Tarantula, SBI, and other similar approaches have been enhanced. Researchers have explored how to cluster test cases in Tarantula to facilitate multiple developers to debug a faulty program in parallel (Jones et al., 2007). Others observed that some groups of statements, referred to as dynamic basic blocks, are always executed by the same set of test cases. To optimize Tarantula, they find a subset of the original test suite that aims to maximize the number of dynamic basic blocks executed (Baudry et al., 2006).

More recent research has focused on identifying the best measure to estimate statement suspiciousness. The explored measures include F^1 measure (harmonic mean of

Suspicious Statement	File	Line #	Rank
if ((dx > 0) && (x ≥ 750)) {	org/newdawn/spaceinvaders/ShipEntity.java	40	0.97
return;	org/newdawn/spaceinvaders/ShipEntity.java	41	0.96
g.setColor(Color.white);	org/newdawn/spaceinvaders/Game.java	304	0.9
g.drawString(message,(800-g.getFontMetrics().stringWidth(message))/2,250);	org/newdawn/spaceinvaders/Game.java	305	0.9
g.drawString("Press any key",(800-g.getFontMetrics().stringWidth("Press any key"))/2,300);	org/newdawn/spaceinvaders/Game.java	306	0.9

FIGURE 12.3 Eclipse plugin for automated debugging based on Tarantula's ranking. Note: the program being debugged is different from the one shown in Figure 12.2.

specificity and sensitivity) (Yu *et al.*, 2008), Tarantula suspiciousness (specificity) (Jones and Harrold, 2005), capture propagation (specificity of basic blocks) (Zhang *et al.*, 2009), and the Ochiai measure (geometric mean of sensitivity and specificity) (Abreu *et al.*, 2007). Research has shown that in practice, the most effective estimate of statement suspiciousness is the Ochiai measure (Abreu *et al.*, 2007).

However, even the Ochiai measure can be further improved. Research has shown that all statement-level suspiciousness estimates contain confounding bias due to subject program control and data flow dependencies and that it can be improved by a causal model at the statement level that reduces these biases (Baah *et al.*, 2010, 2011).

Predicate-Level Approaches Predicate-level statistical debugging approaches represent a class of fault localization techniques that share a common structure. Each approach consists of a set of conditional propositions, or predicates, which are inserted into a program and tested at particular points. A single predicate can be thought of as partitioning the space of all test cases into two subspaces: those satisfying the predicate and those not. Better predicates create partitions that more closely match the space where the fault is expressed. Similar to statements, the predicates are ranked, based on their estimated suspiciousness and guide developers in finding and fixing faults.

As shown in Figure 12.4, the canonical predicate-level statistical debugger Cooperative Bug Isolation (CBI) (Liblit *et al.*, 2003, 2005; Liblit, 2008) computes the proportion of failing executions where a predicate is evaluated (true or false) and the proportion of failing executions where a predicate is evaluated and true. It then calculates the increase from the former proportion to the latter proportion and uses the resulting increase as an estimate of the predicate's suspiciousness. SOBER introduces

Suspiciousness	Predicate
0.769379	(p + passage_index)->last_line < 4
0.686149	(p + passage_index)->first_line < i
0.675982	i > 20
0.671991	i > 26
0.619479	(p + passage_index)->last_line < i
0.600712	i > 23
0.591044	(p + passage_index)->last_line == next
0.567753	i > 22
0.544829	i > 25
0.536122	i > 28

FIGURE 12.4   An example of CBI-ranked predicates from a numeric program.

the concept of statement evaluation bias to express the probability that a predicate is evaluated to be true in an execution (Liu *et al.*, 2005). By collecting such evaluation biases of a statement in all failed executions and those in all passed executions, SOBER compares the two distributions of evaluation biases and accordingly estimates the suspiciousness of the predicate (Liu *et al.*, 2005).

Recent work has improved the effectiveness and efficiency of predicate-level statistical debugging. Adaptive Bug Isolation showed that the number of program points that need to be instrumented and monitored to identify fault predicting predicates can be significantly reduced with adaptive sampling (Nainar and Liblit, 2010). The reduction in the number of instrumented program points improves overall efficiency of predicate-level approaches but not the effectiveness. Compound Boolean predicates combine existing CBI predicates together via Boolean expressions. The result is improved effectiveness but increased overhead (Nainar *et al.*, 2007). In their work with the Holmes debugging tool, Chilimbi *et al.* explored profiling paths instead of predicates to identify sources of program failures (Chilimbi *et al.*, 2009). Profiling paths, as opposed to predicates, require additional instrumentation but can provide more context on how faults are exercised, which can aid developers in debugging. Finally, Zhang *et al.* showed that Boolean expression short circuit rules used in some compilers can cause the effectiveness of predicate-based approaches to vary significantly for some subject programs (Zhang *et al.*, 2010).

State-Altering Approaches

State-altering approaches to deterministic fault localization attempt to identify the cause of program failure by repeatedly altering the program's state and re-executing the program. Here, program state refers to the values of variables, constants, and inputs to the software. As the program is executed, variables can change values, the change in these values reflects a change in the state of program. For example, a control variable in a loop changes the state of the program every time the loop is executed.

Delta Debugging systematically narrows down the variables and values relevant to the failure, by iteratively modifying the program state for each test case and observing if there is any difference in the outcome (Zeller, 2002; Zeller and Hildebrandt, 2002). The process identifies cause–effect chains relevant to a failure and these chains lead to the isolation of the faulty statement (Cleve and Zeller, 2005). Recently, the efficiency of Delta Debugging has been significantly improved by encoding test case inputs into a hierarchy (Misherghi and Su, 2006). Interesting Value Map Pairs (IVMP) is a similar but more aggressive state-altering approach than Delta Debugging. IVMP potentially modifies values at every executed statement in a failing test case. This improves effectiveness but leads to a less efficient approach (Jeffrey *et al.*, 2008, 2009).

Predicate Switching differs from both Delta Debugging and IVMP by only altering the outcomes of branches during the execution of a failing run. As a result, Predicate Switching only provides a subset of the state alterations that are performed in IVMP and Delta Debugging. Moreover, because predicate switching only alters control flow, it may cause the execution of the subject program to enter an inconsistent program state (Zhang *et al.*, 2006).

Other General Approaches

Several fault localization approaches cannot be categorized as statistical or state-altering. These approaches have not been shown to be as effective as the previously described tools; however, they have influenced the design of the most effective fault localization techniques and are useful in preventing errors.

Invariant-based approaches formulate program invariants regarding the proper behavior of a subject program. When the formulated invariants are violated, the behavior is reported to the developer as a possible error. The Daikon tool automatically infers likely program invariants by dynamically analyzing program executions (Ernst *et al.*, 2001). The inferred invariants are inserted into the program and report potential errors when violated at runtime.

Check "n" Crash derives error conditions in a program statically and generates test cases to dynamically determine if the error exists (Csallner and Smaragdakis, 2005). Eclat infers an operational model of the correct behavior of a program and identifies inputs whose execution patterns differ from the model. In Eclat, those inputs with a differing execution pattern are likely to reveal errors (Pacheco and Ernst, 2009). Similarly, the FindBugs tool automatically detects a commonly observed set of predefined error patterns in Java programs (Hovemeyer and Pugh, 2004). The PathExpander tool provides support to increase the path coverage of dynamic error detection tools by executing non-taken paths in a sandbox environment (Lu *et al.*, 2006). This allows for error detection in paths that would have otherwise not been taken and analyzed.

Most fault localization tools identify faulty statements causing an error. However, the goal of the analysis for some tools is to facilitate developers in explaining how an identified fault causes an error. For example, one method uses distance metrics to compare passing and failing program executions and isolate the differences between them (Groce *et al.*, 2006). The identified differences are used to shed light on how a fault causes an error. Ko and Myers developed a debugging tool called The Whyline, shown in Figure 12.5, to help developers better understand program behavior (Ko and Myers, 2008). The Whyline allows developers to select a question concerning the output of a program, and the tool then uses a combination of static and dynamic analysis techniques to search for possible explanations.

Addressing Software with Multiple Faults

Each of the approaches in Isolating the Cause of Deterministic Failure section is also applicable to software with multiple faults. The following algorithm guarantees that each approach gives at least one predication for each fault that is present in a program:

1. Rank each statement in descending order of likelihood that it is faulty.
2. Remove the top-ranked statement and discard all execution traces where the statement is true.
3. Repeat the process until either the set of execution traces or statements is empty.

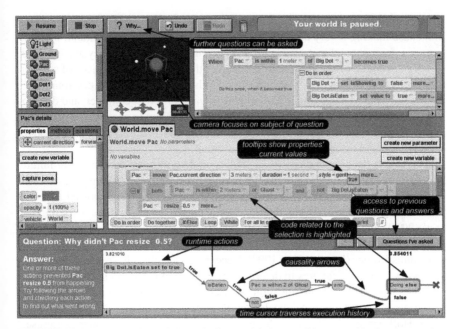

FIGURE 12.5   An overview of the Whyline tool being used in a graphical programming environment.

EMERGENT FAILURES

The previously described approaches assume that the fault in the software is *deterministic*; the activation of the fault is reproduced with a certain set of inputs to the system. These faults are also referred to as Bohrbugs (Grottke and Trivedi, 2005). The term, Bohrbugs, alludes to physicist Niels Bohr and his simple and intelligent atom model. However, for many faults, the assumption that a fault is deterministic is not true. The activation and propagation of the fault are not always reproducible (Munson *et al.*, 2006). These types of faults are *emergent* and are also referred to as Mandelbugs alluding to the "Mandelbrot set" (Grottke and Trivedi, 2005). Emergent faults occur due to:

- Employing stochastics in software: The output of software that employs stochastics can include some natural random variance. In these cases, it is difficult to determine if the software passes or fails a given test case because there is variance in the output.
- The effect of the internal environment of the system on activation conditions: The states of a system (e.g., hardware, operating system behavior, and application) executing the program can impact defect activation. The activation of defects is influenced by states of the system running the program including hardware, software (e.g., operating system and other application programs running on the system), and so on.

In each of these cases, the symptom of the emergent bug is far removed from the root cause making it very hard to detect because the bugs are intermittent, inconsistent, or infrequent and materialize far away from the actual place they are spawned.

As an example of such an episode, consider the case in which a piece of software that works perfectly in one environment, yet fails to work in another environment. If many things have changed (e.g., different hardware, different compiler, and different linker), then there are simply too many degrees of freedom to enable systematic testing under controlled conditions to isolate the bug. Such testing can be done, given enough time and resources, but it is very difficult.

There have been several empirical studies on deterministic and emergent failures in large industrial software systems. The results have been eye-opening. Researchers studied 18 Jet Propulsion Laboratory NASA space missions and showed that even when testing mechanisms were in place, a large number of faults still existed (Grottke *et al.*, 2010). Furthermore, the Tandem system shows that a significant proportion of faults in large and complicated systems are related to emergent failures as opposed to deterministic ones (Lee and Iyer, 1993). Recently, researchers have begun attempting to classify reports of failures to determine if the failure is caused by a deterministic or emergent fault. These classifications have confirmed that emergent failures require more time to fix and more specific debugging strategies to address them (Green *et al.*, 2006; Hoffman *et al.*, 2004). In the remaining sections of this chapter, we present explicit techniques to address emergent faults due to stochastics and emergent faults due to environmental configurations.

STOCHASTICS IN SOFTWARE

The output of software that employs stochastics can include some natural random variance. Frequently these pieces of software are exploratory simulations, where the precise output for a given test case is not known. This occurs in stochastic simulations with test cases where analytical solutions are specified. Despite the existence of an expected value for the test case, the output will include random variance due to the user of stochastics.

In these cases, it is difficult to determine if the software passes or fails a given test case. In the debugging approaches described in Isolating the Cause of Deterministic Failure section, this is required. The labeling of test cases as passing or failing must be a Boolean function. An example helps elucidate this problem.

Consider the pseudocode implementing the Ising model shown in Figure 12.6. The Ising model is used to study ferromagnetism in statistical mechanics. The model consists of discrete variables called spins that can be in one of two states. The spins are arranged in a graph, and each spin interacts with its nearest neighbors (Brush, 1967).

The Monte Carlo implementation of the Ising model shown in Figure 12.6 represents the spin lattice as a two-dimensional array of integers, each valued at positive or negative one. The program makes randomly proposed changes to these spins, using a Metropolis algorithm to enforce the Boltzmann distribution. The output of

ISING(w, h, t, N, β)

```
1   L ← (2-D array of w × h random spins)
2   i ← 0
3   energy ← 0
4   magnetization ← 0
5   while i < N
6       do
7           for (each element in L)    (a sweep)
8               do
9                   x ← (random integer ∈ {0,...w − 1})
10                  y ← (random integer ∈ {0,...h − 1})
11                  ΔE ← (change in energy from flipping L_{x,y})
12                  r ← (random real ∈ [0,1])
13                  if (r < e^{-βΔE}), flip L_{x,y}
14                  if i ≥ t    (the first t sweeps are for thermalization)
15                      do
16                          for (s ∈ L), energy ← energy+ (energy of s)
17                          currentMagnetization ← ∑_{j,k} L_{j,k}
18                          magnetization ← magnetization + |currentMagnetization|
19          i ← i + 1
20  energy ← energy/((N − t)w²h²)
21  magnetization ← magnetization/((N − t)w²h²)
```

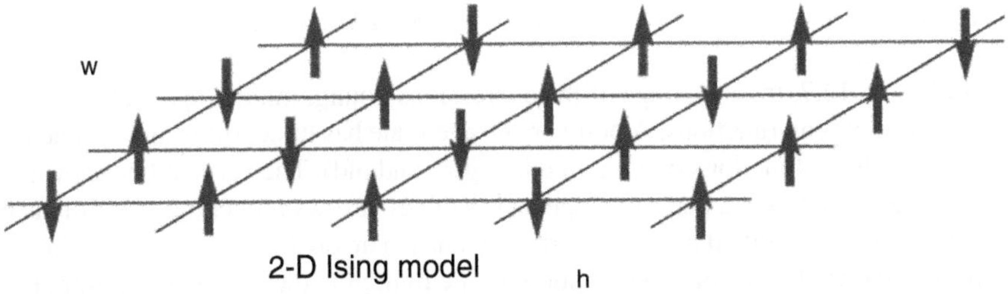

2-D Ising model   h

FIGURE 12.6   A visualization of the Ising model and the algorithm used to implement it.

the simulation is the final value of magnetization – the absolute value of the lattice's total spin per unit volume – averaged over configurations generated by the Metropolis algorithm. For each input, the specified value for the final magnetization is the analytical solution, given by Brush (1967).

Due to the stochastics within the code in Figure 12.6, many test cases will approach their analytical solution but almost none will match it exactly. As a result, almost every test case will be classified as failing. This creates at least two difficult questions for users tasked with localizing emergent faults in this and similar situations: (i) does an output have to match the analytical solution exactly to be considered a passing test case and (ii) if not, how close must an output be to the analytical solution to be considered passing?

To address this problem, researchers employ an approach inspired by fuzzy logic that enables a subject program to both pass and fail a given test case. This approach, fuzzy passing extents, enables users to address the two previous questions for emergent faults caused by stochastics (Gore *et al.*, 2011).

Thus, by employing fuzzy passing extents, debugging approaches built for deterministic software can remain effective for stochastic software, where previously they were in applicable. Overview of Fuzzy Logic section gives an overview of fuzzy logic to facilitate an understanding of fuzzy passing extents. Then Tackling State-Space Explosion section presents fuzzy passing extents.

Overview of Fuzzy Logic

Fuzzy logic is a form of many-valued logic centered around approximate reasoning, as opposed to the exact reasoning featured in traditional logic theory. In traditional logic theory, a variable can have one of the two truth values: true or false. However, in fuzzy logic, variables can have any continuous truth value in the range 0 and 1 ($[0,1]$). This enables fuzzy logic to process incomplete data and provide approximate solutions to problems that traditional logic finds difficult to solve (Zadeh, 2010).

Basic applications of fuzzy logic characterize subranges of a continuous variable. For example, human age can be described with several separate membership functions defining particular age ranges in which one might qualitatively be considered young, middle-aged, and old. Each function maps an age measurement to a truth value in the 0–1 range ($[0,1]$). These truth values are used to determine how one's age might be described.

In Figure 12.7, the meanings of the expressions young, middle-aged, and old are age-scale mapping functions. A point on the age scale has three truth values – one for each of the three functions (young, middle-aged, and old). The vertical line shown in Figure 12.7 represents a particular age and the three arrows reflect the value of each function for the specified age. The dark gray dashed arrow (old) has a truth-value of zero, signifying that the age (35) is not old. The light gray dashed arrow (young) has a truth-value of 0.2 signifying that age 35 is slightly young. Finally, the solid arrow has a truth-value of 0.8 and signifies that age 35 is fairly middle-aged.

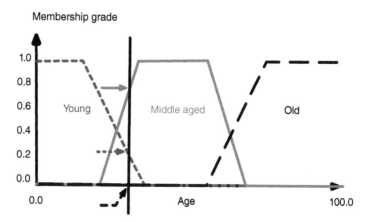

FIGURE 12.7   Fuzzy membership with designations of young, middle-aged, and old.

$$u = \left(\frac{\sum_{i=1}^{|X|} W_i f_i(X_i)}{\sum_{w \in W} w} \right) \text{ where } f_i : \mathbb{R} \to [0,1] \text{ and } |W| = |X|$$

FIGURE 12.8   Equation specifying the passing extent, u, of an output list X.

Fuzzy Passing Extents

Fuzzy passing extents remove the requirement that the outputs of test cases must be either strictly passing or strictly failing. Instead, fuzzy passing extents allow a user to specify a continuous function to classify outputs. Fuzzy passing extents assume that the output of a program is (or can easily be transformed to) some ordered set of real numbers. This is typical of most software that employs stochastics. This set is referred to as the output list, X. The passing extent, u, of the output list, X, is computed using the equation shown in Figure 12.8.

In Figure 12.8, W is an ordered set of weights that focuses attention on particular parts of the output list. The functions f_i encode information about the specified (or passing) output and the tolerance for deviation from that output. The passing extent u reflects the extent to which executing a single test case for a faulty subject program produces the specified (or passing) output. The sum of the passing extent of every test case in the test suite reflects the total number of passing test cases. This sum and similar measures derived from u are used in place of the existing terms in the techniques presented in Isolating the Cause of Deterministic Failure section.

Fuzzy passing extents are a continuous generalization of Boolean pass/fail detection (Harrold et al., 1998; Liblit, 2008; Liu et al., 2005). This means that while they do not guarantee improvements in the effectiveness of predicate-level statistical debuggers, fuzzy passing extents can reproduce the results of each of the suspiciousness estimates previously presented in this chapter.

Ultimately, choosing the functions for the passing extent for the test cases of a faulty subject program is a nontrivial problem, and one that is entirely dependent on the user. However, the fuzzy formalism allows a choice that can enable the Boolean function used to label passing and failing test cases in existing debugging approaches for deterministic software to be applicable to software that employs stochastics.

Fuzzy Passing Extents

An example helps elucidate the utility of fuzzy passing extents. Recall the implementation of the two-dimensional Ising model in Figure 12.6. The implementation represents the spin lattice as a two-dimensional array of integers, each valued at positive or negative one. Random changes are proposed to these spins using a Metropolis algorithm to enforce the Boltzmann distribution. The output of the simulation is the final value of magnetization – the absolute value of the lattice's total spin per unit volume – averaged over configurations generated by the Metropolis algorithm.

As previously discussed, this implementation of the Ising model should approximate the analytical solution, given by Brush (1967).

However, the following fault, based on published mistakes, is injected into the simulation: the absolute value of each sweep's magnetization is not taken while summing magnetizations to compute the average (line 18 of the pseudo-code) (Gore et al., 2011). Over many sweeps, one would expect the faulty behavior to always result in zero magnetization because the algorithm would thoroughly sample states magnetized in positive and negative directions. However, over fewer sweeps, the magnetization will often be close to plus or minus the analytical solution. The reason for this is that both the positive and negative magnetizations are surrounded by relatively probable configurations. If the random walk of the Metropolis algorithm wanders toward one direction of magnetization, it has a low probability of moving to the other within a small number of moves.

The result is an implementation of the Ising model that produces output that approaches the analytical solution at times and output that significantly veers away from the analytical solution at others. To isolate this fault, predicate-level statistical debugging (i.e., CBI) with fuzzy passing extents can be applied to isolate the cause of the failure.

The fuzzy passing function for the model is a curve centered on the analytical solution, $\bar{x}$, for each input. Formally, function is: $e^{\frac{(-(x-\bar{x})^2)}{a^2}}$, where a is the tolerance for deviation. Here, $a = 1.0$. A graph that plots fuzzy passing extents found by applying the fuzzy passing function is shown in Figure 12.9. The divergence between the analytical solution and the expected results for several of the test cases is highlighted by the low fuzzy passing extents shown on the far right of the curve.

Due to the low fuzzy passing extent of these test cases, the suspiciousness of the predicate (magnetization < 0) localizes the values of magnetization (in line 18 of the pseudo-code) that are < 0. The passing extents for this predicate are significantly lower than the other predicates because repeated negative values of magnetization cause the simulation to significantly deviate from the analytical solution. The fuzzy passing extents corresponding to this predicate are shown on the right side of Figure 12.9.

This analysis is not possible without fuzzy passing extents because it is extremely unlikely that any of the test cases will exactly match the analytical solution for the Ising model. This example illustrates how important it is to make debugging approaches applicable to programs that employ stochastics.

FAULTS ACTIVATED BY EFFECTS OF THE INTERNAL SYSTEM ENVIRONMENT

Eliminating faults activated by effects of the internal system environment is not an easy task. Reproducing these types of faults is difficult because the execution sequence of the software is non-deterministic. For example, the multi-threaded program shown in Figure 12.10, contains four lines of code across two threads yielding six unique execution sequences and four unique program end states.

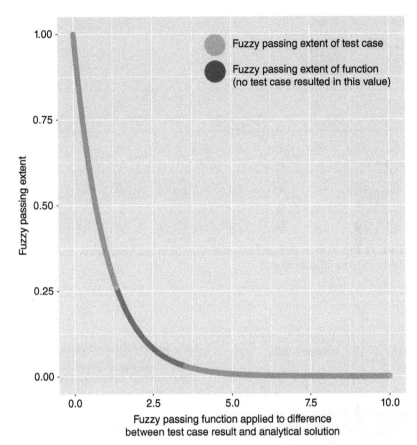

FIGURE 12.9   The result of applying the fuzzy passing function to test cases for the faulty implementation of the Ising model. The center portion of the curve reflects fuzzy passing values that applying the fuzzy function to the test cases did not yield.

Reproducing any one of these six sequences is difficult because threads are scheduled by the operating system, not the software being tested. To debug a fault, the programmer must predict the order of thread execution that caused the fault to gain insight into the error and fix it. Furthermore, many faults activated by effects of the internal system environment, crash the system itself.

To address these emergent faults requires introducing a scheduling software layer to eliminate the non-determinisim within the internal environment. The scheduling software layer, shown in Figure 12.11, is implemented by redirecting calls to the original operating system to alternate implementations where determinism is explicitly scheduled. By capturing and explicitly scheduling the calls related to the non-determinism, the software is able to impose determinism on the system. The result is the capability to organize and explicitly schedule different possible execution sequences. This capability enables the scheduler to explore each possible

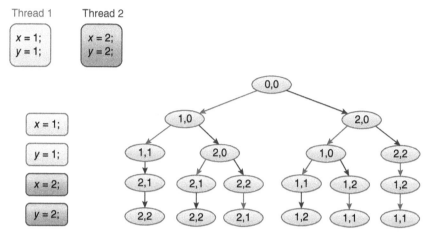

FIGURE 12.10   A multi-threaded program with four lines yielding six unique execution sequences and four unique end states.

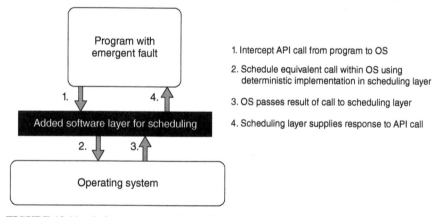

FIGURE 12.11   Software layer and flow diagram for scheduling determinism in programs with emergent faults (Musuvathi *et al.*, 2008).

the execution sequence that could have activated the fault, without repeatedly trying execution sequences that have been previously scheduled. Once the sequence that activates the fault is found, it can be replayed and the debugging methods discussed in Isolating the Cause of Deterministic Failure section can be applied.

An example of this approach is the tool Chess developed by Microsoft for systematic and deterministic testing of concurrent programs (Musuvathi *et al.*, 2008). For multi-threaded program, Chess takes complete control over the scheduling of threads and asynchronous events, thereby capturing all the interleaving non-determinism in the program. This provides two important benefits. First, if an execution results in an error, Chess has the capability to reproduce the erroneous thread interleaving.

This substantially improves the debugging experience. Second, Chess uses systematic enumeration techniques (Clarke and Emerson, 1981; Killian *et al.*, 2007; Godefroid, 1997; Queille and Sifakis, 1981; Musuvathi *et al.*, 2002; Yang *et al.*, 2006) to force every run of the program along a different thread interleaving. Such a systematic exploration greatly increases the chances of finding errors in existing tests without developers having to think of pathological input configurations to stress the system.

In order to build such a software layer, several challenges must be addressed. First, such a software layer should avoid perturbing the system under test and be able to test the code as is. Testers do not have the luxury of changing code. Similarly, testers cannot change the operating system. Therefore, the added software layer should easily integrate with existing test infrastructure with no modification to the system under test.

Second, the layer must systematically accomplish the nontrivial task of capturing and exploring all interleaving non-determinism. For example, concurrency is enabled in most systems via Application Program Interfaces (APIs) that contain a myriad of threading and synchronization functions, many with different options and parameters. The layer must understand the precise semantics of these functions to capture and explore the non-determinism inherent in them. Failure to do so may result in lack of reproducibility, the introduction of false behaviors, and ultimately undermine the credibility of the approach.

Finally, a scheduling layer must explore the space of interleavings intelligently, as the set of interleavings grows exponentially with the number of entities and the number of steps executed by each entity. To effectively search large state spaces, a testing tool must not only reduce the search space by avoiding redundant search (Godefroid, 1996) but also prioritize the search toward potentially erroneous interleavings (Musuvathi and Qadeer, 2007). We address the second and third challenges in more detail next.

Right Layer of Abstraction

A program typically uses system calls implemented in the operating system kernel. In these cases, the scheduler is implemented by redirecting system calls to alternate implementations added on to the operating system. These alternate implementations need to capture and expose the non-determinism of original system calls.

However, for more complex APIs where functionality is implemented outside of the operating system, the design of the scheduler must be adjusted. One acceptable design choice is to include an implementation of these primitives as part of the program. For example, if a third-party library implements locks, the scheduler can treat this library as part of the program under test. Under these circumstances, the scheduler only needs to understand the simpler system primitives that the lock implementation uses.

Although this choice makes the scheduler implementation easier, it also prevents the scheduler from exposing all the non-determinism in the lock implementation. For instance, the library could implement a particular queuing policy that determines the order in which threads acquire locks. Including this implementation as part of the program prevents the scheduler from emulating other queuing policies that future versions of this library might implement. In the extreme, one can include the entire operating system with the program and implement the scheduler at the virtual machine monitor layer.

Ideally, the scheduler should implement a well-documented standard so that the work involved in building the scheduler can be reused across many subject programs.

Tackling State-Space Explosion

Given a program with n threads that execute k atomic steps in total, the number of thread interleavings grows as a scalar factor of n^k. This exponential growth creates an enormous amount of interleavings. To be effective in such a large space of interleavings, it is essential to limit focus on interesting and potentially fault-yielding ones.

One approach to limit focus is to only explore fair interleavings. Systems calls implicitly assume that the underlying operating system schedule is fair. This means that every call will be scheduled eventually. An unfair operating system schedule does not guarantee that every system call will be scheduled.

As a result, the scheduling software layer only has to explore program execution sequences that are fair. Otherwise errors found on execution sequences that are unfair will be useless as the developer would consider these sequences impossible in practice. Finally, fair scheduling is important because it enables the scheduler to return control to the subject program possessing the emergent fault. Unfair schedules result in program sequences that do not terminate. As a result, the scheduling layer never returns control to the subject program.

Other approaches include chunking the code in atomic blocks and only scheduling interleavings of those blocks as opposed to interleavings for every line of code. Identifying where to divide the code into chunks is a difficult process. However, based on the subject program at hand, a number of heuristics exist to facilitate this process (Musuvathi *et al.*, 2008). Employing this approach reduces the number of interleavings for n threads that execute k atomic steps divided into c chunks from n^k to k^c.

The state space can be even further reduced by combining the idea of chunking blocks of code with saving parts of thread interleavings. Given divisions that break the threads into manageable chunks, schedulers can reason that two execution traces with the same interleaving chunks will behave the same. This means that only one of the two threads need to be explored, thus limiting the overall state space.

SUMMARY

In this chapter, we reviewed automated approaches to debugging deterministic and emergent faults. For deterministic faults, a variety of approaches exist for automated debugging including statistical statement-based approaches, statistical predicate-based approaches, and state-altering approaches. Furthermore, a number of other different general approaches to automated debugging exist as well.

Although these approaches are effective, they assume that the fault in the software is *deterministic*. This means that the activation of the fault is reproduced with a certain set of inputs to the system. However, for many faults, the assumption that a fault is deterministic is not true. These types of faults are emergent.

Emergent faults are caused by (i) employing stochastics in software and (ii) the effect of the internal environment of the system on fault activation conditions. In the first case, the output of software that employs stochastics can include some natural random variance. In these cases, it is difficult to determine if the software passes or fails a given test case because there is variance in the output. In the second case, the states of a system (e.g., hardware, operating system behavior, and application) executing the program can impact defect activation.

For each of these cases, a specific approach to account for the non-determinism needs to be introduced to enable existing automated debugging approaches to be applied. Fuzzy passing extents can be employed in software with stochastics to enable a single test case to both pass and fail. Similarly, introducing an additional scheduling layer with alternate implementations of the operating system's system calls can remove the non-determinism in programs where the internal environment of the system impacts fault activation. Both of these techniques, used in combination with traditional automated debugging approaches, represent different methodologies available to handle emergent faults in software.

REFERENCES

Abreu, R., Zoeteweij, P. and van Gemund, A.J.C. (2007) On the accuracy of spectrum-based fault localization. Proceedings of the Testing: Academic and Industrial Conference Practice and Research Techniques – MUTATION. Washington, D.C.: IEEE Computer Society, pp. 89–98.

Baah, G.K., Podgurski, A. and Harrold, M.J. (2010) Causal inference for statistical fault localization. Proceedings of the 19th International Symposium on Software Testing and Analysis (ISSTA '10), ACM, New York, pp. 73–84.

Baah, G.K., Podgurski, A. and Harrold, M.J. (2011) Mitigating the confounding effects of program dependences for effective fault localization. Proceedings of the 19th ACM SIGSOFT Symposium and the 13th European Conference on Foundations of Software Engineering (ESEC/FSE '11), ACM, New York, pp. 146–156.

Baudry, B., Fleurey, F. and Traon, Y.L. (2006) Improving test suites for efficient fault localization. Proceedings of the 28th International Conference on Software engineering (ICSE '06), ACM, New York, pp. 82–91.

Brush, S.J. (1967) History of the lenz-Ising model. *Review of Modern Physics*, **39** (4), 883–893.

Chilimbi, T.M., Liblit, B., Mehra, K., Nori, A.V. and Vaswani, K. (2009) Holmes: effective statistical debugging via efficient path profiling. Proceedings of the 31st International Conference on Software Engineering (ICSE '09), IEEE Computer Society, Washington, DC, pp. 34–44.

Clarke, E. and Emerson, E. (1981) Synthesis of synchronization skeletons for branching time temporal logic. *LNCS: Logic of Programs*, **131**, 52–71.

Cleve, H. and Zeller, A. (2005) Locating causes of program failures. Proceedings of the 27th International Conference on Software Engineering, ACM, New York, pp. 342–351.

Csallner, C. and Smaragdakis, Y. (2005) Check 'n' crash: combining static checking and testing. Proceedings of the 27th International Conference on Software Engineering (ICSE '05), ACM, New York, pp. 422–431.

Ernst, M.D., Cockrell, J., Griswold, W.G., and Notkin, D. (2001) Dynamically discovering likely program invariants to support program evolution. *IEEE Transactions on Software Engineering*, **27** (2), 99–123.

Godefroid, P. (1996) *Partial-Order Methods for the Verification of Concurrent Systems: An Approach to the State-Explosion Problem, Lecture Notes in Computer Science*.

Godefroid, P. (1997) Model checking for programming languages using Verisoft. Proceedings of Principles of Programming Languages, ACM, New York, pp. 174–186.

Gore, R., Kamensky, D. and Reynolds, P.F. (2011) Applying enhanced fault localization technology to Monte Carlo simulations. Proceedings of the 43rd Conference on Winter Simulation (WSC '11), Society for Computer Simulation, San Diego, pp. 2798–2809.

Green, N.W., Hoffman, A.R., and Garrett, H.B. (2006) Anomaly trends for long-life robotic spacecraft. *Journal of Spacecraft and Rockets*, **43** (1), 218–224.

Groce, A., Chaki, S., Kroening, D., and Strichman, O. (2006) Error explanation with distance metrics. *International Journal of Software Tools Technology Transfer*, **8** (3), 229–247.

Grottke, M. and Trivedi, K.S. (2005) A classification of software faults. *Journal of Reliability Engineering Association of Japan*, **27** (7), 425–438.

Grottke, M., Nikora, A.P. and Trivedi, K.S. (2010) An empirical investigation of fault types in space mission system software. Proceedings of IEEE/IFIP International Conference on Dependable Systems and Networks, pp. 447–456.

Harrold, M.J., Rothermel, G., Wu, R., and Yi, L. (1998) An empirical investigation of program spectra. *SIGPLAN Notices*, **33**, 83–90.

Hoffman, A.R., Green, N.H. and Garrett, H.B. (2004, August) Assessment of in-flight anomalies of long life outer planet missions. Environmental Testing for Space Programmes, vol. 558, pp. 43–50.

Hovemeyer, D. and Pugh, W. (2004) Finding bugs is easy. *SIGPLAN Notices*, **39**, 92–106.

Jeffrey, D., Gupta, N. and Gupta, R. (2008) Fault localization using value replacement. Proceedings of the 2008 International Symposium on Software Testing and Analysis, ACM, New York, pp. 167–178.

Jeffrey, D., Gupta, N. and Gupta, R. (2009) Effective and efficient localization of multiple faults using value replacement. 25th IEEE International Conference on Software Maintenance (ICSM '09), IEEE Computer Society, Washington, D.C., pp. 221–230.

Jones, J. and Harrold, M.J. (2005) Empirical evaluation of the tarantula automatic faultlocalization technique. Proceedings of the 20th IEEE/ACM International Conference on Automated Software Engineering, ASE '05, ACM, New York, pp. 273–282.

Jones, J.A., Harrold, M.J. and Stasko, J. (2002) Visualization of test information to assist fault localization. Proceedings of the 24th International Conference on Software Engineering (ICSE '02), ACM, New York, pp. 467–477.

Jones, J.A., Bowring, J.F. and Harrold, M.J. (2007) Debugging in parallel. Proceedings of the 2007 International Symposium on Software Testing and Analysis (ISSTA '07), ACM, New York, pp. 16–26.

Killian, C.E., Anderson, J.W., Jhala, R. and Vahdat, A. (2007) Life, death, and the critical transition: finding liveness bugs in systems code. Proceedings of Networked Systems Design and Implementation, ACM, New York, pp. 243–256.

Ko, A.J. and Myers, B.A. (2008) Debugging reinvented: asking and answering why and why not questions about program behavior. Proceedings of the 30th International Conference on Software Engineering, ACM, New York, pp. 301–310.

Laprie, J.C.C., Avizienis, A., and Kopetz, H. (1992) *Dependability: Basic Concepts and Terminology*, Springer-Verlag, New York, NY.

Lee, I. and Iyer, R.K. (1993) Faults, symptoms, and software fault tolerance in the tandem guardian90 operating system. Proceedings of Twenty-Third International Symposium on Fault-Tolerant Computing, IEEE Computer Society, Washington, D.C., pp. 20–29.

Liblit, B. (2008) Cooperative debugging with five hundred million test cases. Proceedings of the 2008 International Symposium on Software Testing and Analysis (ISSTA '08), ACM, New York, pp. 109–120.

Liblit, B., Aiken, A., Zheng, A.X., and Jordan, M.I. (2003) Bug isolation via remote program sampling. *SIGPLAN Notices*, **38**, 141–154.

Liblit, B., Naik, M., Zheng, A.X., Aiken, A. and Jordan, M.I. (2005) Scalable statistical bug isolation. Proceedings of the 2005 ACM SIGPLAN Conference on Programming Language Design and Implementation (PLDI '05), ACM, New York, pp. 15–26.

Liu, C., Yan, X., Long, F. *et al.* (2005) SOBER: statistical model-based bug localization. *SIGSOFT Software Engineering Notes*, **30**, 286–295.

Lu, S., Zhou, P., Liu, W., Zhou, Y. and Torrellas, J. (2006) Pathexpander: architectural support for increasing the path coverage of dynamic bug detection. Proceedings of the 39th Annual IEEE/ACM International Symposium on Microarchitecture (MICRO 39), IEEE Computer Society, Washington, D.C., pp. 38–52.

Misherghi, G. and Su, Z. (2006) Hdd: hierarchical delta debugging. Proceedings of the 28th International Conference on Software Engineering, ACM, New York, pp. 142–151.

Munson, J.C., Nikora, A.P., and Sherif, J.S. (2006) Software faults: a quantifiable definition. *Advances in Software Engineering*, **37** (5), 327–333.

Musuvathi, M. and Qadeer, S. (2007) Iterative context bounding for systematic testing of multithreaded programs. Proceedings of Programming Language Design and Implementation, ACM, New York, pp. 446–455.

Musuvathi, M., Park, D., Chou, A., Engler, D. and Dill, D.L. (2002) CMC: a pragmatic approach to model checking real code. Proceedings of Operating Systems Design and Implementation, ACM, New York, pp. 75–88.

Musuvathi, M., Qadeer, S., Ball, T., Basler, G., Nainar, P.A. and Neamtiu, I. (2008) Finding and reproducing Heisenbugs in concurrent programs. Proceedings of USENIX Symposium on Operating Systems Design and Implementation, ACM, New York, pp. 267–280.

Nainar, P.A. and Liblit, B. (2010) Adaptive bug isolation. Proceedings of the 32nd ACM/IEEE International Conference on Software Engineering (ICSE '10), ACM, New York, pp. 255–264.

Nainar, P.A., Chen, T., Rosin, J. and Liblit, B. (2007) Statistical debugging using compound boolean predicates. Proceedings of the 2007 International Symposium on Software Testing and Analysis (ISSTA '07), ACM, New York, pp. 5–15.

Pacheco, C. and Ernst, M.D. (2009) Eclat: automatic generation and classification of test inputs. ECOOP 2005 – Object-Oriented Programming, 19th European Conference, Glasgow, pp. 504–527.

Queille, J. and Sifakis, J. (1981) Specification and verification of concurrent systems in CESAR. Proceedings of International Symposium on Programming, Springer-Verlag, New York, pp. 337–351.

Renieris, M. and Reiss, S. (2003) Fault localization with nearest neighbor queries. 18th IEEE International Conference on Automated Software Engineering (ASE 2003), IEEE Computer Society, Montreal, pp. 30–39.

Vessey, I. (1986) Expertise in debugging computer programs: an analysis of the content of verbal protocols. *IEEE Transactions on Systems, Man, and Cybernetics: Systems*, **16**, 621–637.

Yang, J., Twohey, P., Engler, D.R., and Musuvathi, M. (2006) Using model checking to find serious file system errors. *ACM Transactions on Computer Systems*, **24** (4), 393–423.

Yu, Y., Jones, J.A. and Harrold, M.J. (2008) An empirical study of the effects of test-suite reduction on fault localization. Proceedings of the 30th International Conference on Software Engineering (ICSE '08), ACM, New York, pp. 201–210.

Zadeh, L.A. (2010) A summary and update of fuzzy logic. IEEE International Conference on Granular Computing, IEEE Computer Society, Los Alamos, pp. 42–44.

Zeller, A. (2002) Isolating cause-effect chains from computer programs. Proceedings of the 10th ACM SIGSOFT Symposium on Foundations of Software Engineering, ACM, New York, pp. 1–10.

Zeller, A. and Hildebrandt, R. (2002) Simplifying and isolating failure-inducing input. *IEEE Transactions on Software Engineering*, **28**, 183–200.

Zhang, X., Gupta, N. and Gupta, R. (2006) Locating faults through automated predicate switching. Proceedings of the 28th International Conference on Software Engineering (ICSE '06), ACM, New York, pp. 272–281.

Zhang, Z., Chan, W.K., Tse, T.H., Jiang, B. and Wang, X. (2009) Capturing propagation of infected program states. Proceedings of the 7th Joint Meeting of the European Software Engineering Conference and the ACM SIGSOFT Symposium on The Foundations of Software Engineering (ESEC/FSE '09), ACM, New York, pp. 43–52.

Zhang, Z., Jiang, B., Chan, W.K. *et al.* (2010) Fault localization through evaluation sequences. *Journal of Systems Software*, **83**, 174–187.

13

FROM MODULARITY TO COMPLEXITY: A CROSS-DISCIPLINARY FRAMEWORK FOR CHARACTERIZING SYSTEMS

Chih-Chun Chen and Nathan Crilly

Department of Engineering, University of Cambridge, Cambridge, CB2 1PZ, UK

SUMMARY

This chapter introduces a domain-neutral framework and diagrammatic scheme that allow researchers and practitioners from different disciplines to share methods, theories, and findings related to the design and study of different systems. The framework is not tied to any established mode of representation (e.g., networks, equations, and formal modeling languages) nor to any domain-specific terminology (e.g., "vertex," "eigenvector," and "entropy"). Instead, it consists of basic system constructs and three fundamental attributes of system architecture, namely structural encapsulation, function-structure mapping, and interfacing. This allows different aspects of complexity (e.g., degeneracy, multi-structural function realization, emergence, and heterarchy) and different abstractions relating to modularity (e.g., function-driven encapsulation and interface compatibility) to be characterized within a common framework. It also relates these characterizations to existing system characterization schemes (e.g., those based on structure, behavior, and function). Thus, modularity and complexity are seen as two ends of a spectrum of systems possessing the three attributes to different degrees.

Emergent Behavior in Complex Systems Engineering: A Modeling and Simulation Approach,
First Edition. Edited by Saurabh Mittal, Saikou Diallo, and Andreas Tolk.
© 2018 John Wiley & Sons, Inc. Published 2018 by John Wiley & Sons, Inc.

This develops much of the content from Chen and Crilly (2016a), which includes a more accessible treatment of the core concepts for those unfamiliar with complex systems constructs.

INTRODUCTION

In both Engineering and Science, the term "complex system" has been used to refer to systems that exhibit properties that are seen to arise through "self-organization" or where elements and interactions need to be understood at different levels or from different perspectives. When the relationships between these different levels and per-spectives are not well defined or are subject to change, the system can be seen as exhibiting unexpected behaviors, sometimes referred to as "emergent."

Although "complexity" in the design context has traditionally been cast in a rather negative light, attempts have also been made to harness complexity (e.g., as seen in "complexity engineering" (Ottino, 2004; Mittal, 2013; Mittal and Rainey, 2015) or "learning from nature" (Dressler and Akan, 2010)). The goal has been to create more efficient systems with desirable change-related properties, such as adaptability, robustness, resilience, and evolvability (discussions of these properties can be found in Fricke and Schulz (2005), McManus and Hastings (2006), Ross *et al.* (2008), Ryan *et al.* (2013), Schoettl and Lindemann (2014)). In all these cases, concepts of complexity, self-organization, and emergence become central to design practice. Furthermore, a complex systems perspective is becoming increasingly common when tackling design and engineering problems that cut across traditional domain bound-aries and involve both designed and non-designed entities. There are many examples of this:

- distributed computational systems and the internet are studied as natural ecolo-gies (Gao, 2000; Forrest *et al.*, 2005);
- evolutionary design and evolutionary computing study the way selection and diversification mechanisms operate in different environmental conditions (fit-ness landscapes) to give differences in the space of design solutions (Bentley, 2002; de Jong, 2002);
- complex sociotechnical systems are characterized as partially designed and par-tially evolving (de Weck *et al.*, 2011);
- bioengineering seeks to design and manufacture artificial systems from biolog-ical substrates (Endy, 2005; Knight, 2005).

Despite the fact that many disciplines have made significant contributions to addressing complexity, they rarely benefit from each other's methods, tools, or insights due to domain-specific terminology and a lack of explicitness or precision. For example, in Chen and Crilly (2016b), it was found that practitioners within Synthetic Biology and Swarm Robotics sometimes shared more complexity-related engineering issues with each other than they did with other practitioners within the same domain (although commonality within the domains was also identified). Work on rigorously defining complexity-related constructs (e.g., "emergence" and "self-organization") tends to assume a particular representation of the system (e.g., a network) or consensus on the terminology used to describe the system (e.g., what the terms "element," "component," and "subsystem" mean). The lack of an idealized, comprehensive, and consistent representation that generalizes across domains makes it difficult for those working within one domain to have confidence in their interpretation of the solutions proposed within another domain (Goldstone and Sakamoto, 2003). This not only limits the dissemination of useful knowledge but also increases the likelihood that practitioners from different domains will misinterpret or misapply each other's work.

To make the methods, theories, and findings from one domain accessible to other domains, we need to consider different aspects of complexity in domain-neutral terms and how they relate to more general systems characterizations. To provide an accessible means for people working in different disciplines and domains to navigate each other's work, this chapter develops a domain-neutral framework and diagrammatic scheme that relates the notion of "complexity" to more fundamental attributes of system architecture, namely structural encapsulation, function-structure mapping, and interfacing. These three architectural attributes also constitute three core aspects of modularity, which is seen by some as the antithesis of complexity (or as a panacea for complexity). For designers, modular architectures permit a system to be divided into more manageable parts, which can be developed, produced, and modified relatively independently. In other words, modularity is seen as a way of "managing complexity" by containing it within well-defined boundaries. For scientists studying complex systems, modularity offers a way of better understanding the system by conceptually grouping together system elements, states, or behaviors. Relatively strong interactions or dependencies exist within modules, whereas relatively weak interactions exist across them so that different phenomena are modeled as arising through interactions between system elements.

The framework we develop is not tied to any established mode of representation (e.g., networks, equations, and formal modeling languages) nor to any domain-specific terminology (e.g., "vertex," "eigenvector," and "entropy"). However, an equivalent representation can always be found in these alternative representations, providing a bridge between different formal representations, as well as between formal representations and natural language descriptions. The framework also allows more general systems ontologies (e.g., Bunge, 1977, 1979; Gero, 1990; Goel et al., 2009; Tomiyama et al., 1993), formalisms (e.g., discrete event systems in Zeigler (1984), Markov models in Bellman (1957) and Meyn (2007), see also Diallo et al. (2014) and Chapter 3), and systems modeling frameworks (e.g., SysML and CML) to be related to literature on complexity and modularity. Thus, the framework serves as a reference language for the discussion of modularity, complexity, and other systems constructs and the ways in which they are related (as demonstrated in Appendix A).

To ensure conceptual explicitness, we include domain-neutral definitions and diagrammatic representations of the key terms introduced. We also use concrete examples drawn from diverse domains to illustrate the abstract concepts that are developed and the ways in which they relate to actual systems. Although it can be tempting to develop a framework by explicitly drawing on existing frameworks and definitions from the literature, doing so often introduces ambiguities and inconsistencies. This is due to differences in the assumptions that underlie different perspectives and variation in how terminology is used within and between fields. To avoid this, the paper proceeds from first principles and then relates these to those found in the established literature in the appendices. The objective is by no means to comprehensively review the literatures relating to systems, modularity, or complexity and therefore we do not endeavor to cite all the "classic" works from different domains. Instead, we reference other works mainly to illustrate terminological discrepancies or to point the reader to further details on the examples given. For domain-specific reviews, the reader is advised to consult introductory texts, on modularity in design (e.g., Baldwin and Clark, 2000; Gershenson *et al.*, 2003; Ulrich and Eppinger, 2003), modularity in science (Newman, 2006), complexity in design (Luzeaux *et al.*, 2011), complexity in science (Ladyman *et al.*, 2013; Mitchell, 2009), and system characterizations generally (Meadows and Wright, 2008).

The chapter is structured as follows. Characterizing Systems section introduces a framework for characterizing systems, focusing on characterizations that are particularly pertinent to design and scientific domains. The framework also defines composition and classification relationships, which form the basis for levels, hierarchies, and heterarchies. Aspects and Abstractions of Modularity section identifies three core aspects of modularity: structural encapsulation, function-structure mapping, and interfacing. Based on these, two abstractions are introduced: function-driven encapsulation and interface compatibility. Aspects of Complexity section uses the systems characterization framework (introduced in Characterizing Systems section) and the aspects and abstractions of modularity (introduced in Aspects and Abstractions of Modularity section) to characterize different aspects of complexity. Conclusions section concludes the chapter by summarizing the relationships between the different aspects of modularity and complexity.

CHARACTERIZING SYSTEMS

To discuss different aspects of complexity and modularity without being tied to the assumptions made by particular domains about systems, we need to have a set of domain-neutral constructs and terms for talking about systems. We use the term "characterization" to refer to any representation, model, specification, or description of an entity. Indeed, calling an entity a "system" itself assumes a particular kind of characterization, which we call a "systems characterization."

For the purposes of this chapter, we define a *system* as a set of entities and relationships, where the relationships are connections or interactions between the entities. We call the entities in the system the *elements* of the system, which might be considered "components" or "subsystems" with respect to the system, as defined below (of course, these elements might themselves be considered systems in some other characterization). Our definition is therefore broader than the narrower sense in which "system" is employed in General Systems Theory (von Bertalanffy, 1968).

By its very nature, a systems characterization of an entity assumes that it can be characterized in multiple ways, each of which emphasizes different elements or aspects, reflecting different perspectives and purposes. Within a given context, characterizations are often reified by the community who apply them (Whitehead, 1919) so that a particular characterization of an entity is treated as the entity itself or as being inherent to the entity.

In order to avoid confusion between cases where we are referring to an entity "in the world" and cases where we are referring to a *characterization* of an entity in the world, we use the term "instance" to refer to the former and the term "type" to refer to the latter. For example, "Boeing-747 instance" would be used to refer to a particular Boeing-747 aircraft, whereas "Boeing-747 type" would be used to refer to a class of aircraft that is associated with the characterization of Boeing-747 aircrafts. This characterization might include architecture, design specifications, functions, behavior, and so on.

Composition, Classification, and Levels

In terms of the relationships between entities, we can distinguish between two formal relationships, "composition" (part-whole) relationships and "classification" (subtype-type) relationships ("specialization" in Zeigler and Sarjoughian (2013)). These two relationships provide the basis for defining "levels" and "hierarchies" (see Tolk and Muguira, 2003 and Hierarchies and Heterarchies section).

Composition and Classification   A composition relationship implies an entity (the "whole") that can be broken down into a set of further entities (the "parts"). The term "element" itself implies a composition relationship between the element and the system. However, different sets of a system's elements can also have part-whole relationships with each other. We use the terms "subsystem", "component," and "supersystem" to characterize such relationships. These are *relational* terms that only make sense when defined with respect to each other and with respect to a given characterization (see Figure 13.1). With respect to a given system, s:

- A *subsystem* of s is a subset of the entities and relationships in s.
- A *component* of s is an entity in s that is not being further decomposed.
- A *supersystem* of s is a superset of the entities and relationships in s.

Note that when we use the term "system," what we really mean is a system *characterization*; we do not make any metaphysical claims about the decomposability of physical entities. In addition to defining subsystems, components and supersystems, with respect to a given system, we define an *environment* of the system as a set of entities and relationships that are not in the set of entities and relationships constituting the system but that belong to a supersystem of the system. The difference between "the supersystem of s" and "the environment of s" is that the supersystem

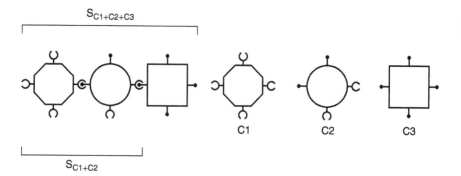

FIGURE 13.1   Composition.

of s includes s, whereas the environment of s does not. In the diagrammatic scheme in Figure 13.1, there are different types of entity (represented by different shapes and interfaces). Here, C1, C2, and C3 represent component types and can be combined to make a system type $S_{C1+C2+C3}$. System type S_{C1+C2} is a subsystem of $S_{C1+C2+C3}$. Entity C3 is a component of $S_{C1+C2+C3}$ but is the environment of system S_{C1+C2} (assuming that no other entities exist, otherwise it is just *part* of the environment). These basic aspects of composition apply both to types and instances of entities.

Entities can be characterized at different levels of *abstraction*. Two elements can be seen to be different to each other at one level but the same as each other at another, more abstract level, where they belong to the same class or "type." Classificatory relationships between characterizations determine which characterizations can be treated as equivalent. We define a *type* as a taxonomic group or "class" associated with a set of subtypes and instances. With respect to a given system type, S (also see Figure 13.2):

- A *subtype* of S is a taxonomic group containing a subset of the entity types, entity instances, and characterizations contained in the set defined by S.
- A *supertype* of S is a taxonomic group containing a superset of the entity types, entity instances, and characterizations contained in the set defined by S.
- An *instance* of S is a concrete realization of S (an entity in the world) that belongs to the set of entities defined by S.

In the diagrammatic scheme in Figure 13.2, component types (outlined shapes) can be represented at two levels of abstraction (although many other levels would be possible): with stars or without stars, where stars represent some feature of the component. These types can also be instantiated (solid shapes). Where components are viewed at a level of abstraction that makes stars visible, there are two options: one star or two stars, representing different variants of the same type of component. Here, two different components are depicted, c2* and c2** (lowercase). Components c2* and c2** are instances of component types C2* and C2** (uppercase), both of which are a subtype of C2. As such, both c2* and c2** are also instances of C2. These basic aspects of classification apply to components, systems, subsystems, supersystems, and environments.

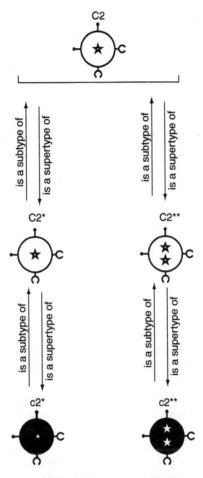

FIGURE 13.2 Classification.

Figure 13.3 shows multiple characterizations of an entity, e, which differ in abstraction and composition. Entity e can be characterized as an instance of the system type $S_{C1*+C2*+C3*}$, composed of C1*, C2*, and C3* but e might also be characterized as an instance of $S_{[C1*+C2*]+C3*}$, composed of $S_{[C1*+C2*]}$ and C3*. It might also be characterized as an instance of the more abstract system type $S_{C1+C2+C3}$ or $S_{[C1+C2]+C3}$. Thus, the compositional hierarchy runs from right to left (from element-level to subsystem-level to system-level characterizations), and the abstraction hierarchy runs from bottom to top, with the bottom characterizations being subtypes of the top characterizations.

Hierarchies and Heterarchies The terms "level" and "hierarchy" are frequently found in systems discourse. The part-whole (composition) and subtype-supertype (classification) relationships defined above give us a means of more precisely understanding these terms. Implicit in the *composition* relationship is what is known

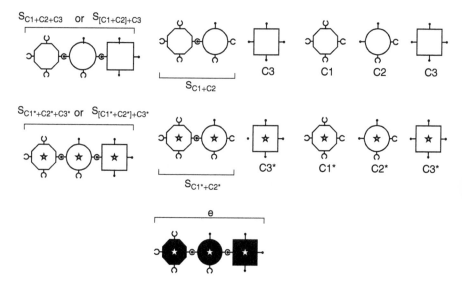

FIGURE 13.3   Multiple characterizations.

as the "scope" of the characterization, which is the set of elements involved. Implicit in the *classification* relationship is the "resolution" of the characterization (also known as "granularity" or "level of abstraction"), which is the set of distinctions that can be made between the elements.

We define *level* as a specification of both the scope and the resolution of a characterization. For example, the level for the system type $S_{C1*+C2*+C3*}$ is defined by the scope of $C1*+C2*+C3*$ and the resolution of $S_{C1*+C2*+C3*}$ as a subtype of $S_{C1+C2+C3}$. Given the definition of "level", a (clean) hierarchy is defined as a set of related characterizations where the levels do not overlap. A *classification hierarchy* is a structure in which if one element is the subtype of another element, it cannot also be its supertype. A *compositional hierarchy* is a structure in which, if one element is the part of another element, that first element cannot also be the whole with respect to that second element. For example, in $S_{C1*+C2*+C3*}$, the component type $C2*$ is related to the system type $S_{C1*C2*C3*}$ in a compositional hierarchy and to the component type $C2$ in a classification hierarchy. Figure 13.4 shows an example of a (clean) hierarchy. $C2$ is related to S_{C1+C2} in compositional hierarchy, and $C2*$ is related to $S_{C1*+C2*}$ in compositional hierarchy. $C2*$ is related to $C2$ in classificatory hierarchy, and $S_{C1*+C2*}$ is related to S_{C1+C2} in classificatory hierarchy.

In the case of complex systems characterizations, multiple hierarchies overlap in a single characterization. This is what is referred to as a "heterarchy" (McCulloch, 1945, Gunji and Kamiura, 2004; Sasai and Gunji, 2008), "panarchy"(Gunderson and Holling, 2001) or "entangled hierarchy" (Palla, 2005) and can be represented by hypernetworks (Johnson, 2007; Chen *et al.*, 2009). Figure 13.5 depicts a heterarchy that contrasts with the hierarchy described above. In Figure 13.5, although there is a classificatory hierarchical relationship between $C2$ and $C2*$, the relationship

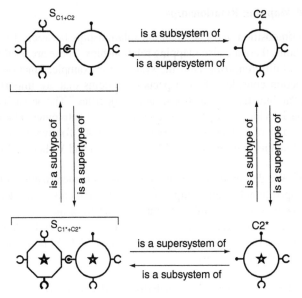

FIGURE 13.4 Clean hierarchy.

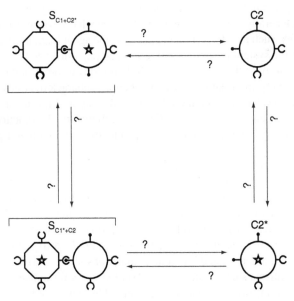

FIGURE 13.5 Heterarchy.

between $S_{C1*+C2*}$ and S_{C1+C2} cannot be characterized by classificatory hierarchy alone. The relationship between C2 and S_{C1*+C2} and between C2* and S_{C1*+C2} cannot be characterized by compositional hierarchy alone.

Aspects and Mapping Relationships

As well as composition and classification relationships between different systems characterizations, there are also mapping relationships. These are used to relate characterizations of different *aspects* of the system, for example, function and architecture. This section considers three aspects of systems that are important in design and science: "architecture," "functions," and "properties." The pervasiveness of these three concepts is evidenced by the existence of several ontologies relating them, both in design domains (e.g., Goel *et al.*, 2009; Gero, 1990; Tomiyama *et al.*, 1993) and in scientific domains (e.g., Bunge, 1977, 1979; Wand and Weber, 1990).

Architecture   We define *system architecture* as a characterization of a system in terms of compositional relationships between its elements. This definition keeps the characterization of a system's structure distinct from the *mapping* relationships between its structure and function and is consistent with several definitions and discussions of architecture in the literature (e.g., IEEE, 2000; Simon, 1962; Alexander, 1964; Maier and Rechtin, 2009).

Functions   The term "function" is much discussed across various literatures on how systems operate (see reviews in Crilly (2010), Erden *et al.* (2008), Houkes and Vermaas (2010), Preston (2009), Vermaas and Dorst (2007)), and it is not always easy to see how a general definition can apply across domains (e.g., to both artifacts and organisms). Generally, however, functions describe what a system *should do* in serving some entity, such as satisfying the goals of some agent (e.g., users and designers) or permitting the system to survive and reproduce (e.g., in an ecosystem or market). We leave debate over the nuances of such definitions to other authors and instead focus on clarifying the relationship that functional characterizations have to other kinds of characterization. Even though the realization or "fulfillment" of a function by an entity is dependent on its properties and architecture, the functional characterization of the entity can be considered independently of these other aspects.

Properties   We use "property" as an umbrella term for anything that can be said to be true of an entity (this might even include having a particular architecture or function). When this is expressed statically (or atemporally, without reference to time or temporal order), we call the property a *state* (Tomiyama *et al.*, 1993). When it is expressed dynamically (in temporally extended terms), we use the term *behavior*, or more precisely, *state transitions* (Gero, 1990; Kam *et al.*, 2001) and *state transition rules* (see also Endogenous and Exogenous Functions section). This can be related to modeling formalisms such as the discrete event systems formalism (Zeigler, 1984), where the state transition rules can be represented by state couplings.

ASPECTS AND ABSTRACTIONS OF MODULARITY

A system characterization with a straightforward compositional hierarchy describes components and subsystems as interacting (or interfacing) with each other in well-defined, well-understood ways and is said to be "modular." Although there exist many different notions of "modularity," they can be understood and distinguished on the basis of three fundamental attributes of system architecture: structural encapsulation, function-structure mapping, and interfacing (Three Core Aspects of Modularity section). Appendix A illustrates how these fundamental attributes can be used to consolidate different definitions of modularity found in the literature. From these three fundamental architectural attributes, we can derive two further abstractions, function-driven encapsulation and interface compatibility (Two Abstractions of Modularity section).

Three Core Aspects of Modularity

The three fundamental attributes of system architecture that we associate with modularity are represented diagrammatically in Figure 13.6. In this chapter, we treat these as the three core aspects of modularity and require that all three of them be satisfied for a set of system elements to be characterized as a *module*:

- Structural encapsulation means that the elements can collectively be treated as a single encapsulated component.
- One-to-one function-structure mapping means that the set of elements collectively map to a particular function.
- Interfacing means that as a collective, the set of elements has well-defined interactions with other system elements.

Figure 13.6 illustrates these three aspects of modularity: structural encapsulation (the module is defined by indifference to the architecture or composition of a set of elements), function-structure mapping (the module is defined by the collective mapping of a structured set of elements to a function – in this case, F_1), and interfacing (the module is defined by how a set of elements interacts with other systems).

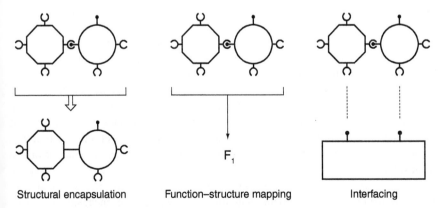

Structural encapsulation   Function–structure mapping   Interfacing

FIGURE 13.6   Three aspects of modularity.

Structural Encapsulation   We use the term *structural encapsulation* to refer to the grouping of related *system elements*, that is, subsystems, into units that can then be treated as component types at some level of abstraction. Structural encapsulation also implies "interface decoupling" as it allows the relationships between a set of related system elements to be considered independently from its interactions with other system elements.

Function-Structure Mapping   We use the term *function-structure mapping* to refer to the mapping between a set of related system elements (i.e., a subsystem) and a function. This structured set of system elements can then be encapsulated into a component type because of its association with the function. We refer to such an encapsulation as "function-driven" (see Function-Driven Encapsulation section).

Interfacing   We define the term *interface* of an element as the aspect(s) of the element that allows it to interact with another element or set of other elements in the same system. For those designing physical products, it might be most natural to think of interfaces in terms of physical structure or geometric fit. However, interfaces can also be realized in nonphysical ways and the interactions need not be determined by geometry. Standards, protocols, agreements, languages, signals, and processes can all be treated as interfaces.

Which aspect(s) of an element is treated as its interface depends on the characterization adopted, which defines the set of elements with which interaction occurs. This might also mean that multiple interfaces are identified for an element.

Two Abstractions of Modularity

From the three core aspects outlined above, we can derive two further abstractions that also pervade the modularity literature: function-driven encapsulation and interface compatibility.

Function-Driven Encapsulation   We use the term *function-driven encapsulation* to describe cases where the criterion for encapsulation is the fulfillment of a function (see Figure 13.7). What connects to each other is that they collectively map to a function, and what makes this set of elements disconnected from other elements is the fact that these other elements do not participate in the fulfillment of that function (being "connected" or "disconnected" might also be a matter of degree, and the mapping to a function is specific to a particular level of abstraction and scope). Function-driven encapsulation can be seen as one of a set of many different forms of encapsulation, each of which is distinguished by the kind of criteria that determines encapsulation. For example, we might also have property-driven or behavior-driven encapsulation where elements are seen as "connected" when they collectively realize a particular property. However, as our definition of modularity is concerned with the relationship between elements and functions, we require encapsulation to be *function*-driven. Figure 13.7 shows an example of function-driven encapsulation

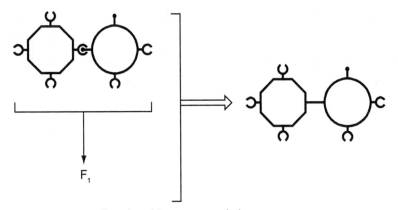

Function-driven encapsulation

FIGURE 13.7 Function-driven encapsulation.

as the structural encapsulation of the module is determined by function-structure mapping.

We say that a system architecture is *completely modular* if every element in the system belongs to a functional group and fulfillment of the system's overall function is completely accounted for by the function-structure mappings to these sets (see Figure 13.7). In the design and management of systems, encapsulation has been said to provide a means of "managing complexity" by hiding the intricacies of certain regions of the system so that characterizations of them can be separated from the characterization of the relationships that exist between them and other regions of the system.

Interface Compatibility Interface compatibility refers to the compatibility between a component and other components of the system. This compatibility might be a matter of degree and characterized as the strength of interaction (e.g., how frequently two entities exchange information and the intensity of attraction). Interface compatibilities between system components determine how different groups of system elements are able to interact with each other, thus providing a characterization of the system's architectural constraints. In a completely modular architecture, as all the elements would be modules or would belong to modules and hence encapsulated in components, interactions between elements in different components would always be via their interfaces. Figure 13.8 shows an example of interface compatibility, leading to a modular architecture. The system type $S_{[C1+C2]+C3}$ has a completely modular architecture as all its elements (both S_{C1+C2} and C3) belong to or constitute modules (M_1 and M_2, respectively). In this case, the modules are defined by function-structure mapping.

If all modules (components mapped to functions) in a system had the same mutually compatible interfaces with each other, there would be no architectural constraints at the module level as any module would be able to interact with any

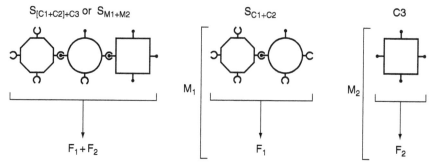

FIGURE 13.8   Interface compatibility.

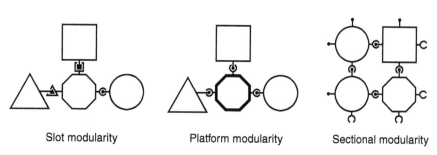

Slot modularity   Platform modularity   Sectional modularity

FIGURE 13.9   Architectural variety from interface compatibility.

other module; that is architectural degrees of freedom would be maximized, and every component could be "repositioned." This is known as "sectional" modularity (Ulrich and Tung, 1991; Ulrich, 1995), where every component in the system has the same set of interfaces. At the other extreme, where interfaces minimize architectural degrees of freedom and each component has a specific "position" or "role" in the system, we have "slot" modularity (Ulrich and Tung, 1991; Ulrich, 1995). In "slot" modularity, each component has a unique set of interfaces, which implies that it has a unique set of interactions with other components in the system and hence can only be located in a single specific position with respect to them. These two extremes are shown in Figure 13.8, together with the intermediate case of "platform modularity," where there is one component that interacts with all the others, and with respect to which the other components can be repositioned as it interacts with them through identical interfaces. Figure 13.9 shows how architectural variety can arise from interface compatibility. In slot modularity, each component type has a unique interface and can hence only connect with a particular set of component types (here, the different interfaces are represented by different-shaped connectors). In platform modularity, a single component type (the octagon in the diagram) connects with all the others with the same kind of interface. In sectional modularity, all components can connect with all the others via the same kind of interface.

Interface compatibilities can provide a means of controlling which parts of the system can vary. In a given system architecture, different elements of different types (possibly mapping to different functions), so long as they have the same interface compatibilities, can interact with the same set of other elements. In a modular architecture (where the system can be decomposed into components mapped to functions), interface compatibilities determine which components can be swapped or substituted for each other. The terms "component-sharing", (Ulrich, 1995) "substitution" (Garud and Kumaraswamy, 1993; Mikkola and Gassmann, 2003), and "standardization" (Miozzo and Grimshaw, 2005) can be found in the literature to refer to cases where, at a particular level of abstraction, different component types have the same interfaces (i.e., they are compatible with the same set of other component types). This "component-sharing", together with overall architectural similarity between products, can be the basis for establishing product "families" (Ulrich, 1995; Galsworth, 1994; Jose and Tollenaere, 2005). The term "component-swapping" (Ulrich, 1995) is used to refer to cases where, at a particular level of abstraction, component types are mapped to different functions but have the same interfaces and therefore can be substituted for each other architecturally (see Figure 13.10). If these differences in function have implications for a product's overall function, they provide the basis for the different product variants in product "families."

It is worth pointing out that "component-swapping" always implies "component-sharing," and vice versa. When we are taking the perspective of a component (the octagon in Figure 13.10) that can interact with a variety of other components, the architecture is characterized as "component-swapping" (different components can be swapped "in or out" of the octagon). When we are taking the perspective of different components that can interact with the same component (the octagon), the architecture is characterized as "component-sharing" (the octagon in Figure 13.10 is a component that can be shared "between" different elements).

The distinction between types and instances introduced in Characterizing Systems section becomes important in discussions of "sharing." Sharing between component *instances* equates to a particular component interacting with several other components (e.g., a USB bus and the devices plugged into it, an organism, and the other

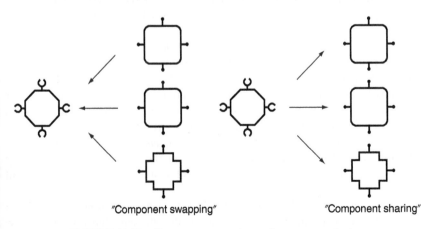

"Component swapping"   "Component sharing"

FIGURE 13.10   Component swapping and component sharing.

organisms in its food web, a protein with many domains engaged in multiple reactions at the same time), whereas sharing between component *types* refers to a particular type of component being able to exist in many different system types, as in different products in a product family or different species in a genus.

Interface compatibility provides us with a systematic means of characterizing and analyzing architectural variety as elements distinguishable by their compatibilities with each other being combined.

ASPECTS OF COMPLEXITY

The term "complexity" is used in different ways in the design literature and is often used interchangeably with "complicated." We treat these as two distinct concepts. A "complicated" system (characterization) is one with many components, subsystems, and interactions but where, at any given level of abstraction, as with a simple system, it is theoretically possible to map functions to the components and subsystems in a one-to-one fashion and describe the interactions between them. By contrast, a "complex" system (characterization) is one for which this is not possible. The three aspects and two abstractions of modularity discussed above can be used to distinguish between different aspects of complexity.

Complexity as non-one-to-one function-structure mappings

Function-driven encapsulation ensures one-to-one mapping between function and architecture. Complexity arises when, at some level of abstraction, the mapping is not one-to-one.

Multi-Structural Function Realization and Architectural Robustness   We use the term *multi-structural function realization* to describe cases where a function maps to more than one architecture (more than one component type). In design and engineering, the possibility of realizing a function with different architectures offers the opportunity for robustness and reduction in cost. Robustness is observed in situations where if one architecture mapping to a function is not realized, others may be able to realize it instead; that is, reduction in cost comes from the fact that the number of components required for a given level of robustness might be lower than if this robustness were achieved through duplication of components. Figure 13.11 shows redundancy through duplicated architectures and distinct architectures. In the top row of Figure 13.11, both the architecture [C2+C3] and the architecture [C1+C3+C1] map to F_X; in the bottom left, redundancy in F_X is provided by an architecture with duplication of [C1+C3+C1]; in the bottom right, redundancy in F_X is provided by two distinct architectural realizations, [C1+C3+C1] and [C2+C3].

In engineering design, the term "principle redundancy" (Pahl and Beitz, 1996) describes cases in which multiple architectures realize the same function. In biology, the term "degeneracy" describes cases where, when a particular element is not able to

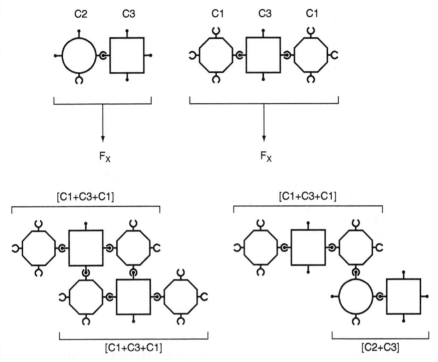

FIGURE 13.11   Redundancy through duplicated architectures and distinct architectures.

fulfill the function, other means of fulfilling that function are possible (Tononi *et al.*, 1999; Edelman and Gally, 2001; Whitacre, 2010; Chen and Crilly, 2014). A function that was previously associated with a single element might also become distributed among multiple elements.

Compared to duplication, multi-structural function realization offers a more robust form of redundancy when the different architectures able to realize the function have different points of fragility and strength. On the other hand, it makes the function-structure mappings more difficult to analyze, and when there is failure, it can be difficult to identify the elements involved. Figure 13.12 shows an example of how multi-structural function realization provides robustness. As in Figure 13.11, both [C2+C3] and [C1+C3+C1] are mapped to F_X but a new component type, C4 is introduced, which prevents the architecture [C1+C3+C1] from realizing F_X. In the bottom left architecture, the presence of C4 prevents $S_{[C1+C3+C1]+[C1+C3+C1]}$ (see Figure 13.11) from performing F_X. In the bottom right architecture, the presence of C4 does not prevent the system type $S_{[C1+C3+C1]+[C2+C3]}$ (see Figure 13.11) from performing F_X because $S_{[C2+C3]}$ is unaffected by C4. The multi-structural function realization architecture of $S_{[C1+C3+C1]+[C2+C3]}$ allows it to be more robust than $S_{[C1+C3+C1]+[C1+C3+C1]}$ with respect to performing F_X, as it can do so in the presence of C4.

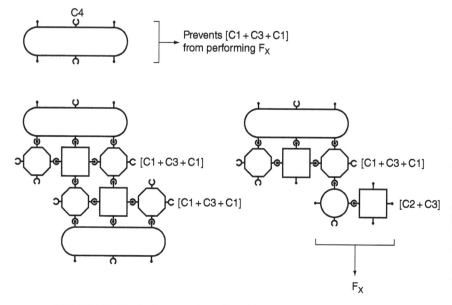

FIGURE 13.12   Robustness through multi-structural function realization.

Context-Dependent Multi-Functionality and Architectural Flexibility

We use the term *context-dependent multi-functionality* to refer to cases where an architecture maps to different functions based on the wider system architecture it is part of. In systems terms, this means a subsystem realizes different functions based on which other systems it is connected to (its environment), that is, the supersystem it is part of. Figure 13.12 shows how C3 can be characterized as context dependently multi-functional. When it is connected to C2, it realizes F_{X1}, and when it is connected to two instances of C1, it realizes F_{X2}.

In design domains, repurposing of products, product parts, and processes is example of context-dependent multi-functionality. For example, a steel rod realizes different functions depending on the wider physical structure it is part of; in software, the same data can have different functions depending on the sections of the program that they flow into; the biochemical function of a protein can depend on the other molecules present; the economic impact of a purchase by a consumer depends on the purchasing activities of other consumers.

When the contexts in which different functions are realized are not well understood, functions may be realized unexpectedly or "emerge" (sometimes resulting in non-fulfillment of other functions). On the other hand, if the context dependencies are well understood, multi-functionality can be exploited to get (desired) functional variety from a given architecture. Figure 13.13 shows an example of context-dependent multi-functionality. The same component (in this case, the square, C3) realizes different functions by participating in different architectures, even if those architectures realize the same function.

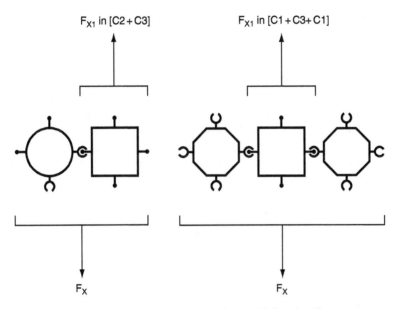

FIGURE 13.13   Context-dependent multi-functionality.

We say that a system is "architecturally flexible" if variety in function is high with respect to architectural variety (in the limit, every architectural variation would be functionally relevant and the ratio would be 1). This has the potential advantage of allowing a system to realize a greater variety of functions with a relatively small number of elements, but also makes it more difficult to analyze.

Complexity as Ill-Defined Interfaces and Shifting System Boundaries

A modular system has subsystems (the modules) with well-defined interfaces, resulting in a perfect compositional hierarchy; each module can be treated as a "closed" system. "Complexity" arises when interfaces are ill defined or changing, and the boundaries between the subsystems are constantly changing so that subsytems are "open" systems. Of course, as with function-structure mappings, this is really a question of characterization.

In a "closed system" characterization where the system has a well-defined boundary, given knowledge of all the possible characterizations within the boundary, it would be theoretically possible to define all the relationships between all the characterizations. However, when the number of characterizations and/or relationships between them is extremely large or not yet known, an idealized "open system" characterization may be used. For example, in design domains, the realization of a product requires the realization of an intricate set of connections between physical components, processes, people, and organizations; in complex systems science domains, models of entities often consist of a web of interdependencies between a large number of system elements. An "open system" characterization of these scenarios would see the system as interacting with itself (as it would with its environment) and would see the interdependencies between the elements of the system as constantly changing.

Complexity as Overlapping Levels

Multi-Level Characterizations and Heterarchy The notion of heterarchy was already introduced in Hierarchies and Heterarchies section. Heterarchical characterizations are ones where several hierarchies *overlap* in a single characterization. These should be distinguished from characterizations that integrate multiple *non-overlapping* hierarchies (e.g., Simon, 1962; Skyttner, 2005). For pragmatic purposes, heterarchies are decomposed into such non-overlapping characterizations, such as in "System of Systems" (SoS) characterizations (Maier, 1998; Rainey and Tolk, 2015), which integrate different resolutions without overlap in scope.

Heterarchies can represent cases where different domains work together to understand a single entity (Alvarez Cabrera *et al.*, 2009; van Beek *et al.*, 2010), as different domains might emphasize different system aspects and consequently "carve up" the entity in ways that overlap. Figure 13.14 shows an example of a complex systems characterization of the entity *e* introduced in Characterizing Systems section based on the heterarchy in Figure 13.5. In Figure 13.14, functions are mapped to architectures specified at different levels. The complexity comes from the fact that in order for the realizations of all the functions to be characterized, different levels of abstraction and scope overlap, that is, heterarchy. In the real world, these different mappings might represent characterizations from different domains, for example, programmers, software architects, and business analysts working on the same software; and cognitive psychologists, neuroscientists, cell biologists, and molecular biologists studying the brain.

Endogenous and Exogenous Functions In both design and scientific domains, the functions being considered in function-structure mapping often relate to different aspects of the system or even to different systems (with different boundaries), resulting in modular architectures that differ substantially from one another (Holtta and Salonen, 2003). For example, in product design, function-structure mappings may be defined with respect to the product's overall function in use (which is typically linked to the satisfaction of user needs and preferences), but they can also be defined with respect to the product's manufacture or contribution to firm strategy. In biology, one set of functions might relate to an organism's survival; another might relate to its development or to its role in evolution.

To generalize, the functions in a function-structure mapping might originate from the consideration of different systems, and we can dissociate (i) the system for which the architecture is defined (e.g., the product and organism) from (ii) the system determining the functions to which this architecture maps (e.g., user and ecosystem). In the case of (ii), we might draw a distinction between "endogenous" functions, which are defined with respect to the system in question, and "exogenous" functions, which are defined with respect to the supersystem of which that system is a part (see Crilly, 2013).

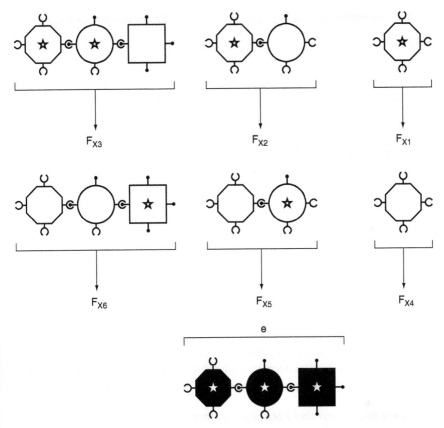

FIGURE 13.14   Functions mapped to architectures defined at different levels.

The distinction between endogenous and exogenous functions is important because they can be associated with different levels of uncertainty. Failure to realize endogenous functions lies in improper realization of the system type (e.g., a system part failing). Failure to realize exogenous functions on the other hand can be attributed to the system's environment, which can change the function-structure mapping. For example, changes in user preferences might mean that elements of the system that previously had the function of satisfying a particular preference no longer can; a new set of conditions in an organism's environment might mean that certain functions of the organism no longer map to the biological elements they were previously mapped to. If knowledge of the system's environment is inferior to knowledge of the system itself, component types mapping to exogenous functions will have higher levels of uncertainty associated with them in terms of function fulfillment (e.g., user preferences compared to product specifications; organism behavior compared to core metabolic functions).

Figure 13.15 shows how different architectures might map to the same endogenous function but to different exogenous functions. The architectures [C1+C3+C1]

System boundary of $S_{[C1+C3+C1]}$ System boundary of $S_{[C2+C3]}$

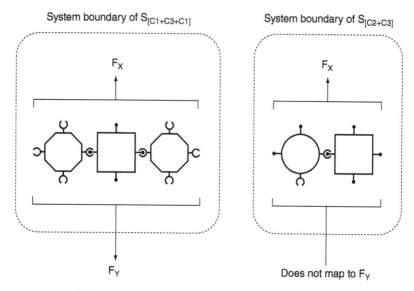

F_Y Does not map to F_Y

FIGURE 13.15   Endogenous and exogenous functions.

and [C2+C3] map to the same endogenous function F_X but only [C1+C3+C1] maps to the exogenous function F_Y. In many cases, endogenous functions and exogenous functions might also be dependent on each other. For example, the realization of the endogenous function F_X might be dependent on the realization of F_Y, or vice versa.

Behavioral Robustness and Flexibility   Although entity change and entity variety can be seen as two distinct concepts, change can also be seen simply as variety observed through time. For example, with an atemporal view, demands to the system due to alterations in physical conditions or consumer preferences (Dahmus *et al.*, 2001) become the same as those made by an environment with a wide range of physical conditions or a market with highly diverse consumer preferences.

Although state *transitions* describe the behavior of a system *instance*, state transition *rules* describe the behavior of a system *type*. State transition rules define the set of state transitions that are realizable (or that must be realized) by instances of the type, thus determining the states that the system can instantiate, its "state space," depending on its initial state, which also determines the behavioral trajectories it can take, as illustrated in Figure 13.16 (see also Chapter 3 on system specifications). The rules mean that in a given system instance, transitions between states can be "guided" and "mutually constraining," so that they follow particular "trajectories" depending on previous states. This can result in *behavioral* "robustness" and "flexibility." For example, the transitions in Figure 13.16 offer more behavioral flexibility than the transitions in Figure 13.16 as for two of the states, there are two possible transitions rather than one. If the choice as to which transition was purely probabilistic, the system would be difficult to predict, whereas if the choice of transition was dependent on some environmental condition, the system would be more adaptive. For example,

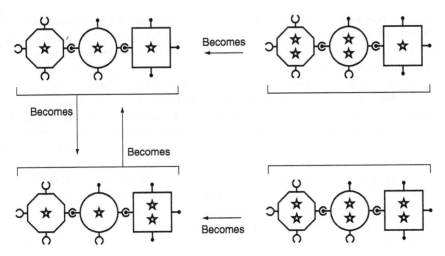

FIGURE 13.16   State transition rules.

in Figure 13.16, the state transition rules result in the system eventually alternating between the two left-hand states, irrespective of the initial state.

In the same way that change can be recast as variety, we can give system behavior (state transitions) an architectural characterization. In the case of "behavioral robustness," it is very difficult to get the system to deviate from a particular behavior, consequently regularities in the architecture of the behavior "emerge." In the case of "behavioral flexibility," there are few constraints on the states that can be realized by the system, and the architecture of the behavior has few regularities. Such a system would be chaotic and difficult to manage, predict, and understand. For example, starting from a particular initial state, the state transition rules in Figure 13.16 would give rise to only one possible behavioral trajectory, whereas the rules in Figure 13.17 could

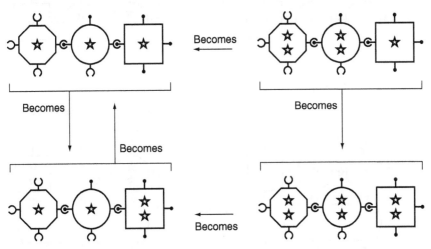

FIGURE 13.17   Different behavioral trajectories depending on initial state.

generate more than one behavioral trajectory. In other words, the state transition rules might generate a system with more behavioral flexibility than the one in Figure 13.16, but if the system is initialized or transitions to the left-hand states, it eventually ends up alternating between the two left-hand states, as shown in Figure 13.16.

Terms such as "positive feedback" and "negative feedback" are used to describe the mechanisms that constrain or "guide" behavior (Ashby, 1962; Heylighen and Joslyn, 2001; Babaoglu *et al.*, 2005; Dauscher and Uthmann, 2005; Yamamoto *et al.*, 2007). In the case of positive feedback, a particular state or behavior increases the likelihood or extent of states or behaviors of the same type, whereas in the case of negative feedback, it diminishes their extent or likelihood. These two mechanisms and interactions between them form the basis for the "emergence" of behaviorally robust self-* properties such as self-replication or self-assembly (Babaoglu *et al.*, 2005), and homeostasis or "autopoiesis," the ability of the system to maintain itself in a viable condition (Maturana and Varela, 1980).

In some complex systems characterizations, the system's *environment* can put the system into a state in which different rules apply or even directly affect which rules apply (see Figure 13.18), thus making different behavioral trajectories available. For example, one of an aircraft's engines could fail as a result of exposure to airborne volcanic ash, which might put a heavier load on the other engines and make them more likely to fail. Similarly, how a stem cell differentiates might be determined by chemical factors in its environment, but once differentiated, it multiplies and "locks" a part of the system into developing in a particular way; that is, it makes the subsystem associated with this developmental trajectory more robust (Bateson

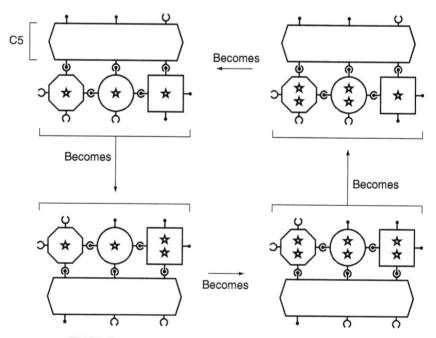

FIGURE 13.18   Environmentally influenced behavioral rules.

and Gluckman, 2012). In Figure 13.18, the presence of C5 changes the state transition rules so that the system cycles through all four states, irrespective of the initial state.

In even more complicated cases, the system can itself influence its environment to make it more likely to realize particular states, which then reinforce the above effect. Identifying such scenarios is a key endeavor in the complex systems sciences.

Architectural robustnesss/flexibility and behavioral robustness/flexibility address different aspects of complexity. In the case of architectural robustness and flexibility, it is the relationship between architecture (which might be the architecture of a system, system type, state, or behavior) and function that we are concerned with. By contrast, in the case of behavioral robustness and flexibility, we are concerned only with the architecture itself (characterized as regularities in behavior).

CONCLUSIONS

This chapter has introduced a domain-neutral framework for understanding the relationships between different aspects of system complexity and modularity in different systems characterizations (see Characterizing Systems section). We defined three core aspects of modularity (structural encapsulation, function-structure mapping, and interfacing) and two further abstractions from them, function-driven encapsulation and interface compatibility (Aspects and Abstractions of Modularity section). These were then explicitly related to different aspects of complexity (Aspects of Complexity section). Table 13.1 summarizes these.

As noted throughout this chapter, the extent to which an entity is considered to be a "complex system" or a "modular system" depends on how the entity is characterized. Systematically relating different aspects of complexity to different aspects of modularity permits complex systems problems to be (re-characterized in different ways to find suitable solutions). It also allows methods from different domains to be applied to similar problems. In particular, we point to the following opportunities for the engineering design community to leverage existing methods (some drawn from the design context, others from scientific contexts):

- Methodologies from design permitting the systematic characterization of the relationship between architectural variety and functional variety in a product family at different levels (e.g., "design for variety," Martin and Ishii, 2002) could be used to analyze the relationship between architectural variety and functional variety of non-designed entities. By generalizing the notion of types, architectures and functions, we would be able to include both designed and non-designed system elements within the same characterization.

- Techniques for exploring system states in the complex systems sciences, such as agent-based modeling (Bonabeau, 2002; Axelrod, 2006), could be used to understand the costs and benefits of different architectures with respect to different functions. When a large number of architectural configurations are possible, being able to simulate them and analyze the functional implications of certain family groupings would provide more solid justification for making architectural decisions at product, product family, and even product

TABLE 13.1 Different Complex Systems Characterizations Related to Different Aspects and Abstractions of Modularity (Non-Complex Systems Characterizations)

Types of Complexity	Section(s)	System Characterization, Aspect(s), and Abstraction(s) of Modularity	Section(s)
Open systems characterization, shifting system boundaries, ill-defined interfaces	Complexity as Ill-Defined Interfaces and Shifting System Boundaries	Structural encapsulation, interfacing, interface compatibility	Structural Encapsulation, Interfacing, Interface Compatibility
Multi-structural function realization, architectural robustness, evolvability	Multi-Structural Function Realization and Architectural Robustness	Function-structure mapping, function-driven encapsulation	Function-Structure Mapping, Function-Driven Encapsulation
Context-dependent multi-functionality, architectural flexibility	Context-Dependent Multi-Functionality and Architectural Flexibility	Function-structure mapping, function-driven encapsulation	Function-Structure Mapping, Function-Driven Encapsulation
Heterarchy, multi-level representations	Hierarchies and Heterarchies, Multi-Level Characterizations and Heterarchy	Composition, classification, levels, hierarchy	Composition and Classification, Hierarchies and Heterarchies
Endogenous and exogenous functions	Endogenous and Exogenous Functions	Composition, classification, levels, hierarchy, function-structure mapping, function-driven encapsulation	Composition and Classification, Hierarchies and Heterarchies, Function-Structure Mapping, Function-Driven Encapsulation
Behavioral robustness, emergence, self-organization	Behavioral Robustness and Flexibility	Composition, classification, levels, hierarchy, architecture, functions, properties, behaviors, and states	Composition and Classification, Hierarchies and Heterarchies, Architecture, Functions, Properties, Endogenous and Exogenous Functions

The relevant sections of the present paper are listed in the columns to the right.

portfolio levels. In addition, for systems with both designed and non-designed elements, we would be able to make better decisions about initialization states and interventions that would help "guide" the system into adopting certain architecturally characterized states with desirable properties.

- Community detection and clustering techniques (Fortunato and Castellano, 2012; Lancichinetti and Fortunato, 2009; Newman, 2006; Palla, 2005) applied in the complex systems sciences could be used to discover different potential "family" groupings with respect to different functions.

- Static architectural "patterns" and dynamic "behavioral motifs" could be shared across domains and application contexts. The domain-neutral nature of our framework would provide a basis for analyzing dynamic architectures structurally to identify further trends and commonalities between them. These could be generalized to higher level design principles and guidelines for designers working on products and problems with complex systems characterizations. In turn, these might be further specialized and adapted for different application contexts.

In both design and scientific contexts, the challenge posed by complex systems problems comes from having to integrate multiple overlapping characterizations. Many emerging technologies are said to involve "complex systems" due to ill-defined mappings between architectures to functionally relevant properties. Similarly, those working in the complex systems sciences often struggle to integrate multiple models of a system with overlapping hierarchies, resulting in "heterarchical"characterizations. The domain-neutral framework we have introduced allows complex systems problems to be expressed in multiple ways so that the insights, methods, and techniques drawn from different domains and application contexts can be appropriately applied to the problems they are most suited to.

APPENDIX A: MAPPING MODULARITY DEFINITIONS TO ASPECTS OF MODULARITY

Table 13.A.1 illustrates how the aspects of modularity introduced in this chapter can be used to characterize different notions of modularity found in the literature. For example, the sources in the first row define modularity in terms of structural encapsulation (or in the case of discourse, this is the aspect emphasized), whereas the sources in the third row emphasize both structural encapsulation and function-structure mapping.

The list below gives the definitions represented in the table, including the domain-specific definitions reviewed in Gershenson *et al.* (2003):

- In Allen and Carlson-Skalak (1998), a module is defined as a component or group of components that can be removed from the product non-destructively as a unit, which provides a unique basic function necessary for the product to

TABLE 13.A.1   Aspects of Modularity Found in Different Definitions of Modularity

	Aspects of Modularity		
Source of Definition	Structural Encapsulation	Function-Structure Mapping	Interfacing
George and Leathrum (1985), Gershenson *et al.* (1999), Jiao and Tseng (1999b), Newman (2010), Ulrich and Eppinger (1995)	X		
Ishii *et al.* (1995), Otto and Wood (2001)		X	
Allen and Carlson-Skalak (1998), Baldwin and Clark (1997), Di Marco *et al.* (1994), Huang and Kusiak (1998), Newcomb *et al.* (1998), Sarkar *et al.* (2013), Spencer (1998), Ulrich and Tung (1991), Ulrich (1995)	X	X	
Galsworth (1994), Walz (1980)			X
Chang and Ward (1995), Chen (1987), Sosale *et al.* (1997)		X	X
Carey (1997), Civil Engineering Research Foundation (1996), Pimmler and Eppinger (1994)	X		X
Hollta and Salononen (2003), Marshall *et al.* (1998)	X	X	X

operate as desired. Modularity is then defined as the degree to which a product's architecture is composed of modules with minimal interactions between modules.

- In Baldwin and Clark (1997), modularity refers to the "building of [a] complex product or process from smaller subsystems that can be designed independently yet function together as a whole."

- In Carey (1997), modularity is defined as design with subsystems "that can be assembled and tested prior to integration … to reduce the time and cost of manufacturing."

- In Chang and Ward (1995), a modular product is "a function-oriented design that can be integrated into different systems for the same functional purpose without (or with minor) modifications."
- In Chen (1987), modularity refers to "tools for the user to build large programs out of pieces."
- In Civil Engineering Research Foundation (1996), modularity is defined as using sets of units designed to be arranged in a variety of ways.
- In Di Marco *et al.* (1994), a clump (module) is a collection of "components and/or subassemblies that share a physical relationship and some common characteristic based on the designer's intent."
- In Galsworth (1994), a module is defined as a group of standard and interchangeable components.
- In George and Leathrum (1985), it is required that for a software module, for a given function, there is no access to, informational flow to, or interactivity between modules.
- In Gershenson *et al.* (1999), it is stipulated that modules contain a high number of components that have minimal dependencies upon and similarities to other components not in the module.
- In Holtta and Salonen (2003), a module is defined as a structurally independent building block of a larger system with fairly loose connections to the rest of the system. It is also required that they have well-defined interfaces which allow independent development of the module as long as the interconnections at the interfaces are retained.
- In Huang and Kusiak (1998), modularity requires similarity of functional interactions and suitability of inclusion of components in a module.
- In Ishii *et al.* (1995), the term "modular" refers to the minimization of the number of functions per component.
- In Jiao and Tseng (1999b), a module is defined as a physical or conceptual grouping of components.
- In Marshall *et al.* (1998), modules are defined as cooperative subsystems which (i) can be combined and configured with similar units to achieve different outcomes; (ii) have one or more well-defined functions that can be tested in isolation from the system and that (iii) have their main functional interactions within rather than between modules.
- In Newcomb *et al.* (1998), a module is described as a set of components grouped together in a physical structure and by some characteristic based on the designer's intent.
- In Newman (2010), a module is defined as a subsystem in which the associations between elements within the subsystem are stronger than the associations between these elements and other elements in the system. This is expressed in network terms. A subsystem is a module when the number of edges within the subsystem is much higher than the expected number of edges derived from

an equivalent random network model with the same number of elements and similar distribution of links between elements with no modular structure.

- In Otto and Wood (2001), "conceptual" modules are introduced. Each of these performs the same functions even if they have different physical compositions.

- In Pimmler and Eppinger (1994), the discussion centers around the interactions between elements. Four types of interaction are identified: (i) spatial, the need for adjacency or orientation between elements; (ii) energy, the need for energy transfer between two elements; (iii) information, the need for information or signal transfer between two elements; and (iv) material, the need for material exchange between two elements.

- In Sarkar *et al.* (2013), a module is defined as a component or subsystem in a larger system that performs specific function(s) and emerges as a tightly coupled cluster of elements sharing dense intra-module interactions and sparse inter-module interactions.

- In Sosale *et al.* (1997), modules are groups of components that can easily be re-used or re-manufactured, also considering material compatibility.

- In Ulrich and Eppinger (1995), the most modular architecture is one in which each functional element of the product is implemented by exactly one chunk (subassembly) and in which there are few interactions between chunks. Such a modular architecture allows a design change to be made to one subassembly without affecting the others.

- In Ulrich and Tung (1991), product modularity is defined in terms of "(i) similarity between the physical and functional architecture of the design and (ii) minimization of incidental interactions between physical components."

- In Ulrich (1995), a modular product or subassembly has "a one-to-one mapping from functional elements in the function structure to the physical components of the product."

- In Walz (1980), modularity is defined as "constructed of standardized units of dimensions for flexibility and use."

REFERENCES

Alexander, C. (1964) *Notes on the Synthesis of Form*, Harvard University Press, Boston, MA.

Allen, R., Douence, R., and Garlan, D. (1998) Specifying and analyzing dynamic software architectures, in *Fundamental Approaches to Software Engineering*, Lecture Notes in Computer Science 1382 (ed. E. Astesiano), Springer-Verlag.

Alvarez Cabrera, A.A., Erden, M.S. and Tomiyama, T. (2009) On the potential of function-behavior-state (fbs) methodology for the integration of modeling tools. Proceedings of the 19th CIRP Design Conference–Competitive Design. Cranfield University.

Ashby, W.R. (1962) *Principles of the Self-Organising System*, Pergamon, New York, NY.

Axelrod, R. (2006) *Agent-based Modeling as a Bridge Between Disciplines. Handbook of Computational Economics 2*, Elsevier, Miamisburg, OHL.

Babaoglu, O., Jelasity, M., Montresor, A. *et al.* (eds) (2005) *Self-Star Properties in Complex Information Systems: Conceptual and Practical Foundations*, Lecture Notes in Computer Science/Theoretical Computer Science and General Issues, Lecture Notes in Computer Science 3460, Springer-Verlag.

Baldwin, C.Y. and Clark, K.B. (1997) Managing in an age of modularity. *Harvard Business Review*, **75**, 84–93.

Baldwin, C.Y. and Clark, K.B. (2000) *Design Rules Vol. 1: The Power of Modularity*, MIT Press, Cambridge, MA.

Bateson, P. and Gluckman, P. (2012) Plasticity, robustness, development and evolution. *International Journal of Epidemiology*, **41** (1), 218.

van Beek, T.J., Erden, M.S., and Tomiyama, T. (2010) Modular design of mechatronic systems with function modelling. *Mechatronics*, **20** (8), 850–863.

Bellman, R. (1957) A Markovian Decision Process. *Journal of Mathematics and Mechanics*, **6**.

Bentley, P.J. (2002) *Digital Biology: How Nature is Transforming Our Technology and Our Lives*, Simon and Schulster.

von Bertalanffy, L. (1968) *General System Theory: Foundations, Developments, Applications*, Braziller, New York.

Bonabeau, E. (2002) Agent-based modelling: methods and techniques for simulating human systems. *Proceedings of the National Academy of Sciences of the United States of America*, **99** (S3), 7280–7287.

Bunge, M.A. (1977) *Ontology I: The Furniture of the World – Treatise on Basic Philosophy*, Reidel, Dordrecht.

Bunge, M.A. (1979) *Ontology II: A World of Systems – Treatise on Basic Philosophy*, Reidel, Dordrecht.

Carey, M. (1997) *Modularity Times Three*, HighBeam Research, Sea Power, Navy League of the United States.

Chang, T.S. and Ward, A.C. (1995) Design-in-modularity with conceptual robustness. Proceedings of the 1995 ASME Design Engineering Technical Conferences, 21st International Conference on Advances in Design Automation, Boston, MA. The American Society of Mechanical Engineers, New York.

Chen, W. (1987) A theory of modules based on second-order logic. Proceedings of the IEEE Logic Programming Symposium, Sendai.

Chen, C.-C. and Crilly, N. (2014) Modularity, redundancy and degeneracy: cross-domain perspectives on key design principles. Proceedings of the Annual IEEE International Systems Conference, Ottawa, ON.

Chen, C.-.C and Crilly, N. (2016a) From modularity to emergence: a primer on the design and science of complex systems. Technical Report CUED/C-EDC/TR.166. University of Cambridge, Department of Engineering. ISSN 0963-5432. 10.17863/CAM.4503

Chen, C.-C. and Crilly, N. (2016b) Describing complex design practices with a cross-domain framework: learning from Synthetic Biology and Swarm Robotics. *Research in Engineering Design*, **27** (3), 291–305.

Chen, C.-C., Nagl, S., and Clack, C. (2009) Complexity and emergence in engineering systems. *Complex Systems in Knowledge-Based Environments: Theory, Models and Applications*, **168**, 99–128.

Civil Engineering Research Foundation (1996) Bridging the globe: engineering and construction solutions for sustainable development in the twenty-first century. Highlight of the *Engineering and Construction for Sustainable Development in the Twenty-first Century: An International Research Symposium and Technology Showcase.*

Crilly, N. (2010) The roles that artefacts play: technical, social and aesthetic functions. *Design Studies*, **31** (4), 311–344.

Crilly, N. (2013) Function propagation through nested systems. *Design Studies*, **34** (2), 216–242.

Dahmus, J.B., Gonzalez-Zugasti, J.P., and Otto, K.N. (2001) Modular product architecture. *Design Studies*, **22** (5), 409–424.

Dauscher, P. and Uthmann, T. (2005) Self-organized modularization in evolutionary algorithms. *Evolutionary Computation*, **13** (3), 303–328.

Di Marco, P., Eubanks, C.F., and Ishii, K. (1994) Compatibility analysis of product design for recyclability. *Computers in Engineering*, **1**, 105–112.

Diallo, S.Y., Tolk, A., Gore, R., and Padilla, J. (2014) Modeling and simulation framework for systems engineering, in *Modeling and Simulation-Based Systems Engineering Handbook* (eds D. Gianni, A. D'Ambrogio, and A. Tolk), CRC Press.

Dressler, F. and Akan, O.B. (2010) A survey on bio-inspired networking. *Computer Networks*, **54** (6), 881–900.

Edelman, G.M. and Gally, J.A. (2001) Degeneracy and complexity in biological systems. *Proceedings of the National Academy of Sciences*, **98** (24), 13763–13768.

Endy, D. (2005) Foundations for engineering biology. *Nature*, **438** (7067), 449–453.

Erden, M.S., Komoto, H., van Beek, T.J. *et al.* (2008) A review of function modeling: approaches and applications. *AI EDAM*, **22** (2), 147–169.

Forrest, S., Balthrop, J., Glickman, M., and Ackley, D. (2005) Computation in the wild, in *Robust Design: A Repertoire of Biological, Ecological, and Engineering Case Studies* (ed. E. Jen), Oxford University Press, Oxford, pp. 207–230.

Fortunato, S., & Castellano, C. (2012) Community structure in graphs, R. A. Meyers, *Computational Complexity: Theory, Techniques, and Applications* 490–512). Springer, Berlin.

Fricke, E. and Schulz, A.P. (2005) Design for changeability (DfC): principles to enable changes in systems throughout their entire lifecycle. *Systems Engineering*, **8** (4), 279–295.

Galsworth, G.D. (1994) *Smart, Simple Design: Using Variety Effectiveness to Reduce Total Cost and Maximize Customer Selection*, Omneo, Essex Junction, VT.

Gao, L. (2000) On inferring autonomous system relationships in the Internet. *IEEE/ACM Transactions on Networking*, 733–745.

Garud, R. and Kumaraswamy, A. (1993) Changing competitive dynamics in network industries: an exploration of sun microsystems' open systems strategy. *Strategic Management Journal*, **14** (5), 351–369.

George, G. and Leathrum, L.F. (1985) Orthogonality of concerns in module closure. *Software Practice and Experience*, **15** (2), 119–130.

Gero, J.S. (1990) Design prototypes; a knowledge representation schema for design. *AI Magazine*, **11** (4), 26–36.

Gershenson, J.K., Prasad, G.J., and Allamneni, S. (1999) Modular product design: a life-cycle view. *Journal of Integrated Design & Process Science*, **3** (4), 13–26.

Gershenson, J.K., Prasad, G.J., and Zhang, Y. (2003) Product modularity: defnitions and benefts. *Journal of Engineering Design*, **14** (3), 295–313.

Goel, A., Rugaber, S., and Vattam, S. (2009) Structure, behavior and function of complex systems: the SBF modelling language. *Artificial Intelligence for Engineering Design, Analysis and Manufacturing*, **23** (1), 23–35.

Goldstone, R.L. and Sakamoto, Y. (2003) The transfer of abstract principles governing complex adaptive systems. *Cognitive Psychology*, **46** (4), 414–466.

Gunderson, L. and Holling, C.S. (2001) *Panarchy: Understanding Transformations in Systems and Nature*, Island Press, Washington, D.C.

Gunji, Y.-P. and Kamiura, M. (2004) Observational heterarchy enhancing active coupling. *Physica D: Nonlinear Phenomena*, **198** (1–2), 74–105.

Heylighen, F. and Joslyn, C. (2001) Cybernetics and second-order cybernetics, in *Encyclopedia of Physical Science and Technology*, 3rd edn (ed. R.A. Meyers), Academic Press, New York, NY, pp. 155–169.

Holtta, K.M.M. and Salonen, M.P. (2003) Comparing three different modularity methods. Proceedings of the ASME 2003 International Design Engineering Technical Conferences and Computers and Information in Engineering Conference, ASME, Chicago, IL, USA.

Houkes, W. and Vermaas, P.E. (2010) *Technical Functions: On the Use and Design of Artefacts*, Springer, Berlin.

Huang, C.-C. and Kusiak, A. (1998) Modularity in design of products and systems. *Systems, Man and Cybernetics, Part A: IEEE Transactions on Systems and Humans*, **28** (1), 66–77.

Software Engineering Standards Committee (2000) *IEEE Recommended Practice for Architectural Description of Software-Intensive Systems*, Technical Report IEEE Std 1471-2000. IEEE Computer Society.

Ishii, K., Juengel, C. and Eubanks, C.F. (1995) Design for product variety: key to product line structuring. Proceedings of the 1995 ASME International Conference on Design Theory and Methodology.

Jiao, J. and Tseng, M.M. (1999b) A methodology of developing product family architecture for mass customization. *Journal of Intelligent Manufacturing*, **10** (1), 3–20.

Johnson, J. (2007) Multidimensional events in multilevel systems, in The Dynamics of Complex Urban Systems: *An Interdisciplinary Approach* (eds S. Albeverio, D. Andrey, P. Giordano, and A. Vancheri), Physica-Verlag, Heidelberg, pp. 311–334.

de Jong, K.A. (2002) *Evolutionary Computation*, A Bradford Book.

Jose, A. and Tollenaere, M. (2005) Modular and platform methods for product family design: literature analysis. *Journal of Intelligent Manufacturing*, **16** (3), 371–390.

Kam, N., Cohen, I. and Harel, D. (2001) The immune system as a reactive system: modelling T cell activation with statecharts. Proceedings of the IEE Symposia on Human-Centric Computing Languages and Environments Visual Languages and Formal Methods, Stresa.

Knight, T.F. (2005) Engineering novel life. *Molecular Systems Biology*, **1** (1), published online.

Ladyman, J., Lambert, J., and Wisener, K. (2013) What is a complex system? *European Journal for Philosophy of Science*, **3** (1), 33–67.

Lancichinetti, A. and Fortunato, S. (2009) Community detection algorithms: a comparative analysis. *Physical Review E*, **89**, 049902.

Luzeaux, D., Renault, J.-R., and Wippler, J.-L. (eds) (2011) *Complex Systems and Systems of Systems Engineering*, Wiley-ISTE, Arlington, VA.

Maier, M.W. (1998) Architecting principles for systems-of-systems. *Systems Engineering*, **1** (4), 267–284.

Maier, M.W. and Rechtin, E. (2009) *The Art of Systems Architecting*, 3rd edn, CRC Press, Boca Raton, FL.

Marshall, R., Leanrey, P.G., and Botterell, O.P. (1998) Enhanced product realization through modular design: an example of product process integration. Proceedings of Third Biennial World Conference on Integrated Design and Process Technology.

Martin, M.V. and Ishii, K. (2002) Design for variety: developing standardized and modularized product platform architectures. *Research in Engineering Design*, **13** (4), 213–235.

Maturana, H. and Varela, F.J. (1980) *Autopoiesis and Cognition: The Realization of the Living*, Reidel, Boston, MA.

McCulloch, W.S. (1945) A heterarchy of values determined by the topology of nervous nets. *Bulletin of Biophysics*, **7** (2), 89–93.

McManus, H.L. and Hastings, D. (2006) A framework for understanding uncertainty and its mitigation and exploitation in complex systems. *IEEE Engineering Management Review*, **34** (3), 81.

Meadows, D.H. and Wright, D. (2008) *Thinking in Systems: A Primer*, Chelsea Green Publishing, White River Junction, VT.

Meyn, S.P. (2007) *Control Techniques for Complex Networks*, Cambridge University Press, Cambridge.

Mikkola, J.H. and Gassmann, O. (2003) Managing modularity of product architectures: toward an integrated theory. *IEEE Transactions on Engineering Management*, **50** (2), 204–218.

Miozzo, M. and Grimshaw, D. (2005) Modularity and innovation in knowledge-intensive business services: IT outsourcing in Germany and the UK. *Research Policy*, **34** (9), 1419–1439.

Mitchell, M. (2009) *Complexity: A Guided Tour*, Oxford University Press, Oxford.

Mittal, S. (2013) Emergence in stigmergic and complex adaptive systems: a formal discrete event systems perspective. *Cognitive Systems Research*, **21**, 22–39.

Mittal, S. and Rainey, L.B. (2015) Harnessing emergence: design and control of emergent behavior in system of systems engineering. Proceedings of the Summer Simulation Multi-Conference, Chicago, IL.

Newcomb, P.J., Bras, B., and Rosen, D.W. (1998) Implications of modularity on product design for the life cycle. *Journal of Mechanical Design*, **120** (3), 482–490.

Newman, M.E.J. (2006) Modularity and community structure in networks. *Proceedings of the National Academy of Sciences*, **103** (23), 8577–8582.

Newman, M. (2010) *Networks: An Introduction*, Oxford University Press, Oxford.

Ottino, J.M. (2004) Engineering complex systems. *Nature*, **427** (6973), 399.

Otto, K.N. and Wood, K.L. (2001) *Product Design: Techniques in Reverse Engineering and New Product Development*, Prentice Hall, Upper Saddle River.

Pahl, G. and Beitz, W. (1996) Developing size ranges and modular products, in *Engineering Design: Systematic Approach* (ed. K. Wallace), Springer-Verlag, Berlin/Heidelberg, pp. 405–453.

Palla, G. (2005) Uncovering the overlapping community structure of complex networks in nature and society. *Nature*, **435**, 814–818.

Pimmler, T.U. and Eppinger, S.D. (1994) Integration analysis of product decompositions. Proceedings of ASME Conference on Design Theory and Methodology, Minneapolis, MN, pp. 343–351.

Preston, B. (2009) Philosophical theories of artifact function, in *Philosophy of Technology and Engineering Sciences* (ed. A. Meijers), Elsevier, Amsterdam, The Netherlands, pp. 213–234.

Rainey, L.B. and Tolk, A. (eds) (2015) *Modeling and Simulation Support for System of Systems Engineering Applications*, John Wiley & Sons, Hoboken, NJ.

Ross, A.M., Rhodes, D.H., and Hastings, D.E. (2008) Defining changeability: reconciling flexibility, adaptability, scalability, modifiability, and robustness for maintaining system lifecycle value. *Systems Engineering*, **11** (3), 246–262.

Ryan, E.T., Jacques, D.R., and Colombi, J.M. (2013) An ontological framework for clarifying flexibility-related terminology via literature survey. *Systems Engineering*, **16** (1), 99–110.

Sarkar, S., Dong, A., Henderson, J.A., and Robinson, P.A. (2013) Spectral characterization of hierarchical modularity in product architectures. *Journal of Mechanical Design*, **136** (1), 011006.

Sasai, K. and Gunji, Y.-P. (2008) Heterarchy in biological systems: a logic-based dynamical model of abstract biological network derived from time-state-scale re-entrant form. *Biosystems*, **92** (2), 182–188.

Schilling, M.A. (2002) Modularity in multiple disciplines, in *Managing in the Modular Age: Architectures, Networks and Organizations*, Blackwell, Oxford, pp. 203–214.

Schlosser, G. and Wagner, G.P. (2004) *Modularity in Development and Evolution*, University of Chicago Press, Chicago, IL.

Schoettl, F. and Lindemann, U. (2014) Design for system lifecycle properties a generic approach for modularizing systems. *Procedia Computer Science*, **28**, 682–691.

Shannon, C.E. (1951) Prediction and entropy of printed English. *Bell System Technical Journal*, **30** (1).

Simon, H.A. (1962) The architecture of complexity. *Proceedings of the American Philosophical Society*, **106** (6), 467–482.

Skyttner, L. (2005) *General Systems Theory: Problems, Perspectives, Practice*, World Scientific Press, London.

Sosale, S., Hashemian, M., and Gu, P. (1997) Product modularization for reuse and recycling. *Design Division publication*, **94**, 195–206. American Society of Mechanical Engineers.

Tolk, A. and Muguira, J.A. (2003) The levels of conceptual interoperability model. Proceedings of the 2003 Fall Simulation Interoperability Workshopvol. 7, Curran Associates, Orlando, FL, pp. 1–11.

Tomiyama, T., Umeda, Y., Ishii, M. *et al.* (1993) A CAD for functional design. *Annals of the CIRP*, **42** (1), 143–146.

Tononi, G., Sporns, O., and Edelman, G.E. (1999) Measures of degeneracy and redundancy in biological networks. *Proceedings of the National Academy of Sciences*, **96** (6), 3257–3262.

Ulrich, K. (1995) The role of product architecture in the manufacturing firm. *Research Policy*, **24** (3), 419–440.

Ulrich, K. and Eppinger, S.D. (1995) *Product Design and Development*, McGraw-Hill, Columbus, OH.

Ulrich, K. and Eppinger, S.D. (2003) *Product Design and Development*, McGraw-Hill Higher Education.

Ulrich, K. and Tung K. (1991) Fundamentals of product modularity. Proceedings of the 1991 ASME Winter Annual Meeting Symposium on Issues in Design Manufacture/Integration, Atlanta, GA.

Vermaas, P.E. and Dorst, K. (2007) On the conceptual framework of John Gero's FBS-model and the prescriptive aims of design methodology. *Design Studies*, **28** (2), 133–157.

Walz, G.A. (1980) Design tactics for optimal modularity. Proceedings of IEEE Autotestcon, Washington, D.C.

Wand, Y. and Weber, R. (1990) Mario Bunge's ontology as a formal foundation for information systems concepts, P. Weingartner & G.J.W. Dorn, *Studies on Mario Bunge's Treatise*, Rodopi, Atlanta, 1990 79–107.

de Weck, O.L., Roos, D., and Magee, C.L. (2011) *Engineering Systems: Meeting Human Needs in a Complex Technological World*, MIT Press, Cambridge, MA.

Whitacre, J. and Bender, A. (2010) Degeneracy: a design principle for achieving robustness and evolvability. *Journal of Theoretical Biology*, **263** (1), 143–153.

Whitehead, A.N. (1919) *An Enquiry Concerning the Principles of Natural Knowledge*, Cambridge University Press, Cambridge.

Yamamoto, L., Schreckling, D. and Meyer, T. (2007). Self-replicating and self-modifying programs in fraglets. Proceedings of the Second International Conference on Bio-inspired Information, Budapest.

Zeigler, B.P. (1984) *Multifaceted Modelling and Discrete Event Simulation*, Academic Press, London.

Zeigler, B.P. and Sarjoughian, H.S. (2013) System entity structure basics, in *Guide to Modeling and Simulation of Systems*, Springer-Verlag, London, pp. 27–37.

14

THE EMERGENCE OF SOCIAL SCHEMAS AND LOSSY CONCEPTUAL INFORMATION NETWORKS: HOW INFORMATION TRANSMISSION CAN LEAD TO THE APPARENT "EMERGENCE" OF CULTURE

Justin E. Lane[1,2]

[1]Institute of Cognitive and Evolutionary Anthropology, Department of Anthropology, University of Oxford, 64 Banbury Road, Oxford, OX2 6PN, UK
[2]LEVYNA, Ústav religionistiky, Masaryk University, Veveří 28, Brno 602 00, Czech Republic

SUMMARY

Social scientists and humanities scholars often treat culture as non-reducible phenomena that are incompatible with scientific study or reductionism (Pyysiäinen, 2001). This has led to claims that culture is an emergent property of human social groups (Sawyer, 2005). This chapter seeks to address these claims using an agent-based model. Particularly, it presents a model of information transmission in human social groups that generates similar patterns of multi-agent consensus from the interactions of individual agents as observed in real-world human groups. As such, it argues that the claims of cultural emergence suggesting sort of strong or "spooky" emergence (Maier, 2014) are unsound and theoretically inoperable in some cases. However, we can reframe cultural emergence as the result of interactions between complex dynamic agents, and we can approximate cultural emergence using

Emergent Behavior in Complex Systems Engineering: A Modeling and Simulation Approach,
First Edition. Edited by Saurabh Mittal, Saikou Diallo, and Andreas Tolk.
© 2018 John Wiley & Sons, Inc. Published 2018 by John Wiley & Sons, Inc.

a lossy intersection of all beliefs held by the agents in a group. Doing so still results in an emergent "culture" that is irreducible to the "sum of its parts" but still on a subset of the "sum of its parts" that are shared by many agents. This chapter presents a new model of cultural information transmission, based on theories from cognitive anthropology (Whitehouse, 2004; Hill and Dunbar, 2003), and discusses its results in relation to cultural dynamics that affect the stability of social groups, potentially resulting in changes in identity, beliefs, and large-scale demographic shifts.

INTRODUCTION

Recent advances in cognitive (i.e., information processing) approaches to complex human social systems have resulted in more opportunities and uses for modeling and simulation. This includes modeling cultural and religious economies (Bainbridge, 2006), social identities (Upal and Gibbon, 2015; Upal, 2015; Schröder et al., 2016), religious ritual transmission (Whitehouse et al., 2012), and social schisms (McCorkle and Lane, 2012), to name a few. In this way, the social sciences and humanities directly engage fields such as engineering and physics. Although these advances have utilized computer modeling and simulation as a bridge between the sciences and humanities, the approach is not without criticism, often resulting from disagreements on ideas of causation and emergence of social phenomena. Scholars in the humanities and social sciences often claim that certain social phenomena such as "cultures" are emergent. Often, they make claims for strong emergence, where cultures are irreducible entities that can "cause" their individual constituents to take on certain properties (i.e., enact certain behaviors or hold specific beliefs).[1] This appears to be backed by many findings in social psychology that suggest that individuals from different cultures exhibit differences on many psychological dimensions (Henrich et al., 2010). However, this conclusion is a logical interpretation of any analysis that treats culture as a categorical independent variable. Nevertheless, most cultural scholars (regardless of department or discipline) are unable to provide a working definition of their causal variable: "culture." Decades of debate has resulted in most scholars using the idea of culture as an explanation for their descriptions of structural or functional differences between groups (Sperber, 1996). Regardless, scholars in all disciplines have noted that members of social groups share some beliefs and that this set of beliefs can become a heuristic for defining cultural boundaries. Although defining a group based on traits such as beliefs and then saying the beliefs cause the group to have certain traits risks tautological reasoning, we can use cultural consensus to refer to a set of mutually held beliefs by members of a social group – and set aside the question of "cultural causation" for the time being.

[1] The idea that cultures can "cause" things is likely held for many reasons; key to this appears to be a rejection of the constructionist hypothesis in conjunction with the rejection of reductionism as an enterprise of explanation; instead, the humanities have rejected explanation and favor description and interpretation as their key role as "researchers." Also, this includes protectionism, as to state that a "culture" is reducible to some other set of rules studied by another discipline calls into question the utility of a separate field dedicated to the study thereof.

In what follows, I present an example of how modeling complex social processes from the perspective of "cultural cybernetics" can result in a form of "emergence" of cultural consensus that can be described as "weak" or "generative." *Cultural cybernetics* is an approach to understanding human cultural systems that focuses on information processing by human cognitive mechanisms and aims to quantify (as best as possible) and predict information changes and interactions between agents and their social and biological environments. First, I discuss the issues of reducing the interactions of agents in an agent-based model (ABM) from their group-level patterns to their individual-level properties and capacities. Then I present a presentation and discussion of an ABM that can generate social consensus in ways that replicates patterns observed by anthropologists, with patterns that are not reducible to the "sum of its parts."

CONSENSUS IN LARGE HUMAN SOCIAL GROUPS

Oftentimes, consensus is studied in small groups with a focus on how groups come to use consensus to make decisions as a social unit. Typically, this is done outside of modeling and simulation by methods such as controlled laboratory protocols (DiFonzo *et al.*, 2013), careful ethnography (Garro, 2000), or field observation (Harton *et al.*, 1998). The formation of "cultural" consensus, among disparate members of large communities is far less understood. Logistically, studying this phenomenon is difficult as many cultural groups are so large that members of the group do not all know one another (Anderson, 2006) and often they are not in a single location where they could be studied. Nonetheless, it is well documented by social scientists that these large "imagined" communities identify as part of a social group because they perceive the other members to share some of their own beliefs (Tajfel and Turner, 1979; Brewer, 1991; Postmes *et al.*, 2005).

This situation presents a thorny issue for social scientists but is well fit to be studied using ABMs. Computationally, the problem can be restated with close isomorphism to the problem in situ:

> can a set of agents with information processing constraints that limit the number of social contacts they can manage sustain similar schemas of information while receiving unique information from other members of their group?

In the model outlined below, this question is addressed by utilizing constraints and information transmission mechanisms that have been studied by cognitive psychologists. Doing so allows the agents to have a "psychological realism" (Sun and Hélie, 2013) that can serve as a basic form of multi-agent artificial intelligence (MAAI).[2] There are several constraints on cultural consensus, which can be used as requirements for our computational model.

---

[2]MAAI is an approach within ABM which seeks to utilize "psychologically realistic agents" in order to develop what can be called multi-agent artificial intelligence. For further discussion of their use in social simulation see Sun and Hélie (2013), Sun (2001), and Lane (2013a).

The first constraint is that of human communicative abilities. Generally, studies show that humans only maintain a limited number of social network links in a conversation, as well as over periods ranging many months and even years (Dunbar, 2008; Roberts *et al.*, 2009; Saramäki *et al.*, 2014; Dávid-Barrett and Dunbar, 2013; Gonçalves *et al.*, 2011; Dunbar *et al.*, 1995). Therefore, information should not simply transfer between any agents in the simulation, as they can only maintain a set number of links at any one point in time, and these links should remain relatively stable. Additionally, individuals do not all have the same ability to transfer information, and some individuals, such as orators and cultural leaders, have the ability to transfer information to others at set intervals of time, such as during rituals or through media dissemination (sometimes referred to as *prestige bias*) (Henrich and Gil-White, 2001). Most importantly, individuals can align themselves with groups through multiple mechanisms: namely, via social relationships and via shared beliefs. When aligning themselves through social relationships, these relationships can be encoded efficiently through memories we have of rare and significant events, ranging from marriages to acts of terrorism and violence. These memories, which psychologists refer to as episodic memories, can become central aspects of one's identity that can promote reflection and serve to "fuse" their identity to that of their social group (Jong *et al.*, 2016). In addition, individuals can utilize shared beliefs to align with a group (Postmes *et al.*, 2005). When this is done, they utilize what psychologists call semantic memories. These memories are socially shared beliefs that individuals have learned over time, but cannot necessarily recall how (unlike episodic memories), such as basic trivia and language. Key to semantic memories is that they are, by definition, consistent of general knowledge that is socially shared (McRae and Jones, 2013). However, if one is basing their alignment with a group on semantic information, they are constrained in that they will not have complete access to the semantic knowledge of all other individuals, only a subset of concepts that should be generally taken as stereotypical beliefs of a group. In this way, the information of a social schema is "lossy." In what follows, I describe how these requirements, when implemented in an ABM, can help us to understand the emergence of social schemas.

First, I will delineate what is meant by emergent in this sense and how this may differ from the traditional social science definition and implicates current approaches to "emergence" in the complexity literature. This is followed by the description of the model and how the parts-to-whole interaction reveals an effect which I will describe as quasi-emergent. I discuss how the calculations used within the model can help to clarify the role of "emergence" in ABMs and leads to a more well-quantified approach to understanding the dynamics of social systems in modeling and simulation generally.

Theoretical Discussion: Reduction and Emergence

Currently, interdisciplinary work in the social sciences generally embraces empirical reductionism to investigate some target phenomenon. When investigating social consensus and group formation, social systems and interactions are reduced to the individual agent level in order to facilitate empirical investigation of the group-level phenomenon. This reductionist approach is extremely useful for empirically investigating the interactions of variables in many systems. However, in social systems, such as cultures or religions, such an approach is insufficient. Even if one can demonstrate that there are statistical relationships between variables across a population, this does not necessarily demonstrate that these relationships generate the target group-level phenomenon when iterated through a heterogeneous population. For example, when a social scientist posits specific claims about how individuals interact with each other and their sociocultural contexts to create social consensus, two levels of analysis are required: (1) the psychological, where individual-level beliefs, motivations, and behaviors will be instantiated or observed, and (2) sociopolitical, where socially shared information can be measured.

Even if empirical studies confirm theoretical relationships proposed at the psychological level, there is no guarantee that these interactions will produce a proposed sociopolitical target phenomenon. Such uncertainty may arise from feedback loops in complex systems. This is a key focus of this chapter as it moves forward. Specifically, the interaction between an individual as part of a group, and how that group is formed conceptually. As such, the feedback loop affects specific changes throughout a population, leading to otherwise unpredictable effects.

Such feedback loops are common in the social systems literature. For example, given the requirements outlined above, frequent participation should result in a feedback loop between learning and identification: if one agent received information from another, this should eventually result in its incorporation into a semantic memory schema. This schema, in turn, affects how the agent aligns themselves with a group; and this, in turn, affects decisions to participate with the group again in the future. Should the social schema change, individuals will be forced to readjust their group alignment. This dynamic is worthy of focus because, if iterated throughout a population, changes in the social schema of the group also affect the alignment between individuals and that group. One issue is that there is a relationship between the social schema and an individual's identification that changes over time and may be unpredictable or irreducible at any one point in time. This could suggest that the social schema of the group may be an "emergent" property of collectives of individuals.

Some researchers in the field of sociology, anthropology, and computer modeling argue that some variables are emergent. This claim requires some unpacking, particularly regarding what could be meant by the term *emergent*. An emergent phenomenon is defined in general as any phenomenon that is not directly reducible to the properties of its constituent parts (Goldstein, 1999). Emergent phenomena come in several varieties, generally broken down into two types of emergence: strong and weak. *Weak emergence* in the social sciences holds that a supervening property of a complex system is related in some way to the collective action of the constituent parts, but in an indirect matter; as such, a macro-level phenomenon could be said to be "materially reducible" but not directly reducible to the system's constituent parts. *Strong emergence* in the social sciences on the other hand posits no material relationship, stating that the supervening property is the result of lower level interactions, but is not directly reducible to these interactions in any material sense.

However, these two categories can be further clarified by considering the engineering literature on modeling systems of systems. Maier discusses four types of emergence relevant for the current discussion: simple, weak, strong, and spooky. Within these types, simple emergence is demarcated by the ability to consistently predict, not just explain, a phenomenon with simplified models of the system's components and the phenomenon can reliably be produced in lower complexity models of the system. Weak emergent phenomena are consistently reproduced in simulations, but the models are not reduced in complexity. Weak emergent phenomenon can be understood through models of reduced complexity after observations are recorded, but cannot be reliability predicted in advance. Strong emergent properties are consistent with the known properties of the system's components, but are not reproducible in simplified models of the system. In direct simulations of the phenomenon, the emergent property is inconsistently reproduced. Models of reduced complexity are also generally unable to reliably predict strong emergent phenomena. Lastly, spooky emergent properties inconsistent with the properties of the system's components are not reproducible in any model of the system, even if the complexity is equal to that of the system itself.

It has been stated that social phenomena are emergent (Sawyer, 2005). This argument is rooted in one of the two schools of thought: social phenomena are emergent because either (1) they are the result of complex interactions among multiple variables at lower levels, thus giving rise to higher level phenomena such as societies, economies, cultures, and religions, but the laws governing the higher level phenomena are reducible to – or demonstrated by – laws at a lower levels; or (2) social phenomena are emergent because they are the result of interacting variables at lower levels, which give rise to higher level phenomena, but the laws governing the higher level phenomena are *not* reducible to – or demonstrated by – laws at the lower levels. That is to say, in position 1, that there is a continuity between principles in the system that are instantiated at all levels of analysis, whereas in position 2, the laws at one level may reduce to the level immediately causing the higher level phenomena, but need not reduce any further – or are unable to be reduced any further. In the psychological literature, Walmsley (2010) refers to position 1 as "dynamic emergence" and position 2 as "nomological emergence."

However, the empirical interactions revealed through experimentation can be utilized within a third option available for ABMs known as *generative emergence* (Epstein, 2006b). This type of emergence is not technically a form of emergence. Rather, it aims to utilize dynamic and complex interactions of lower level variables to reproduce higher level effects of complex systems. It does this by specifying simple rules for agents within the system to follow as these agents interact with each other and their environment. This interaction results in continual changes for variables in the environment, the agents, and groups of agents. However, because these interactions are computationally defined, the generative approach patently denies emergence within these systems. Instead, the generative approach asks the question "how could decentralized local interactions of heterogeneous autonomous agents generate the given regularity" (Epstein, 2006a, 5). The generative school holds that one can only "explain" a complex emergent phenomenon if one can generate some observed higher level regularity (such as a pattern observed in social groups) from the interactions of heterogeneous agents at a lower level (such as interactions among individuals in an environment). Within this approach, it is stated that if you do not "grow" the phenomena, you are unable to explain its emergence (Epstein, 2006a, 8).

When considering the larger issues with approaching complex social systems, the generative approach can circumvent rampant reduction in the nomological position by being uninterested in reducing beyond the level of abstraction necessary to generate some emergent phenomena. It is, in theory, compatible with the dynamic school of emergence but instead of attempting to reduce a phenomenon by laws governing the system beyond the required number of levels, it attempts to utilize laws applicable to the interactions of lower level variables to *generate* (rather than reduce) the higher level phenomenon.

Given the outlines above, we can map the social systems definitions onto those presented by Maier. Generally, weak emergence in the social sciences is poorly defined and can arguably encompass simple, weak, and strong emergence in Maier's typology. Meanwhile, strong emergence in the social sciences maps more closely onto Maier's definition of "spooky" emergence. Maier's typology of emergence also clarifies some – but not all – of the differences between dynamic and nomological emergence used in the social systems and psychological literature. Prima facie, simple, weak, and strong emergence would fall under the category of dynamic emergence, whereas "spooky" emergence would be the only nomological position.

However, understanding these claims within the modeling and simulation literature allows us to be even more specific. Technically, Maier's positions on emergence appear to be making epistemological assumptions that the domain of a model is sufficiently well understood as to potentially offer a model that could reproduce a system (or system of systems). To unpack this, one can draw from the discussion of Mittal and Rainey (2015) that – to a point – further Maier's typology of emergence within the cybernetic perspective of systems of systems modeling. They note that the boundary between strong and weak emergence is really one demarcated by knowledge and that as the knowledge of the underlying dynamics that cause a property to "emerge" become known, the property can move from being strong to weak in its emergence. This is sensible, particularly from a modeling and simulation perspective as more specific or greater knowledge about a system can foster more advanced models. Currently, however, it is important to note that just because a model can be simplified and still offer predictions about a property or phenomenon does not guarantee that the assumptions of the simplified (or original) model do not violate assumptions of laws operating at lower levels of analysis. As such, one cannot rule out that any of Maier's classifications of emergence can be nomological. In this sense, the only way to classify it as nomological or dynamic is if the assumptions of the model are congruent with the laws of lower levels of explanation as well – with the exception of "spooky" emergence, which does not need any such requirement. For succinctness, these are also included in Table 14.1 below, where I list the two types of emergence discussed in the social sciences (weak and strong) and their general mapping to the types of emergence discussed by Maier and others in the modeling and simulation

TABLE 14.1 Mapping Definitions of Emergence in the Social Science and Modeling and Simulation Literatures

Types of emergence from social sciences	Types of emergence from modeling and simulation	Can this emergence be generative?	Dynamical	Nomological
Weak	Simple	Yes	Emergence is dynamical if assumptions of the model are congruent with laws/systems governing lower levels of explanation	Emergence is nomological if assumptions of the model are incongruent with laws/systems governing lower levels of explanation
	Weak	Yes	Emergence is dynamical if assumptions of the model are congruent with laws/systems governing lower levels of explanation	Emergence is nomological if assumptions of the model are incongruent with laws/systems governing lower levels of explanation
Strong	Strong	Yes	Emergence is dynamical if assumptions of the model are congruent with laws/systems governing lower levels of explanation	Emergence is nomological if assumptions of the model are incongruent with laws/systems governing lower levels of explanation
	Spooky	No	The separation between levels of causal systems in spooky emergence does not allow for dynamical emergence	Given that "spooky" emergence is, by its nature, one which does not appear to produce predictions based on known properties of its underlying systems, spooky emergence is only ever nomological. In addition, social scientists rest claims of strong emergence on non-reducibility in such a way that would necessitate new laws to describe the phenomenon that are not congruent with deductive conclusions of lower level explanations, resulting in nomological emergence

literature before listing if it could potentially be classed as a "generative" model and discussing the criteria for each to be models of dynamical or nomological emergence. These positions have serious implications for understanding the dynamics of multi-level systems, such as social systems, scientifically, but – I argue – even more so for modeling and simulation of social systems. The explanation for social systems using a MAAI approach is implicated by the validity of these two assumptions. Take for example, the following definition of religion: Religion is a complex suite of beliefs and behaviors that are supported by – or related to – a belief in a culturally postulated supernatural agent (Lawson and McCauley, 1990). Evidence strongly suggests that these beliefs and behaviors are produced and manipulated by multiple cognitive mechanisms (Barrett, 2004). These mechanisms are produced by the human brain, which is itself an organ that is the result of evolution and selection throughout the ancestral past. We could, in turn, reduce this organ to the interactions between cells (neurons) that use electro-chemical signals (based on complex molecular interactions) to transmit information to one another in response to internal and external stimuli, but such a reduction would likely be unnecessary. Taking for a moment, the observation that religious beliefs and behaviors are the result of the mind we can take for instance a single interaction between two levels (the social and the cognitive) as an example. A great deal of empirical research demonstrates that the mind works in – more or less – mechanistic ways; these mechanisms can be described in multiple ways, such as through the framework of heuristics and biases (Ariely, 2010). However, the laws that govern the interworking of the brain would not be sufficient to instantiate social phenomena as there has been no clear demonstration that any neurophysiological formal interaction gives rise to the cognitive phenomena underlying the decisions of an individual; as such, there is no direct extension (or "covering law") that could sufficiently explain social phenomena (Walmsley, 2010), which are presumably the result of human psychological capacities. Given this, one would seemingly rule out dynamic emergence as a possibility within a framework of scientific reduction and we would then accept nomological emergence and need to look no further than the social context in order to explain the social interactions of individual actors within a complex social system. Although this may be heuristically appealing to some, it runs the risk of committing a common error in the social sciences, which we can call the "Durkhemian fallacy" – that the social can be explained by the social (Lane, 2013a) and that groups can be assumed to have entitativity[3] (Lane, 2017a). These assumptions are problematic because social groups do not behave as singular entities. The history of humanity is riddled with schisms, fusions, and the creation of new groups that cannot be accounted for if one assumes that groups have entitative properties or that social explanations are sufficient for social phenomenon. As such, we can also define our agents in MAAI models at the human (as opposed to group) level.

---

[3]Entitativity is the abstraction of something as a single entity, even if it is a collection of individual parts. It is typically used in psychology to discuss how humans perceive social groups as entities even though they are "imagined" social groups (for further discussions and examples, see Brubaker, 2002; Anderson, 2006; for empirical studies of the subject, see Lickel et al., 2000, 2001).

This approach, it should be noted, is also problematic from an epistemological standpoint as it relies on the assumption (which is currently empirically under-supported) that cognitive functions are the result of neurophysiological forms (such as neural connections). As such, one can argue that any model which assumes that this must exist, at this point is nomological. However, just as Mittal and Rainey (2015) note that strong emergence can become weak emergence as more information about the system is discovered, it is possible that the nomological assumption that cognitive functions that instantiate social behaviors are potentially the result of neurophysiological forms could be shown to be actually the result of dynamically emergent neurophysiological systems (not just potentially or assumed to be); thus, the understanding of the cognitive mechanisms that underlie social behaviors as "nomologically emergent" would become dynamically emergent.

This debate cuts straight to a central issue with social theories that fractionate higher level social phenomena into constituent parts and then claiming that empirically significant interactions explain higher level phenomena. In actuality, what results is that the interaction between variables may be necessary to explain some aspect of a higher level phenomenon, but is not sufficient to explain the phenomenon itself, only a subset of properties related to the target phenomenon. Therefore, the reductive approach employed by psychologists may be necessary but insufficient as an explanation of the formation of social schemas in human groups.

Theoretical Discussion: Method for Computational Testing

To sufficiently demonstrate that a set of properties causes some group-level observation, we should be able to reproduce those group-level patterns by some method. In many ways, controlled lab studies attempt to do just this. They isolate and manipulate the relevant variables to demonstrate that the interactions between them have a desired output. This approach, however, useful for studying many social systems, is not available for understanding large human groups for three reasons: (1) the interactions between variables often have subsequent effects on other variables causing feedback loops that are difficult to measure at the individual level, but are observable at the group level (e.g., social schemas); (2) human groups can be so large that logistical constraints do not allow for empirical investigation; and (3) even if one can overcome methodological and logistical constraints, ethical constraints would not allow for one to experimentally manipulate a large group of people such as a cultural, ethnic, or religious group. Although this may sound common sense to some, these issues create a useful space for the use of computational modeling to better understand these dynamics. Although the hypothetical feedback loops that can exist in human social system can be efficiently studied using system dynamics modeling, the mereological aspects of micro–macro interactions that are of paramount importance to social scientists appear to require an agent-based approach. Furthermore, there is a subsequent issue of validation. Namely, that psychologists have access to empirical studies that can validate causal mechanisms posited within agent architectures. This allows us to address the causal claims of a model through empirical study outside of the model itself as well as establish "mechanistic correspondence" between the mechanisms of the model and mechanisms resulting in the real-world phenomena. Meanwhile, historians and anthropologists have a wealth of data concerning group-level historical trajectories that can be used to validate the model's output as well, demonstrating "variable correspondence" between the model output and data on the target phenomenon. In this way, MAAI models can be useful in a "tripartite" approach that intimately engages computer modeling, history/archeology, and cognitive psychology.

In what follows, I present a brief overview of a MAAI architecture that has undergone validation through comparing the internal mechanisms to new data drawn from psychological studies (seeking to establish mechanistic correspondence) as well as comparing outputs of the model to the historical and ethnographic record (seeking to establish variable correspondence).

OVERVIEW OF MODEL

The objective of the model is to develop an empirically based – and psychologically valid – model of how complex interactions between human memory systems as well as constraints on human sociality (e.g., the number of agents we can actively socialize with) interact at the agent level to study the effects of social stability and the group level. By social stability, it is targeting both the stability of the group (as formalized through a dynamic social network) and that of the information a group holds (as formalized through the semantic networks of interacting concepts held by the agents). Ideally, the model should allow us to answer key questions concerning the stability of large-scale religious groups.

The model focuses on the information identity system (IIS; Lane, 2017b), which proposes that humans have a naturally developed suite of cognitive facilities for using information stored in their memory to draw inferences about their social environment. These cognitive facilities provide both constraints and proclivities for the ways in which humans acquire, store, and process information and how this can result in different ways that we align ourselves socially, such as with large groups or with small numbers of individuals. The IIS draws primarily from the literature in cognitive anthropology, particularly focusing on cognitive mechanisms related to memory consolidation, group alignment, and social network constraints.

The model assumes that there are multiple types of group alignment (in line with recent work in social psychology (Gómez *et al.*, 2011)): one based on aligning oneself with a group of individuals who share a set of beliefs (social identification, where one is tied to a social category – forming a "categorical tie"); another where alignment is based on a visceral sense of oneness with the group (identity fusion). Identity fusion comes in two forms, local and extended. Local fusion is based on the connections one has with other individuals who have mutually experienced some emotional event (as such one aligns with a group based on shared relationships – or "relational ties"). Extended fusion is based on internalizing the social schema of one's group as personally and centrally salient to one's own sense of self (as such, one aligns with a group based on internalizing key beliefs of their group as crucial to their own identity – called *conceptual ties*).

Key to the model is that when agents are "fused" to a group, they are utilizing information stored about rare and emotional events (episodic memory), whereas when agents are identifying with a group, they are looking at how far their beliefs are from that of their social group (i.e., the distance between their own personal schema and their group's social schema). The social schema of a group is calculated using a lossy intersection method. This method, used earlier in cybernetics by researchers such as Carley (1997), approximates the matrix of the "group's" socially shared beliefs by analyzing that of a total population of schemas of a group. First, the belief matrix (a 2×2 sparse matrix) storing connections between the nodes of a network (wherein the network represents different concepts and links represent some relationship between them)[4] of each agent in the group is added together; this results in a union matrix where the presence of a relationship between two concepts is equal to the total number of agents who hold that conceptual relationship in any singe group. The relationships held by greater than t agents in the group – where t is equal to some threshold – are then retained, all other relationships are set equal to 0 as they are not considered to be held by a critical number of agents in the group to be considered "socially shared." For this model, t is set equal to 1/3. This new lossy intersection matrix is considered to represent the social schema of a group in the model. Those agents who share greater than s beliefs with that matrix, where s is equal to some percentage threshold, can be said to identify with that group.[5]

The real-world observation that social identities can change in different contexts (Settles, 2004; Haslam *et al.*, 2011) and that humans can learn as information is re-transmitted (Abel, 2015; Pezzo and Beckstead, 2006) is addressed by utilizing basic epidemiological processes where agents can exchange information along social network ties. At initialization, agents are initialized into a group, which is defined by a social network. There can be multiple groups in the simulation, and the social networks can be constructed to match a range of parameters.[6] What is important is the interaction system that governs how information is exchanged. This is done by having agents cycle through a list of information transmission behaviors: talk to peers, attending frequent events, attend communal gatherings, and reflect. These behaviors are performed by each agent with a frequency rate unique to them based on normal distributions drawn from anthropological field observations and survey data. For this simulation, data were used from Singaporean church groups (Chong and Hui, 2013; Singapore Religion and Spirituality Project, 2016).

At every time step in the model (where each step is 1 day), agents update their beliefs, motivations, and affect level. Their beliefs are represented in an $n \times n$ matrix. Their motivation for the belief is a number greater than 0 stored in a matrix identical to the belief matrix. All values in the matrix decrease at a set (linear) rate for all

---

[4]This way of storing belief schemas is not unlike a primitive adaptation of information schemas as proposed by Quillian (1966) and Minsky (1974).
[5]The model was coded in NetLogo and code for the model is available with documentation from the author upon request.
[6]Allowing for the network to initialize both random networks, scale-free networks, and so on, allows for experimentation and comparison but also facilitates approximations of both small and large group dynamics.

agents at each day. Each agent also decreases their motivation for a belief each time it is presented to them by a group leader. The values governing each rate are global variables that can be set by the user. The only exceptions to this are after a new group leader is introduced into the group, at which point the belief is set at its maximum level (this value is unique for each agent and drawn from a distribution where the minimum, mean, and standard deviation are set by the user), and if the belief is taught during a communal gathering (discussed below).

When the agents talk to peers, an agent chooses n other individuals with whom they have a social network connection and they set one of their beliefs to match that of their peer and they set their motivation for that belief to the average of the two agents. N is equal to an individual-level variable that represents the number of links they aim to maintain in the model (this value is drawn from a gamma distribution discerned through empirical studies of human social networks (Conti et al., 2011; Gonçalves et al., 2011; Dunbar, 2008; Dunbar et al., 1995)). If the number gets too high or low, they create links to agents with whom they share a majority of their beliefs or destroy links to agents with whom they do not share a majority of beliefs.

Agents attend frequent events with an average frequency set at initialization (typically 7 or 14 days depending on the cultural group being simulated). When the behavior to attend frequent events is triggered, agents receive teachings from the leaders of their group. The number of conceptual relationships transmitted during any one attendance session is set by a parameter, but the actual selection of a particular belief for teaching is set by a power law frequency distribution that mimics the general patterns found in religious and cultural materials (Lane, 2013b; Steyvers and Tenenbaum, 2005). When an agent "receives" a teaching, they set the values in their belief matrix to that which are transmitted to them by their leader and they decrease their motivation by a set interval, simulating psychological tedium (or habituation (Whitehouse, 2004)); when motivation reaches 0, the belief is removed from the agent's belief matrix (i.e., forgotten). As such, there is a complex relationship between consensus, which should be effectively maintained through constant contact between individuals, and the extent to which frequent exposure lessens motivation for a belief.

Agents also attend group communal events. These events are held by the group's leaders with a set frequency (e.g., once a year in model time). The agents will attend this event if their attendance frequency has not yet been satisfied for that time period. During the event, agents will exchange beliefs and motivations with agents in their social network that are also in the "attending" state. The agents also receive a special teaching from their group's leader. This teaching consists of the same information as teachings at frequent events but has a special emotional arousal level, which is a random float variable that, when above an agent-level fusion threshold, will cause two things to occur: (1) the agent will set the strength of the social network link to 999, and these links will not deteriorate during the simulation – this is done to simulate the retention of social information in episodic memories; (2) the agent will increase their affect and motivation but will not receive teachings from the groups – also simulating the effects of uncertainty for specific information in episodic memories. Resulting from this experience, agents will also increase their "reflection rate," which is set as an agent-level variable governing how frequently the agent will reflect upon and generate idiosyncratic links in their belief matrix; this is done in accordance with empirical and ethnographic literature.

Reflection is modeled as agents' ability to independently create new links between concepts in their belief network. This is done at a set "reflection rate" for each agent where the agent selects one conceptual link at random from their belief matrix. Then, they will change their belief from one state to another (e.g., 0–1). An agent's reflection rate can increase if they experience some event associated with high arousal, such as the communal events described above. The model assumes that reflection introduces belief idiosyncrasies in the way that untutored reflection on an event could facilitate in actual human populations.

TESTING THE MODEL

To better understand the dynamics of this complex model, a parameter sweep was employed to vary the size of the population, the number of beliefs held in the belief schemas, the rate of information transmission from cultural leaders, and the rate of attending communal events with high sensory pageantry. Varying these variables allows many cultural phenomena to be captured ranging from small groups that are bound by intense rituals to larger groups bound by similar beliefs (these two types of rituals are typically handled separately in the cognitive and anthropological literature, e.g., see the theory of Divergent Modes of Religiosity (Whitehouse, 2004), a theoretical precursor to the IIS). The parameters varied for the discussion here are presented in Table 14.2 (note: for the variable "Retreat-frequency-group-1," the min and max are set so that only annual or triannual communal gatherings are triggered).

RESULTS

The data produced by the parameter sweep were used to test five key hypotheses. Data were collected from each agent at the end of a run for the following variables: identity fusion, social identification, rate of reflection, social consensus with their group, and role (leader or follower). The key hypotheses tested with this data are listed below:

1. Fusion and social identification will be positively correlated
2. Social identification and social consensus will be positively correlated when controlling for fusion
3. Fusion and the rate of reflection will be positively correlated
4. Social identity should not decrease with population size
5. Consensus among the followers in a group will be positively correlated with the proportion of leaders in a group.

The first three hypotheses are confirmed by statistical analysis of the agent-level data. Testing hypothesis 1 revealed a significant correlation between fusion and

TABLE 14.2 Values for Parameter Sweep Experiment

Parameter	Min.	Max.	Intervals
Initial_Population_Level	15	150	15, 50, 150
NetworkType	–	–	"Scale-free"
Num_of_Groups	–	–	1
Num_of_Beliefs	5	25	10
Minfriends	–	–	4
ConnectionProbability	–	–	0.66
Clustering_Exponent	–	–	0.3
the-daily-motivation-level-decay	–	–	0.5
the-motivation-level-decay-per-repetition	–	–	1
the-minimum-motivation-level	–	–	100
the-temple-visit-frequency	–	–	7
the-transmission-rate	3	14	3, 7, 14
toroidal	–	–	FALSE
the-frequency-window	–	–	1000
the-number-of-leaders	–	–	1
the-number-of-new-leaders	–	–	1
Hetrophily_Threshold	–	–	0.5
Homophily_Threshold	–	–	0.79
Retreat-frequency-group-1	365	900	–
Arousal_threshold	–	–	8
new_leader_lag	–	–	35
fusion_threshold	–	–	25

TABLE 14.3 Means for Social Identity by Group Size

Group size	Mean SI	Standard deviation	LLCI	ULCI
15	618.9	474.05	595.66	642.15
50	423.11	423.11	767.3	788.93
150	822.34	392.73	816.48	828.2

social identification ($r_s = 0.99$, $p < 0.01$). Testing hypothesis 2 using a partial correlation revealed that there is a significant correlation between social identification and group consensus when controlling for fusion ($s = 0.926$; $p < 0.01$; $df = 12\,147$). The hypothesis that the relationship between fusion and reflection should be positive (hypothesis 3) was also found to be significant ($r_s = 0.141$; $p < 0.01$). In testing hypothesis 4 above, the model results reveal that social identity does not decrease with population size, showing that large groups can be sustained when not all agents have links to other agents. A one-way ANOVA actually reveals that social identity increases as population sizes go up ($f_{(24742)} = 193.891$; $p < 0.01$; see Table 14.3).

In testing hypothesis 5, the results reveal that there is a significant correlation between the proportion of leaders in a group and the amount of consensus that is sustained ($r_s = 0.161$; $p < 0.01$; 95% CI [0.149–0.173]). Further analysis shows that

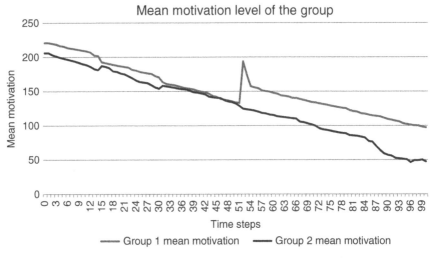

FIGURE 14.1 Mean motivation level for two groups over time.

there is a positive correlation between consensus in the group and the actual number of leaders in the group ($r_s = 0.315$; $p < 0.01$; 95% CI [0.303–0.327]) and a negative correlation between consensus and the number of followers ($r_s = -0.161$; $p < 0.01$; 95% CI [−0.173 to 0.149]).

When viewed over time, trace validation of the model also makes predictions that appear to align with the psychological, historical, and ethnographic data. To demonstrate this, I have recorded data from a simulation that utilizes two groups, one that participates in communal gatherings on time step 50 (group 1) and one that never participates in such an event (group 2). All other parameters are the same between both groups. What we see is that, as expected by the theory, motivation will decrease, and only increases at times of significant social events, such as a communal gathering or when a new leader emerges. The communal event in group 1 is seen to increase motivation at time step 50 (Figure 14.1).

As expected, this abrupt shift in motivation does increase one's social identification with the group. Given that motivation is one of many factors in the calculation of social identification, this is expected. Longitudinal output for the groups' average social identification levels is shown in Figure 14.2.

Intriguingly, this communal gathering increased motivation and social identification but did so at the expense of group consensus. As these events are only attended by some agents, and they all receive teachings, this is somewhat expected. However, it is unexpected to see consensus, which is also a factor in the calculation of social identification, moving in opposite directions at the time of the communal gathering (the longitudinal data for consensus are shown in Figure 14.3). Meanwhile, basic reflection of agents keeps the consensus in group 2 decreasing in a steady state as they interact with the highly motivated individuals in group 1. I will return to this observation shortly.

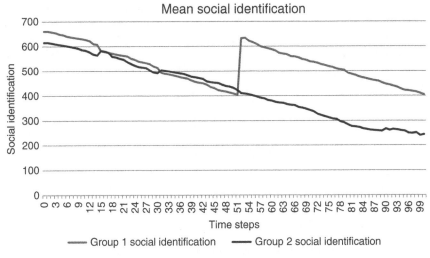

FIGURE 14.2 Mean social identification level for two groups over time.

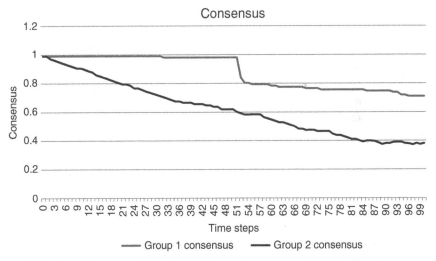

FIGURE 14.3 Consensus for two groups. 1 = total consensus on beliefs among members of a group; 0 = total disagreement among members of a group.

Although it can be noted that attendance at such an event increases the reflection of an agent, and thus their levels of fusion, the levels of fusion were beginning to be manipulated far sooner than time step 50 in group 1, as shown in Figure 14.4. It appears that some highly arousing event, likely the introduction of a new cultural leader, increased the level of fusion among members of group 1. Meanwhile, the interactions between the members of group 2 and those members of group 1 who attended the communal event spread their highly motivated beliefs to the members

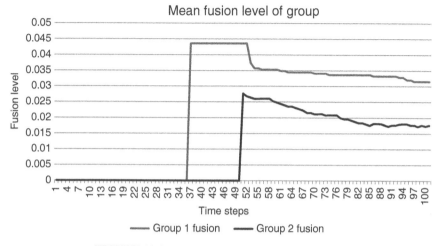

FIGURE 14.4   Average fusion levels for two groups.

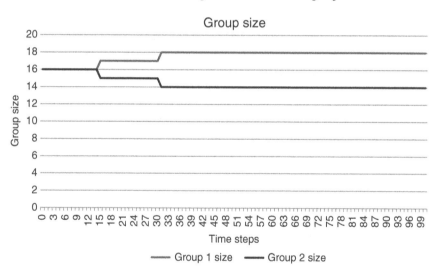

FIGURE 14.5   Group sizes for two groups in the simulation. Note that this only tracks raw group identification.

of group 2, thus showing an effect of cross-contamination between social groups in a single environment.

Lastly, it is worth noting that some individuals converted between groups. Although both groups started with 16 agents at initialization, due to socialization effects and the integration of new beliefs, we observe that multiple individuals leave group 2 to join group 1. This, it would appear, is due to a complex interacting effect of consensus (as no conversations take place after the decrease in consensus occurs in group 1 in conjunction with the communal event performance) and fusion (as no conversions happen after the jump in fusion experienced at time step 36 in group 1, as shown in Figure 14.5).

The results suggest that there are complex non-reducible effects that can be observed at group levels that arise from the complex interactions of "psychological" mechanisms at the individual agent level. The results also suggest many avenues for future directions for exploring this model more fully. Many will inevitably note that the complexity of a model decreases the ease of tractability of the model, and this model is complex. However, as a research tool, a good model should have appropriate correspondence between not only the output of the model and the real-world phenomena that it targets, but also with the mechanisms of the model, including those which it seeks to simulate. In this lies a crucial difference between psychological models that are reliant upon traditional statistics as analytical tools and agent-based simulations. The mechanisms of a psychological model can only be fully described as a dynamic system, and ABMs can facilitate this in ways that other tests of psychological models (i.e., theories, such as those tested with traditional statistics and scanning electron microscopes (SEMs)) cannot. In this regard, I believe that complex models of the psychological foundations of human social systems can be defensible, as the phenomena that they attempt to model are incomprehensibly complex in their totality. Therefore, simplified models (as they will inevitably be) will still be far more complex than some – but not all – people engaged in simulation may be comfortable with.

EMERGENCE OF SOCIAL CONSENSUS IN THE IIS

The system outlined above posits three types of variable interactions: (1) within an individual (e.g., emotional experiences promoting self-reflection upon that experience), (2) between individuals (e.g., a follower learning information from an orator or author), and (3) between individuals and their environment, specifically their social environment as represented by their social group. Statistical procedures drawn from the field of psychology provide several ways in which intra-individual variables can be investigated; these address type 1 variable interactions. However, it is at the type 2 and type 3 levels where potentially emergent effects are likely to be observed. But statistically investigating these interactions is problematic. Interaction between individuals (type 2) is a problem better suited to the tools provided by dynamic network modeling, specifically, social network modeling, and is beyond the scope of traditional linear statistical methods. Additionally, personal-to-group (type 3) interactions can be statistically investigated using methods such as generalized linear modeling, k-means cluster analyses, and multivariate hierarchical modeling. However, these statistical analyses often assume that the interactions occur at a single point in time and those that allow for repeated measures analyses assume that the change functions are the same for all individuals (Hox and Stoel, 2005). This is problematic because the inherent assumption of models of sociopsychological processes is that they occur over time (a point embraced by the identity information system) and that different psychological processes can play different roles. Thus, they assume a diachronic – not synchronic – approach to social stability that can be achieved through more than one change mechanism. Furthermore, the statistical models employed by psychologists cannot address the feedback effects whereby

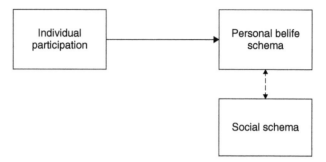

FIGURE 14.6   Individual participation encodes information into personal belief schemas, which in turn are used to align with social belief schemas.

individual interactions create group-level effects and these group level effects – in turn – have an effect on the individuals within the group.

Just such a feedback loop in the IIS model is observed in the way in which individual agent actions and cognitive mechanisms (such as reflection) affect the social schema, which in turn affect the actions of other individual agents. For example, frequently rehearsed information that is presented in contexts of low emotional arousal will encode schemas in semantic memory that are shared throughout a population – a feature common to many large groups such as nations and large "world" religions. Individuals socially identify with a large group by making categorical ties when they feel as if their personal belief schemas align with the social belief schema of a group (depicted by the dotted line in Figure 14.6).

This results in individuals aligning themselves with like-minded individuals, a tendency called *homophily* in the psychological literature. However, when we break down the social schema into more basic components, we see that it is more complex. In the IIS model, as with other cultural theories, the social schema is the information held by other individuals in the social group. This is depicted in Figure 14.7.

This model represents an iterated complex system in the technical sense as the mechanism by which one agent in the system (a person) identifies with their group by means of shared beliefs learned by experience is in itself complex. This process is iterated throughout all the agents in a population at each time period, resulting in groups that are also quasi-recursive in that each member of the group is a partial reflection of their social group as a whole.

The iterated and recursive process described by the IIS results in the apparent emergence of social consensus at the group level. This results in the ability for agents to maintain social identification with a group over time, despite the model allowing for experiences that can codify part of an agent's social network as well as facilitate idiosyncratic reflection that can increase the network distance between the agent's personal schema and the social schema. In this way, the individual stochasticity of a belief (or beliefs) may result in changes at individual and group levels. However, consensus within the group still allows for stable social groups to form. In this way,

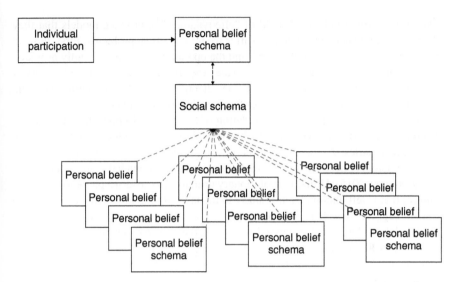

FIGURE 14.7   Consensus visualized as an "emergent" social process.

this model of the cognitive capacities for individual group alignment and information transmission offers an interesting way to simulate the longevity and stability of human social groups while also offering important insights into our understanding of emergence in models of complex social systems.

The takeaway from this model, regarding the emergence of culture, is that the key aspects of culture (a set of beliefs that are socially shared and a structured social group of individuals who share them) can be modeled computationally. In addition, basic validation shows that generative ABMs can capture key aspects of these dynamics. Given the typology of Maier discussed above, it can be stated that culture is (currently) a strong nomological emergent property of human cognitive agents. Although the model generates its target effects from the interactions of lower level components (agents with psychologically realistic cognitive architectures), these architectures are assumed – without direct attachment to their underlying neurophysiological structures – and are therefore nomological. The emergence of culture through this model is a "strong" emergence as the dynamics of the higher level social consensus that anthropologists refer to as culture is the result of under-lying permutations and information processing mechanisms of the agents (based in cognitive psychology) and individual changes can affect the changes of the cultural schema.

This generative approach stands juxtaposed to a top-down cultural approach (i.e., one which assumes culture to be a nomologically emergent causal entity). Such a top-down approach currently lacks any covering law at the cultural level that would govern cultural change that could be empirically verified; as such, it appears that a more simplified (perhaps mathematical or system dynamics) model would be unable to predict cultural shifts and changes. Some scholars, such as those in the field of cultural evolution (and its sub-field of "cliodynamics"[7]), produce models that they claim to capture such laws, but they suffer from two critical problems. First, they do not offer clear definitions of "culture" and the operationalizations currently provided are not generally accepted by subject matter experts. This definitional obscurity is problematic as it is (1) a hallmark of non-scientific research that claims scientific authority (Bunge, 2011) and (2) leaves them unable to measure culture in such a way that could facilitate a data-driven validation of the model. Second, the mechanisms that they do describe are based on stochastic evolutionary processes that have poorly defined boundaries. As such, these models imply that culture is either a strong or spooky emergent phenomena as the models lack a clearly bounded search space (Mittal and Rainey, 2015). Utilizing evolutionary principles to govern a phenomenon (i.e., culture) that is left unbound, particularly to the laws of individuals that comprise the system, suggests that there may be an implicit assumption of "spooky" emergence. The model produced and discussed in this chapter utilized boundaries established (and empirically testable) drawn from empirical studies in cognitive psychology. As such, the model can be said to be a generative model of the weak emergent properties of cultural patterns.

CONCLUSION: EMERGENCE, WHAT IS CLAIMED, AND WHAT SHOULD NEVER BE

Here, I offer some final thoughts on the nature of emergence of cultural patterns in relation to modeling and simulation specifically. There are some in the social sciences that utilize computer modeling, who claim that cultures and human social systems are complex and culture is irreducible. This includes many researchers in the "California School" of cultural evolution. I will refer to those approaches assuming culture to be irreducible as "type 1" approaches. Type 1 approaches treat cultures as emergent aspects of our world, with causal efficacy, that are irreducible to underlying individual properties of the system. In this sense, they are advocating a strong emergence (Bedau, 1997; Kim, 2006) that maps most closely onto Maier's "spooky" emergence (Maier, 2014, 22). Other approaches to human social systems, such as those advocated by evolutionary psychologists, that embrace processes of "evoked culture" (Liénard and Lawson, 2008; Tooby and Cosmides, 1992; Lane, 2013a; Lane, 2017a), rely on claims that cultural patterns, are complex, but the result of dynamic and interacting psychological mechanisms generally assuming a sort of modularity of information processing (Fodor, 1985). This approach will be referred to as type 2. Type 2 approaches often view culture as a supervening property of human cognitive mechanisms in a specific environmental context. These approaches can generally be assumed to represent models that would qualify as simple or weak emergence as they rely on law-like information processes with little or no stochasticity. There is also a third approach (type 3), which could be viewed as an extension

---

[7]Cliodynamics is an attempt to create mathematical models of social system dynamics historical patterns. It is largely an enterprise of non-historians.

of type 2, that holds that culture is the result of complex and interacting cognitive patterns, evoked by information signals input from the environment, and the information produced by these mechanisms creates a "social environment" of information which is subsequently utilized in concert with biological information to evoke cognitive mechanisms. This approach holds that the biological and social aspects of our environment are less separable than the current competing schools from type 1 and type 2 seem to suggest. This implies that the information produced through inter-agent interactions can also influence the elicitation of the cognitive mechanisms that themselves produce cultural information. This interaction results in a feedback loop, which, although potentially tractable within computational systems such as that presented above, constitutes neither support for the idea that "culture" is a "spooky" emergent entity with causal properties nor that it is a basic supervening property of a group of individuals. Instead, it is the structure of inter-agent interactions that defines a group and the information passed therein as well as the elicitation of cognitive mechanisms by biological environments that result in "culture." In this way, type 3 approaches are like the "generativist" approach of ABM advocated by Epstein (Epstein, 1999), who holds that one cannot have spooky emergence in computational systems.[8] Lastly, in type 1 approaches, the cultural group is treated as the appropriate level of explanation, and sometimes the group constitutes its own unit within the explanatory paradigm, whereas in many others, game theoretic or system dynamic models are used, which assume homogeneity of agents.[9] In type 2 approaches, culture is explained via reduction to individual cognitive mechanisms; therefore, the individual is the appropriate unit of explanation. However, while type 3 approaches reject the causal arguments of type 1 on the grounds that the bounds of their group are often undefined (or defined through tautologies)[10] and are insufficient to constitute a well-defined causal entity, type 3 explanations accept that individual cognitive mechanisms can cause "group-level" patterns that have effects on individuals. Therefore, it argues that understanding social information, which can be done by modeling inter-agent interactions by using dynamic social networks, provides a better explanatory level of inquiry. These similarities and differences between the approaches are presented in Table 14.4.

[8] Epstein refers to "strong emergence" specifically in the publication, but his definition of strong emergence does not preclude the "strong" emergence as defined by Maier as utilized in this chapter. It does, however, directly reject the assumptions of what Maier would qualify as Spooky emergence. This is because Epstein is generally looking at emergence through the terminology of social science. See Table 14.1 for clarification.

[9] This is inherently problematic as the experiments they cite as validation of the models utilize statistics from psychology which are totally reliant upon distributions at the individual level in order to reject a null hypothesis. Typically, these models are based in evolutionary theory. For a discussion of the philosophical considerations that should be understood in such an approach, I would point the reader to Knudt (2015) and subsequent discussions (Lane, 2017a). Additionally, readers may be interested in further reading on the "Ludic Fallacy," which is when game theory models are used to understand systems where the mechanisms and decisions are unrelated to the rules of the game utilized in the model (Taleb, 2010).

[10] One could, for example, leverage the social networks described in the model above to define a group based on metrics such as Louvain clustering and still be compatible with the type 3 approach (Blondel *et al.*, 2008), if indeed such an analytical construct was required.

TABLE 14.4   Overview of Approaches to Emergence in Social Systems

	Type 1	Type 2	Type 3
Emergence of culture	Spooky	Weak, simple, strong	Generativist
Informational focus	Social	Biological	Social and biological
Entitativity	Group, culture, trait	Individual	Individual
Explanatory level	Context-context (social)	Context-individual (reductionist)	Context-individual-context (generativist)

The generativist argument, developed by Joshua Epstein, relies on a form of explanation whereby the primary focus is not on reduction but on the ability of general rules and principles (instantiated at a lower level of analysis than the target phenomenon) to produce – or generate – the target phenomenon. In this way, the explanandum[11] is grown from lower level interactions. The cultural cybernetic approach presented here embraces the generativist approach but adds that there are times where generated social phenomena can also result in lower level effects when the generated phenomena is something with which agents can interact. The cultural cybernetic approach, when applied to the sort of consensus between individuals we term *culture*, can appear in a sense to be emergent. However, upon closer inspection – and in line with the generativist research program in ABM – there is no "spooky" emergence. Instead, complex interactions between agents who follow specific rules related to human cognitive facilities can generate (i.e., explain) cultural patterns.

REFERENCES

Abel, T. (2015) Cultural transmission in cycles: the production and maintenance of cumulative culture. *Journal of Cognition and Culture*, **15** (5), 443–492. doi: 10.1163/15685373-12342161.

Anderson, B. (2006) *Imagined Communities*, 2nd edn, Verso, London and New York.

Ariely, D. (2010) *Predictably Irrational: The Hidden Forces That Shape Our Decisions*, Harper Perennial, New York.

Bainbridge, W.S. (2006) *God from the Machine: Artificial Intelligence Models of Religious Cognition*, AltaMira Press, Lanham, MD.

Barrett, J.L. (2004) *Why Would Anyone Believe in God?* Cognitive Science of Religion, AltaMira Press, Walnut Creek, CA.

Bedau, M.A. (1997) Weak emergence, in *Philosophical Perspectives, Mind, Causation and World*, vol. **11** (ed. J. Tomberlin), Blackwell Publishers, Malden, MA, pp. 375–399. doi: 10.1111/0029-4624.31.s11.17.

[11] A term in the philosophy of science means that which is to be explained from the Latin *explanare* – to explain or make clear.

Blondel, V.D., Guillaume, J.-L., Lambiotte, R., and Lefebvre, E. (2008) Fast unfolding of communities in large networks. *Journal of Statistical Mechanics: Theory and Experiment*, 2008 (10), 6. doi: 10.1088/1742-5468/2008/10/P10008.

Brewer, M.B. (1991) The social self: on being the same and different at the same time. *Personality and Social Psychology Bulletin*, **17** (5), 475–482. doi: 10.1177/0146167291175001.

Brubaker, R. (2002) Ethnicity without Groups. *European Journal of Sociology*, **43** (2), 163–189.

Bunge, M. (2011) Knowledge: genuine and bogus. *Science and Education*, **20** (5–6), 411–438. doi: 10.1007/s11191-009-9225-3.

Carley, K.M. (1997) Extracting team mental models through textual analysis. *Journal of Organizational Behavior*, **18**, 533–558. doi: 10.1002/(SICI)1099-1379(199711) 18:1+<533::AID-JOB906>3.3.CO;2-V.

Chong, T. and Hui, Y.-F. (2013) *Different Under God: A Survey of Church-Going Protestants in Singapore*, Institute of Southeast Asian Studies, Singapore.

Conti, M., Passarella, A., and Pezzoni, F. (2011) A model for the generation of social network graphs. 2011 IEEE International Symposium on a World of Wireless, Mobile and Multimedia Networks, WoWMoM 2011 – Digital Proceedings, January 2016. doi: 10.1109/WoW-MoM.2011.5986141.

Dávid-Barrett, T. and Dunbar, R.I.M. (2013) Processing power limits social group size: computational evidence for the cognitive costs of sociality. *Proceedings of the Royal Society B*, **280**, 1–8.

DiFonzo, N., Bourgeois, M.J., Suls, J. *et al.* (2013) Rumor clustering, consensus, and polarization: dynamic social impact and self-organization of Hearsay. *Journal of Experimental Social Psychology*, **49** (3), 378–399. doi: 10.1016/j.jesp.2012.12.010.

Dunbar, R.I.M. (2008) Cognitive constraints on the structure and dynamics of social networks. *Group Dynamics: Theory, Research, and Practice*, **12** (1), 7–16. doi: 10.1037/ 1089-2699.12.1.7.

Dunbar, R.I.M., Duncan, N.D.C., and Nettle, D. (1995) Size and structure of freely forming conversational groups. *Human Nature*, **6** (1), 67–78.

Epstein, J.M. (1999) Agent-based computational models and generative social science. *Complexity*, **4** (5), 41–60. doi: 10.1002/(SICI)1099-0526(199905/06)4:5<41::AID-CPLX9> 3.3.CO;2-6.

Epstein, J.M. (2006a) Agent-based computational models and generative social science, in *Generative Social Science* (ed. J.M. Epstein), Princeton University Press, Princeton and Oxford, pp. 1–43. doi: 10.1002/(SICI)1099-0526(199905/06)4:5<41::AID-CPLX9> 3.0.CO;2-F.

Epstein, J.M. (ed.) (2006b) *Generative Social Science: Studies in Agent-Based Computational Modeling (Princeton Studies in Complexity)*, Princeton University Press, Princeton, NJ.

Fodor, J.A. (1985) Precis of the modularity of mind. *Behavioral and Brain Sciences*, **8**, 1–42.

Garro, L.C. (2000) Remembering what one knows and the construction of the past: a comparison of cultural consensus theory and cultural schema theory. *Ethos*, **28** (3), 275–319. doi: 10.1525/eth.2000.28.3.275.

Goldstein, J. (1999) Emergence as a construct: history and issues. *Emergence*, **1** (1), 49–72. doi: 10.1207/s15327000em0101_4.

Gómez, Á., Brooks, M.L., Buhrmester, M.D. *et al.* (2011) On the nature of identity fusion: insights into the construct and a new measure. *Journal of Personality and Social Psychology*, **100** (5), 918–933. doi: 10.1037/a0022642.

Gonçalves, B., Perra, N., and Vespignani, A. (2011) Modeling users' activity on Twitter networks: validation of Dunbar's number. *PloS One*, **6** (8), e22656. doi: 10.1371/journal.pone.0022656.

Harton, H.C., Green, L.R., Jackson, C., and Latane, B. (1998) Demonstrating dynamic social impact: consolidation, clustering, correlation, and (sometimes) the correct answer. *Teaching of Psychology*, **25** (1), 31–35. doi: 10.1207/s15328023top2501_9.

Haslam, C., Jetten, J., Alexander Haslam, S. *et al.* (2011) 'I remember therefore I am, and I am therefore I remember': exploring the contributions of episodic and semantic self-knowledge to strength of identity. *British Journal of Psychology*, **102** (2), 184–203. doi: 10.1348/000712610X508091.

Henrich, J. and Gil-White, F.J. (2001) The evolution of prestige: freely conferred deference as a mechanism for enhancing the benefits of cultural transmission. *Evolution and Human Behavior : Official Journal of the Human Behavior and Evolution Society*, **22** (3), 165–196, http://www.ncbi.nlm.nih.gov/pubmed/11384884 (accessed 17 November 2007).

Henrich, J., Heine, S.J., and Norenzayan, A. (2010) The weirdest people in the world? *The Behavioral and Brain Sciences*, **33** (2–3), 61–83. doi: 10.1017/S0140525X0999152X.

Hill, R.A. and Dunbar, R.I.M. (2003) Social network size in humans. *Human Nature*, **14** (1), 53–72.

Hox, J. and Stoel, R.D. (2005) Multilevel and SEM approaches to growth curve modeling, in *Encyclopedia of Statistics in Behavioral Science* (eds B.S. Everitt and D.C. Howell), John Wiley & Sons, Ltd, Chichester, pp. 1296–1305.

Jong, J., Whitehouse, H., Kavanagh, C., and Lane, J.E. (2016) Shared trauma leads to identity fusion via personal reflection. *PLoS ONE*, **10** (12), e0145611. doi: 10.1371/journal.pone.0145611.

Kim, J. (2006) Emergence: core ideas and issues. *Synthese*, **151** (3), 547–559. doi: 10.1007/s11229-006-9025-0.

Knudt, R. (2015) *Contemporary Evolutionary Theories of Culture and the Study of Religion*, Bloomsbury Academic, London.

Lane, J.E. (2013a) Method, theory, and multi-agent artificial intelligence: creating computer models of complex social interaction. *Journal for the Cognitive Science of Religion*, **1** (2), 161–180.

Lane, J.E. (2013b) Semantic network mapping of religious material. Society for Complex Systems in Cognitive Science, Berlin, Germany.

Lane, J.E. (2017a) Contemporary evolutionary theories of culture and the study of religion. *Journal for the Cognitive Science of Religion*, **3** (2), 210–221. doi: 10.1558/jcsr.30498.

Lane, J.E. (2017b) The evolution of doctrinal religions: using semantic network analysis and computational models to examine the evolutionary dynamics of large religions, University of Oxford.

Lawson, E.T. and McCauley, R.N. (1990) *Rethinking Religion: Connecting Cognition and Culture*, Cambridge University Press, New York.

Lickel, B., Hamilton, D.L., and Sherman, S.J. (2001) Elements of a Lay theory of groups: types of groups, relational styles, and the perception of group entitativity. *Personality and Social Psychology Review*, **5** (2), 129–140. doi: 10.1207/S15327957PSPR0502_4.

Lickel, B., Hamilton, D.L., Wieczorkowska, G. *et al.* (2000) Varieties of groups and the perception of group entitativity. *Journal of Personality and Social Psychology*, **78** (2), 223–246. doi: 10.1037/0022-3514.78.2.223.

Liénard, P. and Lawson, E.T. (2008) Evoked culture, ritualization and religious rituals. *Religion*, **38** (2), 157–171. doi: 10.1016/j.religion.2008.01.004.

Maier, M.W. (2014) The role of modeling and simulation in system of systems development, in *Modeling and Simulation Support for System of Systems Engineering Applications* (eds L.B. Rainey and A. Tolk), John Wiley & Sons, Inc, Hoboken, NJ, pp. 11–44.

McCorkle, W.W. and Lane, J.E. (2012) Ancestors in the simulation machine: measuring the transmission and oscillation of religiosity in computer modeling. *Religion, Brain & Behavior*, **2** (3), 215–218. doi: 10.1080/2153599X.2012.703454.

McRae, K. and Jones, M.N. (2013) Semantic memory, in *The Oxford Handbook of Cognitive Psychology* (ed. D. Reisberg), Oxford University Press, Oxford, pp. 1–24. doi: 10.1093/oxfordhb/9780199988693.013.0017.

Minsky, M. (1974) A framework for representing knowledge. The Psychology of Computer Vision, Artificial Intelligence, Cambridge, MA, https://web.media.mit.edu/~minsky/papers/Frames/frames.html (accessed 17 November 2007).

Mittal, S. and Rainey, L. (2015) Harnessing emergence: the control and design of emergent behavior in system of systems engineering. Proceedings of the Conference on Summer Computer Simulation, Society for Computer Simulation International, Chicago, pp. 1–10, http://dl.acm.org/citation.cfm?id=2874979&CFID=982354657&CFTOKEN=72808548 (accessed 17 November 2007).

Pezzo, M.V. and Beckstead, J.W. (2006) A multilevel analysis of rumor transmission: effects of anxiety and belief in two field experiments. *Basic and Applied Social Psychology*, **28** (1), 91–100.

Postmes, T., Spears, R., Lee, A.T., and Novak, R.J. (2005) Individuality and social influence in groups: inductive and deductive routes to group identity. *Journal of Personality and Social Psychology*, **89** (5), 747–763. doi: 10.1037/0022-3514.89.5.747.

Pyysiäinen, I. (2001) *How Religion Works: Towards a New Cognitive Science of Religion*, Brill Academic Pub, Leiden.

Quillian, M.R. (1966) Semantic memory. Scientific Report No. 2. Project No. 8668, Bolt Beranek and Newman Inc, Cambridge, MA.

Roberts, S.G.B., Dunbar, R.I.M., Pollet, T.V., and Kuppens, T. (2009) Exploring variation in active network size: constraints and ego characteristics. *Social Networks*, **31** (2), 138–146. doi: 10.1016/j.socnet.2008.12.002.

Saramäki, J., Leicht, E.A., López, E. *et al.* (2014) Persistence of social signatures in human communication. *Proceedings of the National Academy of Sciences of the United States of America*, **111** (3), 942–947. doi: 10.1073/pnas.1308540110.

Sawyer, R.K. (2005) *Social Emergence: Societies as Complex Systems*, Cambridge University Press, Cambridge.

Schröder, T., Hoey, J., and Rogers, K.B. (2016) Modeling dynamic identities and uncertainty in social interactions: Bayesian affect control theory. *American Sociological Review*, **81** (4), 828–855. doi: 10.1177/0003122416650963.

Settles, I.H. (2004) When multiple identities interfere: the role of identity centrality. *Personality and Social Psychology Bulletin*, **30** (4), 487–500. doi: 10.1177/0146167203261885.

Singapore Religion and Spirituality Project (2016) Singapore Religion and Spirituality Project. Data. https://singaporereligion.wordpress.com/ (accessed 17 November 2017).

Sperber, D. (1996) *Explaining Culture: A Naturalistic Approach*, Blackwell Publishers, Oxford.

Steyvers, M. and Tenenbaum, J.B. (2005) The large-scale structure of semantic networks: statistical analyses and a model of semantic growth. *Cognitive Science*, **29** (1), 41–78. doi: 10.1207/s15516709cog2901_3.

Sun, R. (2001) Cognitive science meets multi-agent systems: a prolegomenon. *Philosophical Psychology*, **14** (1), 5–28. doi: 10.1080/09515080120033599.

Sun, R. and Hélie, S. (2013) Psychologically realistic cognitive agents: taking human cognition seriously. *Journal of Experimental and Theoretical Artificial Intelligence*, **25** (1), 65–92. doi: 10.1080/0952813X.2012.661236.

Tajfel, H. and Turner, J.C. (1979) An integrative theory of intergroup conflict, in *The Social Psychology of Intergroup Relations* (eds W.G. Austin and S. Worchel), Brooks-Cole, Monterey, CA, pp. 33–47.

Taleb, N.N. (2010) *The Black Swan: The Impact of the Highly Improbable*, 2nd edn, Random House Trade Paperbacks, New York.

Tooby, J. and Cosmides, L. (1992) The psychological foundations of culture, in *The Adapted Mind: Evolutionary Psychology and the Generation of Culture* (eds J.H. Barkow, L. Cosmides, and J. Tooby), Oxford University Press, Oxford and New York, pp. 19–136.

Upal, M.A. (2015) A framework for agent-based social simulations of social identity dynamics, in *Conflict and Complexity: Countering Terrorism, Insurgency, Ethnic and Regional Violence* (eds P.V. Fellman, Y. Bar-Yam, and A.A. Minai), Springer, New York, pp. 89–109. doi: 10.1007/978-1-4939-1705-1.

Upal, M.A. and Gibbon, S. (2015) Agent-based system for simulating the dynamics of social identity beliefs. Proceedings of the 48th Annual Simulation Symposium, Society for Computer Simulation International, pp. 94–101.

Walmsley, J. (2010) Emergence and reduction in dynamical cognitive science. *New Ideas in Psychology*, **28** (3), 274–282. doi: 10.1016/j.newideapsych.2009.09.003.

Whitehouse, H. (2004) *Modes of Religiosity: A Cognitive Theory of Religious Transmission*, Cognitive Science of Religion, AltaMira Press, Walnut Creek, CA.

Whitehouse, H., Kahn, K., Hochberg, M.E., and Bryson, J.J. (2012) The role for simulations in theory construction for the social sciences: case studies concerning divergent modes of religiosity. *Religion, Brain & Behavior*, **2** (3), 182–201. doi: 10.1080/2153599X.2012.691033.

15

MODELING AND SIMULATION OF EMERGENT BEHAVIOR IN TRANSPORTATION INFRASTRUCTURE RESTORATION

Akhilesh Ojha[1], Steven Corns[1], Tom Shoberg[2], Ruwen Qin[1], and Suzanna Long[1]

[1]Department of Engineering Management and Systems Engineering, Missouri University of Science and Technology, Rolla, MO 65401, USA
[2]U.S. Geological Survey, CEGIS, Rolla, MO 65409, USA

SUMMARY

Extreme events such as earthquakes, hurricanes, and the likes result in mass destruction leading to partial or total disruption of various infrastructure and supply chain systems. This causes substantial economic loss. The damaging effects of an extreme event last well after the termination of the emergency response system, and therefore, the development of efficient restoration and disaster management strategies warrants a thorough cost analysis of the critical infrastructure disrupted and their interdependencies. The economic analyses must account for both direct and indirect losses associated with infrastructure system failure and thus the need to model the supply chain interdependent critical infrastructure. The objective of this study is to understand how an extreme event affects the road transportation network. In this study, a system dynamics approach is used to model the transportation road infrastructure system to evaluate the different factors that render road segments inoperable and calculate economic consequences of such inoperability.

Emergent Behavior in Complex Systems Engineering: A Modeling and Simulation Approach,
First Edition. Edited by Saurabh Mittal, Saikou Diallo, and Andreas Tolk.
© 2018 John Wiley & Sons, Inc. Published 2018 by John Wiley & Sons, Inc.

INTRODUCTION

Economic losses from infrastructure and supply chain failure that result from extreme events such as earthquakes, hurricanes, or the like are considerable. These losses continue to amass well after emergency response has terminated. To ameliorate the losses from large-scale disasters, it is important to understand the critical infrastructures damaged and to analyze the various interdependencies among them in order to design efficient restoration strategies.

Defining and modeling supply chain interdependent critical infrastructure (SCICI) is a complex problem (Ramachandran *et al.*, 2015, 2016) as disruption of one infrastructure network can produce a ripple effect of failure through other infrastructure networks. This will potentially result in large economic losses. Therefore, understanding the interdependencies between various infrastructure systems is critical to a cost analysis for an infrastructure network failure in the aftermath of a disaster. There are two types of economic losses that result from infrastructure disruption: direct losses and indirect losses. Direct losses include the costs of rebuilding or repairing damaged property, whereas indirect losses include losses due to changes in demand and supply behavior. For instance, if a bridge is damaged by an earthquake, direct loss would include the cost of rebuilding the bridge, whereas indirect cost would include the costs associated with the extra distance and delays that vehicles must endure over a period of time until the damaged bridge is restored. Such indirect losses result, in part, to emergent behavior within the system. The highly interdependent nature of infrastructure elements makes a system dynamics approach ideal for studying these complex infrastructure networks. A system dynamics approach has the capability to effectively incorporate a large number of variables in its algorithm and model the complex nature of interactions between these variables efficiently. A system dynamics approach uses decision trees and cause-and-effect relationships among different variables to understand the behavior of a variable due to changes in other variables. This extends to the detection of emerging dependencies. An emergent property of a system is a property that is possessed by the system as a whole but is not possessed by any components of the system individually (Maier, 2014). Analyzing traffic patterns due to a major disruption in the transportation infrastructure is a complex problem. For example, if a road segment becomes damaged to the point where at least a part of the traffic flow must be diverted to different routes, this diversion will lead to an increase in the travel costs per vehicle that can depend on flow rate, volume, topography, route mileage, and so on. As the traffic is redirected to alternate routes, the road capacities of these alternates are utilized, which leads to reduced speeds, increases in travel time, and traffic flow congestion. The increased travel costs, travel time, reduced speed for the traffic flow, and congestion constitute emergent behavior within the transportation system due to a disruption in one or more road segments.

In this study, a system dynamics model is applied to the transportation network for estimation of traffic disruption costs in the aftermath of a disaster. The causal loop diagram (CLD) used in system dynamics closely models system behavior. The system dynamics approach is used to model the interdependencies between system variables. In a dynamic system, the value of the variable changes with time and a system dynamics approach makes it possible to update these values accordingly and hence capture these interdependencies. Mittal and Rainey (2015) state that any complex system that exists in the space and time domain demonstrates emergent behavior. The transportation infrastructure system is a complex system that has a spatiotemporal character. A system dynamics approach can be used to study the spatial as well as the temporal nature of the system making system dynamics methodology a good fit to understand the emergent behavior of the system. By means of a CLD, a visual framework depicting the interdependent nature of the infrastructures involved in the network is presented. Estimations of these costs may serve as an important tool in decision-making processes of policy makers for disaster restoration and recovery plans. The degree to which these ripple effects are being realized and the economic losses in which they result are calculated. These ripple effects are ascribed to the emergent behavior of the system as described above. Rerouting vehicles to alternate paths cause decreases in available road capacity that slows down the overall traffic, which, in turn, leads to increases in travel times and congestion. Such emergent behavior can be understood by analyzing the overall speed of traffic flow post-disruption and comparing it with the overall traffic speed before a disruption in the transportation network. The drivers, when given information about the cost and travel time for each alternate route, can make informed decisions to avoid congestion. This study discusses different scenarios where the cost and time for different alternate routes are calculated. Results from this research will help in understanding the costs of infrastructure failure from how traffic patterns are altered due to a disruption in the transportation network.

The following section gives insight into system dynamics methodology and its applications. In Methodology section, the model methodology is explained, first with a discussion of how different factors affect available road capacity in general, and second, how a system dynamics approach can be used to construct a road transportation disruption model that calculates available road capacity in the aftermath of infrastructure failure. Then, an illustrative example is used to demonstrate the model. Finally presented are conclusions, limitations of this model, and future work.

SYSTEM DYNAMICS APPROACH

System dynamics is a methodological approach to study the dynamics of complex systems involving a large number of variables (Coyle, 1996). System dynamics methodology has a qualitative part that visually represents cause-and-effect relationships between different variables and a quantitative part that parameterizes these relationships. In this methodology, feedback loops describe the parameter interactions within the model. Feedback loops are either positive or negative. Positive loops, also known as reinforcing loops, are ones in which a change (positive or negative) in one variable induces a similar change (positive or negative) in another variable, whereas negative or balancing loops are ones in which a change (positive or negative) in one variable induces an opposite change (negative or positive) in another variable.

Extra distance to be ———————————————$\xrightarrow{+}$ Travel costs per
travelled per vehicle   vehicle

FIGURE 15.1   Cause-and-effect relationship.

System dynamics modeling invokes a four-stage developmental process. The first stage requires a qualitative analysis of the different variables involved in the problem and identifying the cause-and-effect relationships between these variables. The second stage involves building a CLD that describes the variables under consideration. These variables are connected and typically arrow diagrams are used to describe the cause-and-effect relationships among each other. Each arrowhead will have either a positive or a negative sign. A positive sign on the arrowhead means that an increase in the value of the variable at the tail of the arrow will lead to an increase in the value of the variable at the arrowhead and a decrease in the value of the variable at the tail of the arrow will lead to a decrease in the value of the variable at the arrowhead. A negative sign on the arrowhead means that if the value of the variable at the tail of the arrow decreases, then the value of the variable at the arrowhead increases, and if the value of the variable at the tail of the arrow increases, then the value of the variable at the arrowhead decreases. For instance, in Figure 15.1, the positive sign on the arrow connecting extra distance to be traveled per vehicle and travel costs per vehicle means that an increase in the distance to be traveled per vehicle leads to an increase in travel costs per vehicle and vice versa. The third stage in a system dynamics approach involves constructing a model, before finally testing this model in the fourth stage.

System dynamics finds wide application in economic, business, ecological, and population systems due to its ability to model simple, linear, as well as complex nonlinear systems (Sha and Huang, 2010; An and Jeng, 2005; Sterman, 1992). It is also a useful tool to study complex systems involving a large number of variables as well as nonlinear feedback loops otherwise considered unmanageable by the conventionally used algorithms such as the critical path method (CPM) and the program evaluation and review technique (PERT) (Sterman, 1992). The nonlinearity in the complex systems can be attributed to the emergent behavior of the system. The feedback loops in the system dynamics methodology models the dynamic patterns in a complex system and maps out these patterns in terms of their structural relationships. In complex systems, as new information becomes available, the behavior of the system might change. A CLD depicts the cause-and-effect relationships between different variables to show the complex interactions among these variables. The presence of decision trees and cause-and-effect relationships in system dynamics models makes them a popular choice in analyzing social and economic systems (Lyneis *et al.*, 2001). There is a tendency for the users to include more variables than required because of the ease of how cause-and-effect relationships are mapped in a CLD. To avoid incorporating excess variables in system dynamics modeling, Li *et al.* (2009) advocated dividing every model into four subsystems - project, profit, resource, and knowledge - and allocate variables to these categories, eliminating all variables that do not belong to these subsystems. Alasad *et al.* (2013) advised using expert knowledge and perceptions of stakeholders to create realistic system dynamics models.

The ability of a system dynamics approach to incorporate different aspects of a problem (economic, infrastructure, etc.) makes it a good fit for this study. System dynamics models have been applied to study many different systems and subsystems. Qing and Mingchao (2011), for example, applied the system dynamics approach to study the economy-environment-resource system in Jiangxi, China, to analyze the sustainability of the current development mode and the substitution rate of technology for natural resources. Liu *et al.* (2011) integrated the transportation systems to improve capital-use efficiency and economic development, and Zheng *et al.* (2009) integrated metrics such as infrastructure, foreign trade, regional logistics cost, and growth rate of foreign trade to conclude that investment in aviation logistics is a good way to promote trade and economic development. System dynamics approaches have also been applied to complex construction projects that contain multiple independent systems and highly nonlinear feedback loops (Lyneis *et al.*, 2001) and to port operation systems to improve service time and cost of service (Gui *et al.*, 2005). Researchers have also combined policy decisions with practical operations to understand and analyze an area's logistics system (Li *et al.*, 2009) and to identify key factors for promoting regional logistics hubs formation (Zhao *et al.*, 2011). System dynamics models have been integrated with business process simulation model to evaluate, design, and optimize the business process and study the evolution of business over long periods of time (An and Jeng, 2005), and with a project management software tool to track project abilities in terms of budget, schedule, and rework hours and improve planning (Sycamore and Collofello, 1999). To evaluate unanticipated problems associated with the emergency medical service system, Su *et al.* (2008) supplemented their discrete event emergency medical services simulation model with a system dynamics model to account for the feedback effects of human decisions. Mittal explains how any complex system model is guaranteed to show some emergent behavior for any system that exists in space and time (Mittal and Rainey, 2015). To conclude, system dynamics methods have been used in the fields of logistics, economy, business processes, and construction projects just to name a few. The ability of a system dynamics approach to model the spatiotemporal character of a system generates a greater understanding of the emergent behavior arising out of interdependencies within a complex system, in this case, the effects of disruption in a transportation network and its associated indirect costs.

METHODOLOGY

Disruption in one part of a transportation system creates a ripple effect throughout other parts of the system, as well as other critical infrastructure systems linked to it. It is therefore necessary to categorize and parameterize the different factors that result from such a disruption. A system dynamics approach can be used to understand the effects of disruption in the transportation system. The qualitative part of system dynamics, that is, constructing the CLD, helps to visually depict the causes as well as the effects of disruption in the transportation network. The quantitative part of this approach helps to study the magnitude of the disruption and thereby helps in calculating the economic losses due to the disruption. In this study, the available road capacity is the metric used to quantify the change in traffic patterns due to a disruption and estimate the costs or losses associated with it. The following section explains how different factors affect the road capacity.

Factors Affecting Available Road Capacity

The quantitative part of system dynamics deals with parameterizing the relationships between different variables. These relations are defined by a set of equations. Available road capacity refers to the length of the road that is accessible to the vehicle transport. A number of factors affecting the road capacity must be considered when calculating the total road capacity. Table 15.1 includes the various factors that affect the available road capacity along with the magnitude of the effect. In this section, it is explained how different factors affect the available road capacity.

1. *Connectivity Issue (T_{ci})* – Figure 15.2 gives the road capacity lost due to connectivity issues. The length of the road capacity lost due to road closure is denoted by T_{ci}. Here, T_{ci} is the product of length of closure, T_{cil}, and the number of lanes closed, T_{cin}.

$$T_{ci} = T_{cil} * T_{cin} \tag{15.1}$$

2. *Road Maintenance (T_{rm})* – refers to the length of road capacity lost due to ongoing maintenance (Figure 15.3). The road length used for maintenance is denoted by T_{rm}. Here, T_{rm} is equal to the length of the ongoing road maintenance (T_{rml}) multiplied by the number of lanes closed (T_{rmn}).

$$T_{rm} = T_{rml} * T_{rmn} \tag{15.2}$$

3. *Traffic Jams and Accidents (T_{tj})* – This is essentially the same parameterization as required for connectivity issues with the major difference being the amount of time for which a segment of road is closed. The road length closed to use by a traffic jam or accident covering all lanes of a road is denoted by T_{tjl}. Here, T_{rc} refers to the number of lanes (T_{tjn}) on closed road times multiplied by the length of the closed segment.

$$T_{tj} = T_{tjl} * T_{tjn} \tag{15.3}$$

4. *Regulatory Enforcement event (T_{re})* - This has the same parameterization as required for road maintenance with the main difference being that the typical length of the lane or partial lane closure is a little over one or two car lengths. The road length of capacity used for regulatory enforcement is denoted by T_{re}. Here, T_{re} is equal to the length of closure (T_{rel}) multiplied by the number of lanes closed (T_{ren}).

$$T_{re} = T_{rel} * T_{ren} \tag{15.4}$$

5. *Road Construction Transit (T_{rc})* – the road length used by each road construction vehicle (T_{rcl}) in transit multiplied by the total number of road construction vehicles (T_{rcn}) in transit affects the available road capacity. These road construction

vehicles can be further divided into the road lengths used by graders, bulldozers, flatbed semi-tractor trailers, asphalt removers, and so on. To calculate the length used by road construction transit, the following equation is used.

$$T_{rcl} * T_{rcn} = T_{grcl} * T_{grcln} + T_{brcl} * T_{brcn}$$
$$+ T_{febrcl} * T_{febrcn} + T_{artl} * T_{artn} + \ldots \quad (15.5)$$

In the above equation, the road length used by each road construction vehicle is denoted by T_{rcl} and the total number of such vehicles is denoted by T_{rcn}.

6. *Emergency Vehicles* – the road length used by each emergency vehicle (T_{el}) in transit multiplied by the total number of emergency vehicles in transit affects the available road capacity. The length of the road used by emergency vehicles can be further subdivided into the length of the road used by police cars (T_{pc}), ambulances (T_a), fire trucks (T_{fe}), and tow trucks (T_{tt}) separately. The road length used by emergency vehicles is defined by the equation below:

$$T_{el} * T_{en} = T_{pcl} * T_{pc} + T_{al} * T_a + T_{fel} * T_{fe} + T_{ttl} * T_{tt} \quad (15.6)$$

In the above equation, T_{el} is the length of road required by vehicle for safe transit, and T_e is number of emergency vehicles operating on roads in a given area. Road capacity used equals the length (T_{el}) between the forward and rear buffer zone (the closest distance that the emergency vehicle can approach another vehicle and the closest approach another vehicle can safely have behind the emergency vehicle, respectively). Figure 15.4 shows the area occupied by an individual emergency vehicle.

TABLE 15.1 Factors Affecting Available Road Capacity

Factors Affecting Available Road Capacity	Road Capacity Lost per Factor
Connectivity issue	$T_{ci} = T_{cil} * T_{cin}$
Road maintenance	$T_{rm} = T_{rml} * T_{rmn}$
Traffic jams and accidents	$T_{rc} = T_{tjl} * T_{tjn}$
Regulatory enforcement	$T_{re} = T_{rel} * T_{ren}$
Road construction transit	$T_{rcl} * T_{rcl} = T_{grcl} * T_{grcln} + T_{brcl} * T_{brcn} + T_{febrcl} * T_{febrcn}$ $+ T_{artl} * T_{artn} + \ldots$
Emergency vehicles	$T_{el} * T_{en} = T_{pcl} * T_{pc} + T_{al} * T_a + T_{fel} * T_{fe} + T_{ttl} * T_{tt}$

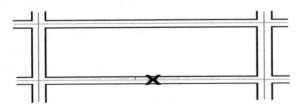

FIGURE 15.2 Road capacity lost due to connectivity issue.

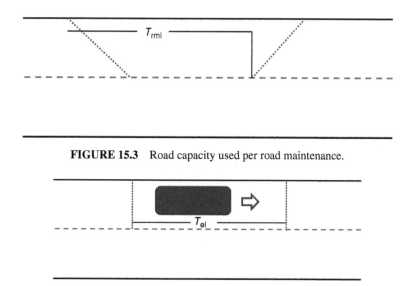

FIGURE 15.3   Road capacity used per road maintenance.

FIGURE 15.4   Road capacity used per emergency vehicle.

Model Explanation

After identifying the various factors affecting the available road capacity, a CLD (Figure 15.5) is created to visually represent the causes leading to a change in the available road capacity and the effects on travel costs when the available road capacity changes. The equations have parameterized the relationship between these variables. Figure 15.5 shows the different variables affecting available road capacity and their interrelationships. A change in the factors affecting the available road capacity may lead to some degree of inoperability of the road segment. If available road capacity decreases, the average speed per vehicle may decrease, which would increase the travel time per vehicle. This leads to an increase in the travel costs per vehicle. For example, if a bridge becomes completely inoperable, there is no capacity available on the stretch of road going to the bridge in both directions and, therefore, traffic must be rerouted, which increases the distance traveled per vehicle and hence increases travel times and costs. In another example, if a segment of the road is under construction leading to some loss of capacity (Figure 15.5) that may decrease the average speed which, in turn, increases the travel time per vehicle and hence the travel costs. Such changes in the capacity of one road segment may also affect the traffic patterns on the other road segments acting as alternate routes leading further complications to calculating the average cost per vehicle.

To estimate the maximum number of vehicles that can be at the road segment at a given time, capacity of the road has to be calculated. Length of a vehicle is used to calculate the road capacity occupied by a vehicle on the road. Two types of vehicles are considered: cars and trucks. A buffer length is included in the length of each vehicle to accommodate the minimum safe distance between each vehicle so as to avoid collisions and maintain the advisory speed limit. Cars are denoted by c and the

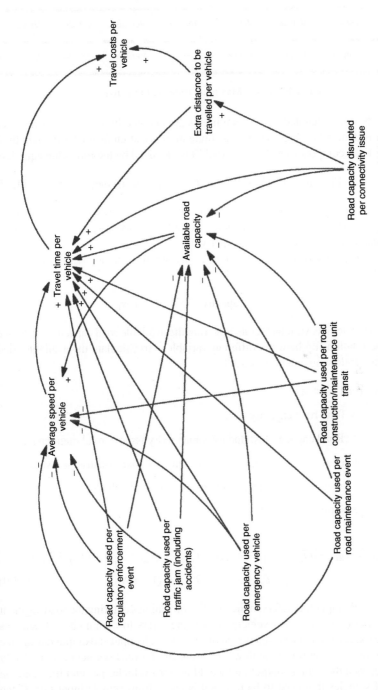

FIGURE 15.5 Effect of transportation network disruption on travel costs per vehicle.

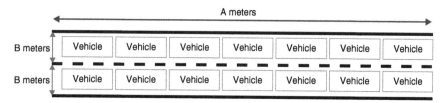

FIGURE 15.6   Maximum capacity of the road.

length of the car is denoted by C_L, and trucks are denoted by t and the surface area of a truck is denoted by T_L. To obtain a single multiplication factor for the length of a vehicle, V_L, a composite car/truck ("cruck") is idealized by the following equation.

$$V_L = ((c\% * (C_L + B_C)) + (t\% * (T_L + B_T))) \text{ meters} \tag{15.7}$$

Here, B_C and B_T are the buffer length for cars and trucks, respectively. The maximum capacity of a road could be depicted using Figure 15.6, where each vehicle is maintaining a safe distance from the other vehicle.

The total capacity of the road is equal to the length of a lane (L) multiplied by number of lanes (N_L) as given in Eq. (15.8).

$$\text{Total Road Capacity,} \quad T_{RC} = L * N_L \tag{15.8}$$

Figure 15.7 depicts various factors affecting the available road capacity and hence defines the relationship between different variables. To calculate the available road capacity, the following equation is used.

Available road capacity

$\qquad$ = Total road capacity

$\qquad$ − (Road capacity lost due to(connectivity issue + road maintenance

$\qquad$ + regulatory enforcement + traffic jams + road construction vehicles

$\qquad$ + emergency vehicles)) − (Average length of vehicle)

$\qquad$ * Number of vehicles

$$T_{ARC} = T_{RC} - ((T_{cil} * T_{cin}) + (T_{rml} * T_{rmn}) + (T_{tjl} * T_{tjn}) + (T_{rel} * T_{ren})$$
$$+ (T_{rcl} * T_{rcn}) + (T_{el} * T_{en})) - (V_L * N_V) \tag{15.9}$$

Here, vehicle input rate refers to the number of vehicles entering the road segment in a given period of time and vehicle output rate refers to the number of vehicles exiting the road segment in a given period of time. Vehicle input and output rate are the variables that most control traffic flow. For instance, if a road segment is completely inoperable, then the vehicle output rate would be zero vehicles per unit time and the number of vehicles that need to be rerouted is taken from vehicle input rate. If the

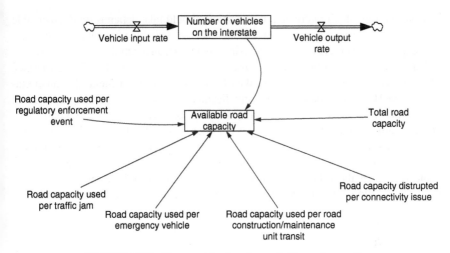

FIGURE 15.7   Factors affecting the available road capacity.

available capacity of the road is reduced due to a disruption, some of the traffic needs to be diverted to alternate feasible routes. Depending on the amount of traffic being diverted, the available road capacity on the alternate routes may also be affected as the number of vehicles on alternate routes increase. This methodology can be applied to different road segments to study the effect on their available capacity due to an increase in the number of vehicles. The next section includes an example using this methodology and calculating the indirect economic costs.

ILLUSTRATIVE EXAMPLE

For this study, a single bridge (the Eads Bridge over the Mississippi River) in St. Louis, Missouri metro area, is considered damaged. For simplicity, only the east-bound traffic flow on the bridge is modeled. Eads Bridge has two east-bound lanes, and it is asserted that both these lanes are closed due to road maintenance. With this inoperability of the road, vehicles have to be rerouted. Alternate paths (two neighboring bridges) are chosen for these vehicles. The alternate paths are prioritized based on the minimum indirect costs. The alternate paths selected for the traffic to flow from the west side of the Mississippi river to the east side are by using the adjacent bridges to the north (The Martin Luther King Bridge (alternate path 1)) and south (The Poplar Street Bridge (alternate path 2)) of the Eads Bridge. The length of the Eads Bridge is approximately 1600 m (1 mile). Both the alternate paths have two lanes for the traffic going from the west side of the Mississippi river to the east side. Using the methodology above, indirect economic loss associated with a change in traffic pattern due to disruption in a road segment is calculated. The main objective of this illustrative example is to use the methodology to determine the alternate path the vehicle (cruck) will be rerouted to and average cost per vehicle for covering the extra distance.

Determining which alternate path a vehicle should take depends on the available road capacity of each alternate path. To calculate the available road capacity, the average length of the vehicle is calculated using Eq. (15.7), an example of which is shown in Eq. (15.10). Here, out of total traffic, 83.33% are cars and 16.67% are trucks. This case is based on representative data from Missouri Department of Transportation (MoDOT, 2017). The safe following distance for a car is one car length for every 5 miles per hour and the safe following distance for a truck is two and a half times of the length of the car for every 5 miles per hour. This velocity dependence requires that a change in the speed of the vehicle leads to a change in the buffer length of the vehicle. Therefore, the average length occupied by a vehicle changes with the speed. Table 15.2 gives the value for the average length of a car and a truck and the value for the safe following distance when the vehicles are traveling at 55 miles per hour.

$$V_L = ((83.33\% * (4 + 44) + (16.67\% * (16 + 110)))) \text{ meters}$$

$$V_L = 61 \text{ meters} \tag{15.10}$$

The average length occupied by a vehicle when traveling at 45 miles per hour and 32.5 miles per hour is calculated and the results are presented in Table 15.3.

Once the length of the cruck is calculated, the next step is to calculate the available road capacity for the alternate routes. The available road capacity is calculated, given that there are 30 vehicles already present on alternate route 1 and 45 vehicles already present on alternate route 2. The values for the length of the alternate paths and the number of vehicles already on the alternate paths are also given in Table 15.4. It is assumed that there is no disruption on either of the alternate paths and the only factor affecting the available road capacity is the number of vehicles that are originally on that route.

TABLE 15.2 Average Length and Buffer Length for Vehicles Traveling at 55 mph

Vehicle Type	Average Length (m)	Buffer Length (m) When Traveling at 55 mph
Car	4	44
Truck	16	110

TABLE 15.3 Average Length of Vehicle Based on the Speed at Which the Traffic is Flowing

	When the Vehicle is Traveling at 45 mph	When the Vehicle is Traveling at 32.5 mph
Buffer length for car	36 m	26 m
Buffer length for truck	90 m	65 m
Length of cruck, V_L	51 m	38.5 m

TABLE 15.4 Number of Vehicles on Alternate Paths and Length of Alternate Paths

Route	Number of Vehicles Already on the Route	Distance
Path 1 (Martin Luther King Bridge Route)	30	4200 m (2.6 miles)
Path 2 (Poplar Street Bridge Route)	45	7100 m (4.4 miles)

Based on Eq. (15.9), available road capacity for alternate path 1 and path 2 when the vehicle is traveling at 55 miles per hour is calculated using Eq. (15.11). As both the alternate paths have two lanes, the available road capacity for the two alternate paths will be calculated as below.

For alternate path 1,

$$T_{ARC} = (4200 * 2) - (61 * 30)$$

$$T_{ARC} = 6570 \, \text{meters} \qquad (15.11)$$

Similarly, the available road capacity for alternate path 2 is calculated as shown in Eq. (15.12).

For alternate path 2,

$$T_{ARC} = (7100 * 2) - (61 * 45)$$

$$T_{ARC} = 11\,455 \, \text{meters} \qquad (15.12)$$

Similarly, the available road capacity for alternate path 1 and alternate path 2 is calculated when the vehicle is traveling at 45 miles per hour and 32.5 miles per hour and the results are presented in Table 15.5.

After calculating the available road capacity for the alternate paths, the next step is to calculate the number of vehicles that can be rerouted to these alternate paths using Eq. (15.13).

$$\text{Number of vehicles/crucks that can be rerouted to the path} = \frac{T_{ARC}}{V_L} \qquad (15.13)$$

Using Eq. (15.13), the number of vehicles that can be rerouted to alternate path 1 and alternate path 2 is calculated for the vehicles traveling at 55 miles per hour,

TABLE 15.5 Available Road Capacity on Alternate Routes for Vehicles Traveling at Different Speeds

	Alternate Path 1	Alternate Path 2
Available road capacity when the vehicle is traveling at 55 mph	6570 m	11 455 m
Available road capacity when the vehicle is traveling at 45 mph	6870 m	11 905 m
Available road capacity when the vehicle is traveling at 32.5 mph	7245 m	12 467.5 m

TABLE 15.6 Available Road Capacity

Average Speed of the Vehicle	Number of Vehicles that can be Rerouted to Path 1	Number of Vehicles that can Rerouted to Path 2
55 mph	107	187
45 mph	134	233
32.5 mph	188	323

45 miles per hour, and 32.5 miles per hour. The available road capacity for each alternate path and the number of vehicles that can be rerouted to that alternate path are presented in Table 15.6.

The travel costs per mile due to rerouting are calculated as shown in Eq. (15.14).

$$C = (c\% * G) + (t\% * D) \tag{15.14}$$

Here, C denotes the average cost per mile per vehicle, G denotes the fuel price per mile per car, and D denotes the price of fuel per mile per truck. Given a gasoline price per gallon of \$2.08, diesel price per gallon of \$2.18, average miles per gallon (mpg) for a truck is 6 miles per gallon, and average mpg for a car is 23.6 miles per gallon, then the average cost per mile per vehicle is calculated using Eq. (15.14) as follows:

$$C = \left(83.33\% * \left(\frac{\$2.08}{23.6\,\text{miles}}\right)\right) + \left(16.67\% * \left(\frac{\$2.18}{6\,\text{miles}}\right)\right)$$

$$= \$0.13 \text{ per mile per vehicle} \tag{15.15}$$

Indirect costs due to rerouting would include the cost incurred due to extra miles traveled and the extra time a cruck has to travel. The total indirect costs are given by Eq. (15.16).

$$\text{Total Indirect cost per cruck} = \text{Cost due to extra miles travelled per cruck}$$

$$+ \text{Costs due to extra time a cruck must travel} \tag{15.16}$$

The cost incurred due to extra miles traveled per cruck is given by Eq. (15.17).

$$\text{Indirect cost incurred due to extra miles travelled per cruck}$$

$$= C * \text{Extra distance travelled} \tag{15.17}$$

As the extra distance traveled per cruck is equal to the difference between the length of the alternate path and the length of the original path that a cruck would follow if there is no disruption. Hence, Eq. (15.17) can be rewritten as Eq. (15.18).

$$\text{Indirect cost incurred due to extra miles travelled per cruck}$$

$$= C * (\text{Length of alternate route} - \text{Length of original route}) \tag{15.18}$$

Using Eq. (15.18), the indirect cost incurred due to extra distance traveled by a cruck for the two alternate routes are calculated in Eqs (15.19) and (15.20).

Indirect cost per cruck due to extra distance travelled using alternate path 1

$$= \$0.13 * (2.6 \text{ miles} - 1 \text{ mile}) = \$0.20 \text{ per cruck} \qquad (15.19)$$

Indirect cost per cruck due to extra distance travelled using alternate path 2

$$= \$0.13 * (4.4 \text{ miles} - 1 \text{ mile}) = \$0.44 \text{ per cruck} \qquad (15.20)$$

Similarly, the indirect costs incurred due to the extra time a cruck takes due to rerouting can be calculated using Eq. (15.21).

Indirect cost due to extra time a cruck must travel

$$= C_T \left(\left(\frac{\text{Distance of alternate route}}{\text{Speed on the alternate route}} \right) - \left(\frac{\text{Distance of original route}}{\text{Speed on the original route}} \right) \right)$$
$$(15.21)$$

Here, C_T is the cost factor of travel time. For this example, cost factor of travel time is considered to be the minimum wage in St. Louis, Missouri, which is \$7.70/h. However, the cost factor of travel time can be varied depending on the traveler's destination, field of work, and so on. The speed on the original route is considered to be 55 mph.

For this illustrative example, it is assumed that 140 vehicles have to be rerouted from Eads Bridge at a given moment of time. For this example, we have three cases, that is, vehicles traveling at 55, 45, and 32.5 mph. These cases are explained below.

Case 1 When the vehicles are traveling at 55 miles per hour on alternate path 1.

From the results presented in Table 15.6, alternate path 1 has a capacity to accommodate 107 more vehicles traveling at 55 miles per hour. This implies that other 33 vehicles will have to be rerouted to alternate path 2.

Using Eq. (15.18), the indirect costs due to the extra distance traveled per cruck for rerouting 107 vehicles through alternate path 1 and 33 vehicles through alternate path 2 are \$21.4 and \$14.52, respectively. The indirect costs due to extra time these crucks must travel are \$23.97 for 107 crucks on alternate route 1 and \$15.71 for the 33 crucks on alternate route 2. Therefore, the total indirect cost due to rerouting these 140 vehicles would be \$75.60.

Case 2 When the vehicles are traveling at 45 miles per hour on alternate path 1.

Table 15.6 shows that alternate path 1 has a capacity to accommodate 134 vehicles traveling at 45 miles per hour. This implies that the other 6 vehicles will have to be rerouted to alternate path 2. Using Eq. (15.18), the indirect cost due to the extra distance traveled per cruck for rerouting 134 vehicles through alternate path 1 is \$26.80 and 6 vehicles through alternate path 2 is \$2.64. Indirect costs due to extra time a cruck must travel are found out to be \$40.86 and \$3.68 for alternate route 1 and 2, respectively, using Eq. (15.21). This implies that the total indirect costs due to rerouting 140 vehicles would be \$73.93.

Case 3 When the vehicles are traveling at 32.5 miles per hour on alternate path 1.
From the results presented in Table 15.6, alternate path 1 has a capacity to accommodate 188 more vehicles traveling at 32.5 miles per hour. This implies that all 140 vehicles will be rerouted to path 1. Using Eq. (15.18), the indirect cost due to the extra distance traveled per cruck for rerouting all 140 vehicles through alternate path 1 is equal to $28. The indirect cost due to extra time a cruck must travel is found out to be $66.64 for 140 crucks on alternate route 1, using Eq. (15.21). This implies that the total indirect costs due to rerouting 140 vehicles would be $94.64.

After analyzing the results from the above three cases, case 2 (vehicles traveling at 45 mph) is preferred to be the best case as the indirect costs are minimum for this case. Even though the result in case 3 shows that the vehicles will have to follow the shortest distance, it is not a preferred option as the time penalty associated with this methodology makes case 3 one of the most expensive options.

This methodology has been applied for rerouting 140 vehicles, but the methodology is flexible and scalable. As more vehicles and more alternate routes are added, the equations can simply be adjusted. The indirect economic losses for a large number of vehicles can be calculated using the results from Eqs (15.16), (15.19), and (15.21) depending on the alternate route that is followed by the vehicle. Figure 15.8 is a speed versus cost graph that shows the cost of rerouting 140 and 280 vehicles along the two alternate routes considered in the above example. The same procedure is followed to calculate the indirect economic loss for rerouting 280 vehicles, as shown in the above example.

The indirect cost of rerouting 280 vehicles is calculated using the same methodology used in the example. For the case when the vehicular traffic is flowing at a speed of 55 mph, the indirect cost due to the extra distance traveled per cruck for rerouting

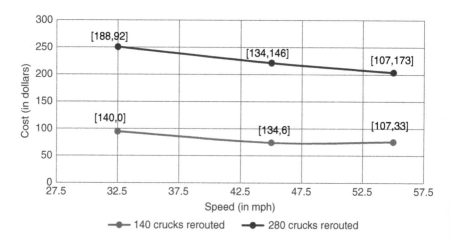

FIGURE 15.8 Speed versus cost graph when 140 and 280 vehicles are rerouted on the two alternate paths. The light gray line shows the results when 140 crucks are rerouted and the dark gray line shows the results when 280 crucks are rerouted. The first and second numbers in brackets on the graph are the number of vehicles on alternate route 1 and 2, respectively.

107 vehicles through alternate path 1 is \$21.4 and 173 vehicles through alternate path 2 is \$76.12. The indirect costs due to extra time a cruck must travel are found out to be \$23.97 and \$82.5 for alternate route 1 and 2, respectively. This implies the total indirect cost due to rerouting 280 vehicles when the traffic is flowing at a speed of 55 mph is equal to \$203.84. Similarly, when the vehicular traffic is flowing at a speed of 45 mph, the indirect cost due to extra distance traveled and extra time added for rerouting 280 vehicles is \$91.04 and \$130.34, resulting in a total indirect cost of \$221.38. For the case when the vehicular traffic is flowing at a speed of 32.5 mph, the indirect cost due to extra distance traveled and extra time added for rerouting 280 vehicles is \$78.08 and \$172.51, respectively, resulting in a total indirect cost of \$250.59. From the results of this example, the best scenario for rerouting 280 vehicles would be the case when the traffic is flowing at 55 mph as it is the least expensive option. As seen from the two examples, the amount of added travel time influences the decision along with the extra distance that needs to be traveled. As number of vehicles keep increasing, it will be necessary to add additional alternative routes so as to accommodate them. The benefit of this approach lies in its ability to account for lost time while selecting the most cost-effective alternative.

CONCLUSION AND FUTURE WORK

The objective of this study is to create a methodology to model the emergent behavior during a disruption in the transportation system and that calculates economic losses due to such a disruption. A CLD visually represents the different factors that affect available road capacity and travel costs. A CLD mapping the interdependencies between system variables provides greater insight into the spatiotemporal character of the transportation network system. This model also posits equations that allow the user to calculate available road capacity and to determine the number of vehicles that need to be rerouted to alternate paths. This, in turn, allows for the calculation of indirect losses associated with that traffic being rerouted. These indirect costs are, in part, due to emergent behavior within the alternate transportation system, include costs due to extra distance traveled per vehicle as well as costs due to the extra time a vehicle had to travel due to the disruption in the transportation network. With the traffic being rerouted to alternate routes, the available road capacities on these routes are reduced as more and more vehicles utilize them. This, in turn, affects traffic speed and causes congestion, thereby increasing the indirect costs due to extra travel time each vehicle must endure. To demonstrate the methodology, an illustrative example based on bridges crossing the Mississippi River in St. Louis is used where the two east-bound lanes of the Eads Bridge are under maintenance. Two alternate paths are examined and the extra cost per vehicle is calculated for these alternate paths. This methodology calculates the most cost-efficient traffic reorientation scenario.

This methodology can be applied to other transportation networks with alternate paths added as needed. Care should be taken when increasing the number of paths, as this will likely result in a nonlinear increase in the number of options evaluated. This could be alleviated either through the application of heuristics or a self-organizing approach. The cost per vehicle per alternate path can be calculated and multiplied by the number of vehicles going through those alternate paths to calculate the indirect economic losses. This research can further be extended to estimate the extent of disruption of the transportation network that will not only necessitate a higher freight transportation load on rail, water, and air networks but also make them a more viable option by minimizing economic losses.

This approach could be modified to investigate the factors leading indirect costs due to the inoperability for other critical infrastructure systems such as power, water, and communications. A system dynamics model is advantageous for determining the factors that render such infrastructure systems inoperable. Understanding these factors allows the design of strategies and solutions to abate the economic losses owing to the inoperability of the infrastructure system. Systems dynamics methods also allow the modeling of the spatiotemporal character of a system and therefore yield a greater insight to the emergent behavior arising out of interdependencies within a complex system. By using a common method to evaluate indirect losses, it can simplify the integration of the data into a larger evaluation framework.

The example evaluated in this study is a steady-state representation of the number of vehicles that are present on each bridge at any particular point in time. This methodology can model different states and time steps to map the emergent behavior arising out of the transportation system. Expanding this to include a discrete event simulation (Zeigler and Muzy, 2016) would allow for capturing some of the dynamic effects of the traffic building up to reach capacity. This model assumes that the information about rerouting is shared with individual drivers, thereby guiding emergent behavior to minimize congestion. Introducing human behavior effects into the model will allow the exploration of the willingness of drivers to accept different routes. This study is focused on a particular area of a particular transportation system. This work will be expanded to include the other infrastructure elements mentioned into a holistic representation to give decision makers better and more representative information regarding how best to restore critical infrastructure systems.

REFERENCES

Alasad, R., Motawa, I., and Ougunlana, S. (2013) A system dynamics-based model for demand forecasting in Ppp infrastructure projects – a case of toll roads. *Organization, Technology & Management in Construction*, **5**. doi: 10.5592/otmcj.2013.2.4

An, L. and Jun-Jang J. (2005) On developing system dynamics model for business process simulation. Proceedings of the Winter Simulation Conference, 2005, DOI: 10.1109/WSC.2005.1574489.

Coyle, R.G. (1996) *System Dynamics Modelling: A Practical Approach*, Chapman & Hall, London.

Gui, S., Zhu, Q., and Lu, L. (2005) Area logistics system based on system dynamics model. *Tsinghua Science and Technology*. doi: 10.1016/S1007-0214(05)70065-1

Li, L., Zhang, C. and Li, H.M. (2009) Application of system dynamics to strategic project management. 2009 First International Conference on Information Science and Engineering, DOI: 10.1109/ICISE.2009.337.

Liu, Y., Yu, S., Liang, X. and Guo, T. (2011) System dynamics model for structure configuration of transportation corridor. Proceedings 2011 International Conference on Transportation, Mechanical, and Electrical Engineering (TMEE), DOI: 10.1109/TMEE.2011.6199617.

Lyneis, J.M., Cooper, K.G., and Els, S.A. (2001) Strategic management of complex projects: a case study using system dynamics. *System Dynamics Review*, **17** (3), 237. doi: 10.1002/sdr.213

Maier, M. (2014) The role of modeling and simulation in system of systems development. Chapter 2, in *Modeling and Simulation for Support for System of Systems Engineering* (eds L.B. Rainey and A. Tolk), Wiley, pp. 11–41.

Mittal, S., and Rainey, L. (2015) Harnessing emergence: the control and design of emergent behavior in system of systems engineering. SummerSim'15 Proceedings of the Conference on Summer Computer Simulation. Chicago, Illinois – July 26–29, 2015, Society for Computer Simulation International, . San Diego, CA, USA.

MoDOT (2017), http://www.modot.org/safety/documents/2015_Traffic_SL_06212016.pdf (accessed 1 May 2017).

Qing, Z. and Mingchao, W. (2011). Simulation of economy-resource-environment system of Jiangxi Province based on system dynamics model. 2011 International Conference on Business Management and Electronic Information, DOI: 10.1109/ICBMEI.2011.5920505.

Ramachandran, V., T. Shoberg, S. Long, S. Corns, and H. Carlo. 2015. Identifying geographical interdependency in critical infrastructure systems using open source geospatial data in order to model restoration strategies in the aftermath of a large-scale disaster. *International Journal of Geospatial and Environmental Research* 2 1, Article 4, http://dc.uwm.edu/ijger/vol2/iss1/4

Ramachandran, V., Long, S.K., Shoberg, T. *et al.* (2016) Post-disaster supply chain interdependent critical infrastructure system restoration: a review of data necessary and available for modeling. *Data Science Journal*, **15**. doi: 10.5334/dsj-2106-001

Sha, M. and Huang, X. (2010) A system dynamics model for port operation system based on time, quality and profit. 2010 International Conference on Logistics Systems and Intelligent Management (ICLSIM), DOI: 10.1109/ICLSIM.2010.5461258.

Sterman, J.D. (1992) *System Dynamics Modeling for Project Management*, Sloan School of Management, MIT.

Su, Y., Yang, L. and Jin, Z. (2008) Simulation and system dynamics models for transportation of patients following a disaster. 2008 International Workshop on Modelling, Simulation and Optimization, DOI: 10.1109/WMSO.2008.109.

Sycamore, D. and Collofello, J.S. (1999) Using system dynamics modeling to manage projects. Proceedings. Twenty-Third Annual International Computer Software and Applications Conference (Cat. No.99CB37032), DOI: 10.1109/CMPSAC.1999.812703.

Zeigler, B.P. and Muzy, A. (2016) Some modeling & simulation perspectives on emergence in system-of-systems. Spring Simulation Multi-conference (SpringSim'16), April 2016, Pasadena, CA, USA, http://www.scs.org/springsim, hal-01315199.

Zhao, D., Sun, D., Li, Y. and Zetong L. (2011) Research on formation mechanism of the modern regional logistics hub based on system dynamics. 2011 International Conference on System Science, Engineering Design and Manufacturing Informatization, DOI: 10.1109/ICSSEM.2011.6081229.

Zheng, C., Liu, Z., Wang, C., Wang, X. and Xu, B. (2009) A system dynamics model of the interaction of aviation logistics with regional economy development in Guangxi faced to CAFTA. 2009 International Conference on E-Business and Information System Security, DOI: 10.1109/EBISS.2009.5137909.

SECTION IV

RESEARCH AGENDA

16

RESEARCH AGENDA FOR NEXT-GENERATION COMPLEX SYSTEMS ENGINEERING

Saikou Diallo[1], Saurabh Mittal[2], and Andreas Tolk[3]

[1]*Virginia Modeling, Analysis & Simulation Center, Old Dominion University, Suffolk, VA, USA*
[2]*The MITRE Corporation, McLean, VA, USA*
[3]*The MITRE Corporation, Hampton, VA, USA*

SUMMARY

In this chapter, we attempt to set the research agenda on emergence in the medium and long term. We first summarize the view of emergence from the authors in this book and conclude that almost all of them are focused on epistemological emergence, which results in better understand systems but fails in explaining the emergence of new categories as we seem to observe in the real world. Although the epistemological perspective is essential, the community must also focus on ontological emergence. We propose a simulation experience approach (SEA) based on the mix of live–virtual and constructive (LVC) simulation. We demonstrate that the research in employing emergence in complex systems (CSs) engineering must be transdisciplinary and propose a set of grand challenges that must be tackled to move forward.

ON ENGINEERING EMERGENCE: LEARNING FROM OTHERS

What are the reasons for emergence? Before we can begin to sketch an answer to this question, we must first determine what we mean by the term "emergence."

Emergent Behavior in Complex Systems Engineering: A Modeling and Simulation Approach,
First Edition. Edited by Saurabh Mittal, Saikou Diallo, and Andreas Tolk.
© 2018 John Wiley & Sons, Inc. Published 2018 by John Wiley & Sons, Inc.

An investigation into the existing literature results in equal parts frustration and curiosity. That is because as engineers and scientists, we seek well-defined and unambiguous concepts and emergence is nothing but as showcased in this volume. The most fundamental source of frustration is the elusive nature of the concept as illustrated by the multitude of closely related but not quite equivalent definitions used to describe it. On the one hand, we have formal approaches to explaining emergence, which yield very crisp simulations. However, these simulations are rooted in a computational view of the world and therefore fail to satisfactorily describe phenomena that are said to be emergent in nature or in social contexts. For instance, a computational view of emergence is limiting when explaining the dynamics of human migration that include social, political, cultural, and human dimensions at the local and global level. Such complexities have to be greatly reduced and purposefully bounded to fit a model that can be executed on a digital computer. As the resulting model is a purposeful abstraction of reality, emergent behavior observed in a computational environment may be completely different from the behavior observed in the real world.

On the other hand, more sophisticated and context-sensitive definitions of emergence found in Social Sciences and the Humanities capture the complexity of socio-cultural phenomena but lack the rigor required to be interpretable by a computational model and the follow-on simulation. Thus, one can only make sense of these phenomena post-facto or by being immersed in them with a full understanding of all the actors and variables while events are unfolding in the real world. The variable list captured in such a world model will never be complete and consistent and the resulting model will (i) only reflect the worldview of the person experiencing the event and the stakeholders who participated in the modeling process and (ii) not be implementable on a digital computer because of the human, cognitive, and emotional processes involved in the process.

The study of emergence is made even more challenging because, even though, the concept appears in almost all scientific and engineering domains, it often does so under a different name (error, transformation, adaptation, fusion, etc.) and therefore lessons learned from one field are not carried to others. We suspect that emergence has underlying commonalties or a set of universal properties but so far, we have not been able to formulate and verify those properties in a satisfactorily manner to engineers and scientists. That is not to say that we should give up on understanding emergence. Rather, it means that we must take a multidisciplinary and transdisciplinary approach to the study. A multidisciplinary approach involves redefining problems based on inputs from different academic disciplines and reach solutions based on new understanding, whereas a transdisciplinary approach refers to a research strategy wherein inputs, concepts, and methods from one discipline are used across other disciplines in a systematic way.

For engineers and scientists, it is the predicted universality of emergence that makes it attractive and the subject of our curiosity. After all, emergence is a powerful concept that has the potential to provide insights and explanations where uninformed conjecture is the norm today. Can we explain the rise of organized societies with technology and connectivity as an emergent phenomenon? Or better yet, why does the concept of emergent societies persist even when their original actors are no longer present? Is the concept persistent or always in flux? With emergence, we can potentially ask and attempt to answer grand societal questions and explore alternative realities in unpredictable ways and are doing so in the context of religion and social integration. From an engineering standpoint, understanding emergence can lead us to design smarter and more resilient systems while at the same time furthering our understanding of natural and sociocultural phenomena.

From a philosophical standpoint, Wildman and Shults explain that emergence has an ontological and an epistemological component. To understand ontological emergence, one must imagine the world as a layered set of entities interacting with one another. A property P is *ontologically emergent* at layer L if it occurs as the results of the combination of the interactions at the layers below it and an emergent law that is irreducible to the laws at the lower levels (Van Cleve, 1990; McLaughlin, 1997; Wilson, 1999, 2002). In this view of emergence, it is a consequence of interactions at lower levels at some time t and the emergent layer is said to be *supervenient* to the layer below. Ontological emergence is extremely hard (or impossible) to reproduce for engineers because it implies the appearance of new levels of causality (the irreducible laws) at every layer that cannot be explained by the subsequent layers. From a system standpoint, ontological emergence implies the emergence of new structures with unique properties that are discernible from any preexisting ones, that is, the appearance of new categories.

Epistemological emergence on the other hand defends the notion that emergence is a result of our limited ability to predict the future behavior of a system given our current understanding of it. An essential feature of epistemological emergence is the notion of a complex system. A system is a set of interconnected things that work together to perform a function. A complex system is one that exhibits nonlinear behavior that can be termed *emergent* but only because of our epistemological limitations. Therefore, emergence adds to our knowledge of the system. However, ascribing emergence to complexity leads to a tautological stance where complex systems are defined as being emergent and emergence is described as a defining property of complex systems.

Both the ontological and the epistemological components of emergence pose problems for modelers and simulation engineers. The ontological view requires a sort of sui-generis that digital computer simulations are incapable of, whereas the epistemological view results in computer models that are

- *Purely exploratory* and potentially useless when one is looking for emergent behavior because any unexpected behavior is potentially emergent or

- *Purely explanatory*, meaning one seeks to explain a macro-level behavior from micro-level interactions, which implies that we must know the emergent behavior a priori. As a result, we are robbed of any predictive capability and are stuck providing alternative explanations without any plausible way to claim validity.

TABLE 16.1   Sample of Fields that Study Emergence

Field of Study	Definition
Computer Science, Artificial Intelligence, Artificial Systems	Formation of global patterns from local interactions (Holland, 1998; Mittal, 2013)
Cognitive Science, Linguistics	Formation of categories and rule-like behaviors due to the interactions of individual items and experiences (McClelland, 2010)
Physics, Biology, Ecology. Systems Science	A property of a unit that is unpredictable from an observation of its components (Salt, 1979)
Economics, Social Sciences, Evolution, Game Theory	Appearance of a system that cannot be explained by its preceding instantiation

To illustrate, Table 16.1 lists a sampling of disciplines that study the concept of emergence with a capture of the essence of the term in those fields without claiming completeness or exclusiveness.

The disciplines listed in Table 16.1 seem to embrace the epistemological stance that emergence is a property of a system that results from a dynamic interaction of its subparts. It is important to note that some disciplines focus on the outcome or the result (system and property), whereas others are more interested in the process or the relationship between the emergent structure and the preexisting one. Table 16.1 also reveals deep contradictions as to what constitutes emergence. Depending on the field:

- *Emergence is the Result of an Evolutionary Process*: The source of emergence is the need for the system/organism to adapt to either survive or grow and thrive. This evolutionary view is most often found in the computational sciences (Computer Science, Artificial Intelligence, etc.). It implies some form of collective intelligence that cannot be ascribed to any component/organism (no central control) and can only exist under the right conditions. In this case, emergence is persistent and depends on the environment in which the system exists. In terms of systems thinking, this type of emergence can be loosely mapped to Complex Adaptive Systems (CAS).

- *Emergence is Not Predictable*: The source of emergence is the nonlinear combination of components at different states over time. Emergence, in this case, is one of the possible states a system might take even if the observer or designer of the system is not aware of the possibility (Mittal and Rainey, 2015). This lack of awareness might be due to limitations in knowledge or the size of the state space. In contrast to the evolutionary view, this school of emergence does not require for systems to be goal-seeking; they simply act and react based on either (seemingly) randomly or following a set of causal rules. In this case, although what emerges is unpredictable and not always persistent, we can reengineer it, once we know the conditions under which it appears. This is the view that is most prevalent in the Sciences (Physics, Biology, etc.), and in terms of systems thinking, this type of emergence can be ascribed to the broad category of Complex Systems (CS).

- *Emergence is Not Explainable by Past Information*: The source of emergence is the result of interactions between parts of the systems that results in novel, more complex structures that, in turn, generate new interactions. The system is characterized by delays and has feedback mechanisms that make it dynamic and nonlinear (complex). This form of emergence is often used in Social Sciences and Economics and is usually classified under the umbrella of Complex Dynamic Systems.

- *Emergence is Recurrent and Reproducible*: The source of emergence is a combination of components interacting together and with the environment over time to form new patterns. The process of forming patterns is a function of simple (or complex) rules inherent to the system and the interactions between the system and the *environment* in which it lives. This view of emergence is favored in Cognitive Science and Linguistics and from a system thinking viewpoint corresponds to a Complex System of Systems.

This book purposefully sought out multiple perspective of emergence including how to define, engineer, and explain emergence. Each group of authors reflects the meaning of emergence from their discipline, background, and application area. As mentioned earlier in this chapter, it is not a surprise to see the same contradictions found in the emergence literature be reflected in this volume. Let us attempt to briefly summarize the concept of emergence as we understand it from the perspective of the contributors.

Zeigler and Mittal propose a formal system specification that accounts for emergence. They offer a layered approach to describing systems and argue that emergence is the result of a combination of interactions within components and between layers. Further, they propose a general description of properties that a goal-oriented system must possess to show emergence when operating in a resource-constrained environment. In fact, they describe being "constrained" is pursued as a necessary condition for emergence to occur. They argue that the proposed Resource-Constrained Complex Intelligent Dynamical System (RCIDS) can be used to show how language emerges through the purposeful use of communication between two agents. The underlying mechanism employs an alternating active and passive mode of the agent with each agent giving feedback to another agent as they alternate their mode. For a language to emerge, both agents cannot be active or passive at the same time. It is interesting to note that Zeigler and Mittal do not reject the idea of layered emergence and increasing levels of complexity, which is in line with the original view of *emergentist* in the eighteenth century.

Yilmaz views models (and simulations) as emergent artifacts that represent a viewpoint and act as proxy for the data and theories used to construct it. The chapter proposes an approach to contrast, compare, and select models based on studying how they emerge. Emergence is defined as the apparition of patterns due to the purposeful transformation a system undergoes to adapt and evolve. The chapter brings together essential elements for generative science through the proposed generative parallax simulation (GPS) framework that links a particular domain of interest, a model generator that generates multiple models, a simulator that performs experimentation that facilitates model evaluation, and a creative cognitive process that ties this all together. The chapter also provides a reference architecture to construct such a system.

Mittal, Diallo, and Tolk discuss the challenge of emergence in complex systems engineering. They contrast the landscape of traditional systems engineering and how it is being transformed with the concept of multi-mode emergence: simple, weak, strong, and spooky. They associate the emergent behavior taxonomy with the complex systems engineering through the lens of modeling and simulation (M&S). Rouse defines emergence in its most colloquial sense of appearing or becoming visible or noticeable. The author distinguishes emergence that can be explained or should have been expected in hindsight from emergence that cannot be explained by preexisting knowledge of the system under study. Rouse focuses on complex enterprises and demonstrates that they are prone to emergence due to their goal-seeking nature and interactions with a dynamic environment.

Frydenlund and Earnest describe the quest for emergence as an epistemological journey to unlock the relationship between system-level behavior and component-level behavior. They illustrate this notion of emergence by modeling the emergence of preference for technologies in Western societies using information economics and computational social science. Johnson, Sousa-Poza, and Padilla model emergence using factors derived from observations of emergence in physics and chemistry. They use analogical reasoning to establish parallels between the properties of physical and chemical systems and complex system of systems in general. The resulting conceptual model of emergence is described as a complex dynamical system with feedback and loops. In this view, emergence is akin to potential energy that can transform into kinetic energy as a result of interactions between the system and its environment over time.

Ören, Mittal, and Durak introduce the notion of induced emergence as a form of engineered (for good or bad) behavior that can be observed in society. From this chapter's perspective, emergence is not due to an epistemic failure to fully understand the system due to its size, interconnectivity, or dynamics. Rather, it is a purposeful steering of a system toward a desired outcome. The chapter introduces the notion of multi-modeling and proposes it as an approach for studying induced emergence in computational social sciences. Norman, Koehler, and Pitsko discuss the emergence of swarms in Unmanned Autonomous Systems with a focus on making future systems resilient. The chapter provides an overview of real-world examples of emergence from this domain. They conceive emergence as a feature of the system that is novel but ultimately traceable to lower level behaviors, although not trivially, through a scientific or "reductionist" approach.

Szabo and Birdsey propose, implement, and test an architecture that helps engineers and subject matter experts detect emergence, which they define as a behavior that is found in any component of the system but can be explained analytically once observed. The chapter demonstrates how a combination of M&S, statistical, and exploratory analysis can be applied to detect the presence of a potentially emergent behavior. Gore explores the nature of emergent failure in software code. The chapter defines emergence as an unpredictable and unforeseen behavior in a software program. The chapter discusses both deterministic and nondeterministic faults and various causes for both of these categories. The chapter discusses the reasons for emergent failures and proposes a fuzzy passing extent approach. This approach helps create test cases that both pass and fail to handle the non-deterministic behaviors. They also propose an additional scheduling layer with alternate implementation of the underlying operating system's system-calls to get a handle on the thread interleaving and control the multi-threaded execution environment. The proposed approach facilitates detecting and isolating the sources of failure when it occurs.

Chen and Crilly propose an architecture and language neutral framework, which they use to unify system descriptions across disciplines. The chapter argues that emergence, at least in software-systems engineering, is due to an epistemological gap in the relationship between the components of a system and our understanding of whole system functions. Using the framework, the authors explain emergence as the result of the interactions between the reinforcing and balancing behaviors of a system. Lane focuses on the emergence of culture and argues that a computational social science approach is sufficient to explain the emergence of culture as opposed to claims that culture spontaneously emerges. For Lane, emergence is the appearance of new behavior that can be explained by our knowledge of the system. Although the emergent behavior is new, it is a combination of behaviors and interactions at lower levels.

Ojha, Corns, Shoberg, Qin, and Long do not address emergence explicitly. Rather, the chapter focuses on the use case of transportation infrastructure restoration which can be described as a complex dynamic system with feedback. Based on the choice of formalism to describe the process, the authors abide by the definition emergence as a novel behavior that cannot be ascribed to the component parts of the system.

Table 16.2 summarizes the broad category of emergence to which a chapter seemingly belongs to. In addition, we provide how emergence is modeled and attempt to describe whether authors believe that their model aims to contribute to further our understanding of epistemic emergence (epistemic stance) or whether they are contributing to our understanding of emergent categories (ontological stance).

Table 16.2 confirms the assertion by Wildman and Shults that computer and systems engineers in general and simulation engineers, in particular, are making a great contribution to epistemic emergence. Nonetheless, when exposed to the mix of philosophical and engineering views, one recurring question we have is the following: If we can build a simulation that produces exactly the observed emergent behavior, is it still truly emergent? Or, did we just gain additional understanding that we did not have before we built the simulation?

One thing that is apparent from working on this book is that engineers need to engage more with philosophers and humanists to clarify what exactly they are contributing when it comes to the subject of emergence. Table 16.2 also shows that at least some engineers are interested in ontological emergence, although they realize that a pure computational approach is insufficient for many reasons, some of which

TABLE 16.2 Summary of Chapter's View of Emergence

Authors	Evolutionary	Predictable	Explainable	Recurrent	Model Type	Stance
Frydenlund and Earnest	Yes	No	Yes	Yes	Agent-based model	Epistemic
Lane	Yes	No	Yes	Yes	Agent-based model	Epistemic
Norman, Koehler and Pitsko	Yes	No	Yes	Yes	Agent-based model	Epistemic
Szabo and Birdsey	Yes	No	No	Yes	Formal, agent-based modeling, statistics	Epistemic
Chen and Crilly	Yes	Yes	Yes	No	Formal, system diagrams	Epistemic
Yilmaz	Yes	No	Yes	Yes	Graph theory, system diagrams	Epistemic
Ören, Mittal and Durak	Yes	Yes	No	No	Formal, DEVS	Epistemic
Zeigler and Mittal	Yes	Yes	Yes	Yes	Formal, DEVS	Epistemic
Johnson, Sousa-Poza, Padilla	No	No	No	No	Causal-loop-diagram	Ontological
Rouse	No	Yes	Yes	No	System diagram	Epistemic
Gore	No	No	No	No	Formal methods, algorithms	Epistemic

are explained in Chapter 3. In the next section, we discuss how M&S can be used to explore ontological emergence.

MAKING SENSE OF EMERGENT BEHAVIOR: A MIXED METHOD APPROACH

Wildman and Shults pose an interesting challenge to simulation engineers. Stated briefly, how can we design models and simulations that account for the type of emergence that we see abundantly around us? Mittal, Diallo, and Tolk have boldly stated that "Ontological (spooky) emergence marks the upper bound of system thinking." In other words, system thinking may not be the right apparatus for engineers to engage with ontological emergence, and Yilmaz points in the direction of "Model Thinking" as a way to embrace the messy process of people and objects interacting with a changing environment. We also have the increased use of the concept of a System of Systems to account for the interaction of humans, engineered physical systems, and the natural (or social) environment in which they exist (Mittal and Zeigler, 2017). Although we understand and recognize the limits of systems thinking, let us ironically apply it to systematically study the problem of modeling ontological emergence by conceptualizing a system with a set of inputs, outputs, and states (discrete or continuous). The question is, can we generate a model of this system such that the model generates truly new categories of things with their own fundamental laws? We have already stated without proving it that a computable model alone would not be able to achieve this feat and a mixed approach might be necessary. In this section, we propose a strategy to model and simulate ontological emergence.

Live, Virtual, and Constructive as a Generator of Ontological Emergence

In M&S, we are not limited to only computational implementations of models. We distinguish between *live* simulations in which the model involves humans interacting with one another (role playing, play acting, etc.), *virtual* simulations where the model is simulated by a fusion of humans and computer-generated experiences, and *constructive* simulations where the model is entirely implemented in a digital computer and may have increased levels of abstraction. Increasingly, we are mixing the three forms of simulation in what is commonly known as live–virtual–constructive (LVC) simulation (Hodson and Hill, 2014). LVC simulations are used mainly for training but they can be adapted for the type of experimentation/exploration needed to investigate ontological emergence.

Figure 16.1 shows the LVC continuum where each component can meaningfully contribute to ontological emergence. The live component brings in the richness of real-world data, human decision-making, inconsistency, and real conflict. The virtual component provides a stimulating environment that engages the human in a way that makes her respond or react, whereas the constructive environment embodies our current knowledge and understanding of the world. During an LVC simulation,

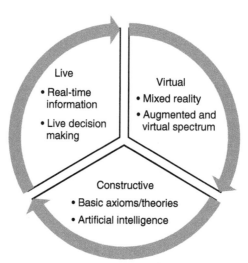

FIGURE 16.1   Experimental LVC approach for generating emergence.

responses and reactions are recorded and both the virtual and constructive components are responding and adapting to the human input as the simulation evolves over
time. At the end of the LVC simulation, we should have information about the final
state of the model comprised of all three components and access to both qualitative
and quantitative data about the whole and the parts. The analysis of such an LVC
experiment will require a holistic understanding rather than a piecemeal decomposition (or reduction) of the components. In other words, the best way to analyze the
experiment is to experience it with the goal of gaining some insight that is necessarily
tailored to the subject. From a design standpoint, Figure 16.2 shows a flow diagram
of an LVC experiment geared toward ontological emergence.

The simulation experience approach has eight steps that are not necessarily linear
but we feel are necessary to create an environment conducive to the effective study
of ontological emergence. Before we describe the intent of each step, we first make
some basic observations:

- *Open-Ended Environment*: As participants can draw on their experiences and
  we encourage observers to participate as they see fit, the number of variables
  that can be introduced is very large. Thus, we have an experiment where literally anything can happen and the goal of the experiment is not to reveal some
  truth (absolute or relative). Rather, we are focused on generating insights, new
  connections, or a new environment, that is, one that is shaped by the interactions
  between people (real and artificial) and their environment. The artificial agents
  can thus change or be shaped in ways that we could not imagine or even become
  something that was not programmed; virtual and augmented objects can take on
  different meaning or purpose and people can relate in new ways. We also have a

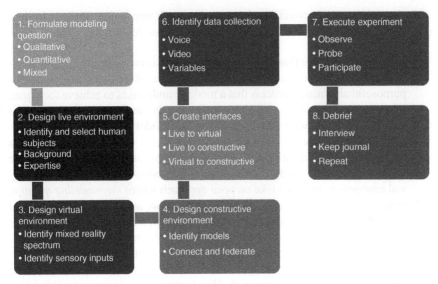

FIGURE 16.2   Simulation experience approach (SEA) to design an LVC experiment to investigate ontological emergence.

very large number of states and outputs because each participant holds a mental model that includes their physical and emotional state as well as their beliefs and understanding (or lack thereof) of what is occurring.

- *Rich Data*: The focus of the framework is on the nuances and richness of the LVC experiment. Although a large amount of qualitative and quantitative data is generated, the simulation is to be experienced preferably while it unfolds or later by watching recordings of it. Traditional analytical methods can be used to further investigate some question a researcher might have, but it is understood that any conclusions drawn from such analysis are necessarily reductionist. Outputs are expressed physically, verbally, physiologically, and mentally.

- *Delay and Misunderstanding*: With this approach, we understand that (i) insights might not be generated or experienced during the simulation, (ii) there is a strong possibility that misunderstanding and bias will cloud what is experienced, and (iii) the experience produces nothing emergent. The frameworks are a reflection of the nonlinear, holistic real-world interactions that occur in the real world as opposed to the controlled laboratory experiment where everything is scripted.

- *Embrace Transdisciplinarity*: The framework we propose relies on a mix of methodologies, worldviews, and bodies of knowledge that transcends disciplines. To achieve ontological emergence, we must embrace the difference between artists, engineers, humanists, and neuro-atypical thinkers. Each step relies on the strengths of one or more disciplines but the framework emphasizes diversity in teams.

Having laid out basic principles, let us explore each step and attempt to map it to what we currently know from the state of the art in other domains:

1. *Formulate Modeling Question*: The goal of M&S is to answer a modeling question. The modeling question defines the boundaries of the model and guides the purposeful abstraction process that a modeler undertakes to achieve their goal. Formulating a modeling question usually entails identifying (i) a hypothesis that will be tested with the simulation of the model in a confirmatory study, or (ii) a formal description of the phenomenon that we wish to explore in an exploratory study. In both cases, we assume an epistemological stance because we have a priori knowledge of the question we wish to ask. To achieve ontological emergence, we must take an open approach where *the modeling question is specified as a rich description of the situation we wish to explore*. We do not assume that there is a problem to be solved. Rather, we abide by the fact that we are in a problem situation (Avison and Taylor, 1997) and want to create an environment that facilitates the emergence of a problem. It is possible that each person (or artificial intelligence) perceives the problem differently and that some might not perceive a problem at all. The observer effect in second-order cybernetics (Heylighen and Joslyn, 2001) explores this issue and its relationship to M&S is explored by Mittal and Zeigler (2014). As the modeling question is a description of the situation, we need to involve participants who are trained in soft system methodology, capable of elicitation and good at facilitating a conversation between people with different worldviews. The specification of the modeling question also involves setting up the virtual environment through the selection/creation of virtual worlds. These worlds might be preexisting or are created by all participants as a first step of the LVC experiment. It is possible for each participant to live in their world and share it with others or to create a shared world where they agree to live. The constructive component is set up by the selection of models or groups of models that best match each person's worldview and understanding of the aspect they want modeled. This also includes selecting input variables, ranges for values and outputs of interest. As with the virtual environment, each participant can create their own version of the LVC experiment with their world and their simulations or create a shared world with shared models. Throughout the setup of a modeling question, facilitators help foster a conversation between participants and can also create their own version of the LVC experiment from which they can share. At the end of this process, we should have an LVC experiment that is uniquely tailored to the participants and their understanding of the situation that they wish to explore.

2. *Design Live Environment*: In this phase, we need to identify a mix of participants that are susceptible to gain insight or cause ontological emergence. The selection process involves interviews or might simply target a community of people bound by some characteristic (race, gender, profession, etc.). Consideration should be given to the diversity (or uniformity) of the group under study. This goes beyond statistical sampling and having a representative subset of

the population; it involves a deeper understanding of the participant's thinking, background, and worldview. Designing a live environment requires an understanding of the role space, time, movement, and the five senses play in how we interact with one another and our environment. In this phase, we need to determine how movement occurs (physically or not), what senses (sight, sound, taste, smell, and touch) we need to stimulate, and how much. We also need to determine how much we want to augment the live environment and what impact it has on creating a stimulating experience. The focus is on presenting options for participants to design their world while recognizing that we are limited in time and resources and cannot satisfy the needs of everyone.

3. *Design Virtual Environment*: The virtual environment embodies a sensory representation of the environment a participant wants to experience. We are interested in creating worlds that (i) engages a mix of all five senses, (ii) is intuitive and somewhat familiar in that it fits a worldview, and (iii) is dynamic, responsive, and adapts to interactions with the participant. In a sense, the virtual environment should be an extension of the live environment with a mix of augmented and fully virtual components.

4. *Design Constructive Environment*: The constructive environment captures closed-world computational components of the participant's world in the form of computational models. Constructive components can be representations of physical phenomena, social dynamics in groups or crowds, cultural evolution, cognition, decision-making, and any other aspect a participant is interested in interacting with or using. The constructive environment can consist of independent models working separately in sequence or parallel or be made of a composition of models working together in an orchestrated fashion. Regardless of how the constructive environment is constituted, its role is to further extend the live environment and therefore must be adaptable and evolutionary.

5. *Create Interfaces*: Interfaces embody how a participant interacts with their world. Beyond linking Live to Virtual, Live to Constructive, and Virtual to Constructive, we need to think through how interfaces combine to provide a seamless experience. Here, we need to explore the use of voice, sight, brainwaves, and any other means of input to determine the best way a participant experiences their world. These interfaces should be selectable during the first step as the participant is designing their world.

6. *Identify Data Collection*: Data collection starts in step 1 and ends well after the experiment is completed. In this phase, we need to determine what will be recorded and how. We are concerned with the technical aspects of tracking enough information without affecting the experience of participants. We want to be careful in making sure that there a sensible mix of data and an organized way to relate it. We know that we will deal with both qualitative and quantitative data, but we are also interested in recording the experiment, what is said, how it is said. The vastness of the data we could collect need to be balanced with the richness of the experience.

7. *Execute Experiment*: When conducting a scientific experiment, it is always recommended to separate the observer from what is being observe, although we know that to be problematic in some cases. We advocate that observers in this case act more like anthropologists who need to be part of the experiment to fully understand it. Otherwise, we fear that even when we observe ontological emergence, we will be unable to explain it. In addition, observers who participate are able to shape their lines of inquiry as the LVC experiment evolves allowing them to capture the richness of the experiment while certainly affecting participants through their probing questions. The dynamic of observer and participant blurs the line between open and closed worlds and constantly engages participants in a dialog that might lead to the generation of a spark. We understand that each LVC experiment is different and the extent to which an observer might inject themselves into the experiment might vary. Application of second-order cybernetic principles such as having a panel of subject matter experts may help address the observer effect.

8. *Debrief*: The experiment can end at a set time of at the discretion of participants. In either case, we do recognize that a conversation is required as a closing step. This conversation might take the form of a formal survey or an open discussion/forum. It is also important to note that recency bias might skew the opinion of participants and that some participants might need time to process what they experienced. Participants might be encouraged to keep a journal to record their thought and some might be brought back for follow up. Although ontological emergence is possible throughout the process, we can only fully attempt to explain it when we combine our findings in step 8 with all the information we have. What might emerge is anyone's guess and we can use this step to reconfigure the world, open lines of deeper investigations into relationships that we would not have suspected before the experiment. Simply stated, even in cases when do not have ontological emergence, we might still learn a great deal about the situation we were exploring.

The SEA framework demonstrates the complexity associated with investigating the next frontier of emergence, which is ontological emergence. It also highlights the need for a diverse, transdisciplinary team of investigators. In Table 16.3, we list some disciplinary skills required to meaningfully achieve that step.

The disciplines listed can apply to more than step. The point is that it is impossible for a single discipline to successfully implement the SEA framework. If the technical component dominates, we might miss important human dimensions, and if the technical component is missing, we will not achieve the virtual and constructive components. The mix of people involved in the research team will necessarily dictate the outlook of the framework and the type of situations that we can investigate. In other words, the SEA framework is a biased representation of the SEA team. For that reason, it must be chosen carefully and more importantly be balanced. M&S has been applied to multiple disciplines and there are guiding principles on how to apply it to

TABLE 16.3   SEA Framework as a Transdisciplinary Approach

Step	Disciplines
Formulate modeling question	Psychology, Anthropology, Social Studies, Performing Arts, Visual Arts
Design live environment	Engineering, Theater, Scenography, Education, Special Education
Design virtual environment	Film, Language and Literature, Mathematics, Computer Science
Design constructive environment	Computer Engineering, Social Sciences, History, Geography, Physics, Biology, Geology
Create interfaces	Neuroscience, Humanities, Physiology, Linguistics, Systems Engineering
Identify data collection	Statistics, Psychology, Sociology
Execute experiment	Sociology, Philosophy
Debrief	Ergonomics, Communication, Criminal Justice

specific problems for a multidisciplinary and transdisciplinary undertaking (Mittal et al., 2017).

GRAND CHALLENGES OF EMERGENCE RESEARCH: A RESEARCH AGENDA

We attempt to approach grand challenges of emergence research by proposing (i) that we need to turn our attention to ontological emergence and (ii) by proposing the SEA framework as an initial framework to potentially achieve that goal. The challenges in implementing the SEA approach fall in the category of grand challenges because they take a long horizon to tackle and require a community level of effort to overcome. In this section, we present the grand challenges of emergence research that have to be addressed in the research agenda not only of academia, but in research efforts of industry and practitioners as well:

- *Architecture Methods in Open-Ended Environments*: Systems engineering – and in particular model-based systems engineering – successfully developed systems architecture methods and tools that have been proven to successfully support the design and engineering of systems (Maier, 2009; Weilkiens *et al.*, 2015). However, such architecture methods are focusing on the system and do not allow the sufficient specification of an open-ended environment with the possibility to create a system of systems by being coupled with other systems. Such methods, however, will be pivotal to better understand and manage emergence. Open-ended environments can draw from system of systems and enterprise approaches, but they will likely require a new set of methods and tools.

- *Natural Modeling Methods for Humanities*: Scientific work requires to comprehend, share, and reproduce research results to increase knowledge over time. Currently, the use of simulation-based solutions is often overshadowed by the lack of comprehensibility, shareability, and reproducibility and is limited to a subgroup of simulation experts. New modeling methods that support the world view of humanities, but are combined with the rigor of a model that can be implemented as a simulation, are going to help to bridge the gap. A common, mathematically consistent, representation of such methods may provide a new methodology that fuses the expressiveness of art with the rigor of engineering. The grand challenge of "modeling for everyone" with the necessary rigor is an open task.

- *Meta-Formalisms Supporting the Transition from Multi- to Inter- to Transdisciplinary Research*: Multidisciplinary describes the loosest coupling of disciplines to solve a common problem. Each discipline remains sovereign. Juxtaposing disciplines and their methods, the team coordinates and sequences the contributions of various disciplines. Interdisciplinary creates a closer linkage, as the focus is on overlapping domains of knowledge, building permanent bridges that link the domains. Transdisciplinary represents the strongest coupling, as new disciplines are created by transcending, transgressing, and transforming the contributing disciplines and specialties. To move from juxtaposing, sequencing, and coordinating vie integrating, interacting, and linking finally to transcending, transgressing, and transforming, meta-formalism have to ensure the transition of knowledge in a logically consistent way. First approaches are discussed in Tolk (2016), and some interesting work has been published by Aliyu *et al.* (2016), but the "great unifying theory of complexity engineering" remains unformulated so far.

- *Robust Simulation Environments*: Simulation of a complex adaptive system model that manifests accurate emergent behavior must be supported with a robust simulation infrastructure. As Mittal and Martin (2017) point out, we must rule out the emergent behavior resulting out of the computational implementation of the model. Assuming we have a valid conceptual model that is implemented correctly in a modeling workbench, that is, it is a valid digital model, it still does not guarantee that the execution of this model by any simulator will yield the correct dynamic behavior. Formal approaches to modeling and system theoretic approaches to simulator engineering in a distributed environment must be robust enough that time management is transparent and the temporal behavior is strictly a function of the model and not of the simulation infrastructure. Complexity research is needed in M&S of multi-level concurrent systems.

- *Experiencing Complexity and Emergence*: Benjamin Franklin is attributed with the quote "Tell me and I forget, teach me and I may remember, involve me and I learn." In particular for decision makers, this insight is directly applicable and derives the necessity for better training and education by immersive technology and means. The military is well known for their exercises that actually motivated that SEA approach. Something similar is needed for the general complexity and emergence field. The same technology may actually be used to bring research teams closer together, as they can experience the effects together, and then apply their individual strength to better understand and manage what they experienced. So, the grand challenge is to make emergence immersive.

- *Beyond Systems Thinking*: Systems thinking replaced the traditional reductionists approach by holistically analyzing the way systems' constituent components interrelated and cooperate to make the system work within the context of larger systems over time. At several points in this book, the approach has been shown to possibly reach its limits, as open environments and applications in unforeseen contexts are no longer the exception, but the rule in complex systems engineering applications. Similarly to the quantum leap, we experienced as a community when moving from reductionism to system thinking, a new set of methods that allows to better address holism of open systems in open environments may result in another quantum leap of understanding. Model thinking beyond bounded models under closure may help, or another radically different approach may be needed. The collection of methods and tools enumerated in Table 5.2, as compiled by Sheard *et al.* (2015), is a step into the right direction, but it is just a step, not the end of the journey, of which we are not sure what the destination is.

This short list of topics is intended to spawn a discussion and kick-off a research agenda. These topics are mainly driven by the group of experts behind this book, so it needs to be extended and discussed in the broader audience as envisioned in this chapter.

Complexity and emergence will require new methods and tools, and maybe even different structures. After years of specialization and focusing on more and more details, which helped us gain a tremendous amount of knowledge and understanding and led to breakthroughs in so many disciplines and domains, a new set of research characteristics may "emerge" that takes the opposite approach. After we focused on understanding the last detail for years, we may now need Universalists with an eye for the whole, without sacrificing the details. We want to see the forest as well as the trees, and as complex systems engineering introduced the metaphor of the engineer as the gardener instead of the watchmaker, this may be the right metaphor to close this last chapter with.

DISCLAIMER

REFERENCES

Aliyu, H.O., Maïga, O., and Traoré, M.K. (2016) The high level language for system specification: A model-driven approach to systems engineering. *International Journal of Modeling, Simulation, and Scientific Computing*, **7** (01), 1641003.

Avison, D.E. and Taylor, V. (1997) Information systems development methodologies: a classification according to problem situation. *Journal of Information Technology*, **12** (1), 73–81.

Heylighen, F. and Joslyn, C. (2001) Cybernetics and second order cybernetics. *Encyclopedia of Physical Science & Technology*, **4**, 155–170.

Hodson, D.D. and Hill, R.R. (2014) The art and science of live, virtual, and constructive simulation for test and analysis. *The Journal of Defense Modeling and Simulation*, **11** (2), 77–89.

Holland, J.H. (1998) *Emergence: From Chaos to Order*, Oxford University Press.

Maier, M.W. (2009) *The Art of Systems Architecting*, CRC Press.

McClelland, J.L. (2010) Emergence in cognitive science. *Topics in Cognitive Science*, **2** (4), 751–770.

Mittal, S. (2013) Emergence in stigmergic and complex adaptive systems: a formal discrete event systems perspective. *Cognitive Systems Research*, **21**, 22–39.

Mittal, S. and Martin, J.L.R. (2017) Simulation-based complex adaptive systems, in *Guide to Simulation-Based Disciplines: Advancing Our Computational Future* (eds S. Mittal, U. Durak, and T. Oren), Springer, UK, pp. 127–150.

Mittal, S. and Rainey, L. (2015) Harnessing emergence: the design and control of emergent behavior in system of systems engineering. Summer Simulation Multi-conference.

Mittal, S. and Zeigler, B.P. (2014) Context and attention in activity-based intelligent systems. ITM Web of Conferences, 3(1).

Mittal, S. and Zeigler, B.P. (2017) Theory and practice of modeling and simulation in cyber environments, in *The Profession of Modeling and Simulation* (eds A. Tolk and T. Oren), Wiley & Sons.

Salt, G.W. (1979) A comment on the use of the term emergent properties. *The American Naturalist*, **113** (1), 145–148.

Sheard, S., Cook, S., Honour, E., Hybertson, D., Krupa, J., McEver, J., McKinney, D., Ondrus, P., Ryan, A., Scheurer, R. and Singer, J. (2015) A complexity primer for systems engineers. INCOSE Complex Systems Working Group White Paper, http://www.incose.org/docs/default-source/ProductsPublications/a-complexity-primer-for-systems-engineers.pdf (accessed September 2017).

Tolk, A. (2016) *Multidisciplinary, Interdisciplinary, and Transdisciplinary Research*, The Digital Patient: Advancing Healthcare, Research, and Education, pp. 225–240.

Weilkiens, T., Lamm, J.G., Roth, S., and Walker, M. (2015) *Model-Based System Architecture*, John Wiley & Sons.

Van Cleve, J. (1990) Mind--Dust or Magic? Panpsychism Versus Emergence. *Philosophical Perspectives*, **4**, 215–226.

McLaughlin, B. (1997) Emergence and supervenience. *Intellectica*, **25** (1), 25–43.

Wilson, J. (1999) How superduper does a physicalist supervenience need to be? *Philosophical Quarterly*, **49**, 33–52.

Wilson, Jessica (2002). "Causal Powers, Forces, and Superdupervenience," *Grazer Philosophische Studien*, **63**: 53–78.

Mittal, S.; Durak, U.; and Oren, T. (editors): *Guide to simulation-based disciplines: advancing our computational future. Simulation Foundations, Methods and Applications Series*; Springer International Publishing: Cham Switzerland, 2017.

Index

abstractions, 285, 288, 294–295, 391
acyclic multimodel, 193
agent
  behavior, 39, 193, 232
  interactions, 133, 234, 238, 343
  level, 47, 174, 324, 331, 339
  types, 243
agent-based model (ABMs), 130, 134,
  144, 234, 321, 323, 378
anti-emergentist, 13
AOR Simulation, 193–194
approaches
  generative, 15–17, 59, 61–62, 65, 68,
    74, 131, 323, 326–329, 341–342, 376
  statistical, 280
architectural variety, 298, 300, 303,
  309
architecture, 30, 37, 42–45, 82, 102–103,
  112–115, 150, 176, 209, 230, 234–236,
  241, 254, 258–259, 285, 287, 289,
  294–295, 297–302, 304, 307, 309–310,
  312, 314, 376–377, 385
autonomous agents, 326

behavioral trajectories, 306–308
behaviors, 231, 286
belief matrix, 332–334
beliefs, 70, 107–108, 322–325, 329,
  331–335, 337–338, 340–341, 381
birds model, 230–231, 234, 239, 241, 251,
  254, 256

characterizations, 152, 285, 287–292, 294,
  297, 303–304, 308–311
claims, 11, 15–16, 18, 166, 289, 321–322,
  325, 327–328, 330, 342, 377
CLD (causal loop diagrams), 159, 166,
  350–353, 356, 365
cognitive agents, 341
cognitive mechanisms, 323, 329–331, 340,
  342–343
coherence models, 73–74
communication networks, 129, 132,
  134–135, 137, 139
complex enterprises, 99–100, 102–104, 106,
  108, 110, 112, 114, 116, 118, 120,
  122–124, 376

Emergent Behavior in Complex Systems Engineering: A Modeling and Simulation Approach,
First Edition. Edited by Saurabh Mittal, Saikou Diallo, and Andreas Tolk.
© 2018 John Wiley & Sons, Inc. Published 2018 by John Wiley & Sons, Inc.